Statistical Methods and Analyses for Medical Devices

Scott A. Pardo

Statistical Methods and Analyses for Medical Devices

Springer

Scott A. Pardo
Global Medical & Clinical Affairs
Ascensia Diabetes Care
Monsey, NY, USA

ISBN 978-3-031-26138-1 ISBN 978-3-031-26139-8 (eBook)
https://doi.org/10.1007/978-3-031-26139-8

This Springer imprint is published by the registered company Springer Nature Switzerland AG
The registered company address is: Gewerbestrasse 11, 6330 Cham, Switzerland

Preface

Medical devices are many and diverse; everything from cotton swabs to MRI machines. Some require little or no supporting data for regulatory registration, and some require multiple clinical studies. Inasmuch as anything that is manufactured by people require data and associated analyses in order to insure quality, all medical devices can and should be associated with data gathering and analyses (even cotton swabs). As such, it behooves those who are involved in the design, manufacture, dispersal, and usage of medical devices to know something about statistical methods; what they are, when to use them; assumptions they require; and even how to apply them using software.

The aim of this book is not to turn everyone working in the medical device industries into mathematical statisticians. Rather, the goal is to provide some help in thinking statistically, and knowing where to go to answer some fundamental questions, such as justifying a method used to qualify/validate equipment, or what information is necessary to support the choice of sample sizes. There are no statistical methods specifically designed for analysis of medical device data. However, there are some methods that seem to appear regularly in relation to medical devices. For example, the assessment of receiver operating characteristic curves is fundamental to development of diagnostic tests, and accelerated life testing is often critical for assessing the shelf life of medical device products. Many of the chapters, especially Chaps. 5, 6, 7, 8, 9, 10, 11, 12, 13, 14, 15, 16, and 17, can be (and are) books unto themselves. This treatment is meant to give the reader a place to start. As in the case of all written material, this work is not a place to finish.

For those who are in fact statistically knowledgeable, the hope is to provide some insight into what methods are likely to help provide rationale for choices relating to data gathering and analysis activities for medical devices.

Monsey, NY, USA Scott A. Pardo

Acknowledgments

I owe gratitude to many, many individuals who have broadened my knowledge of the many applications of statistics in the worlds of medical devices. First, my friends and colleagues, Dr. Joan Parkes and Dr. Rezi Zawadzky, who provided me with inspiration and encouragement. I would like to acknowledge the input, inspiration, and thought-provoking problems and questions posed by Dr. Michael (Mickey) Pardo, Dr. Yehudah (Yudi) Pardo, and Dr. Jeremy Pardo. You guys always made me think hard, and I learned so much because of you. I would also like to acknowledge and thank, for the support, inspiration, and input, my wife, who has put up with me, and provided me with thoughts and questions that tremendously improved this and other works. Te amo Palomika mia.

Contents

Chapter 1
Some Fundamentals: Probability

Abstract Basics of probability theory are covered, in the context of random variables. Expectation, variance, covariance, correlation, and moments are described. Parametric distributions are introduced.

1.1 The Basics

The primary probability concept we will use is called a random variable (Meyer, 1970). A random variable is actually a function. It is generally sufficient to think of a random variable as a number whose value cannot be predicted exactly before any observation is made. So, gravity is not usually thought of as a random variable. On earth, gravity is 9.8 m/s^2, roughly. We do not need to measure it in order to know it. However, the flight velocity of the laden European swallow is not a known value, per se. Some swallows may be faster than others, even when laden with the same weight and volume.

Random variables have ranges of values, and the ranges are associated with probabilities. So, for example, tensile strength of some object is a force that may have some variation from part to part, even though the parts are essentially "identical." Thus, we cannot say a priori exactly what the tensile strength of a particular part might be. Rather, we might be able to say that the tensile strength of a particular part will fall between 1.000 and 1.010 lbs with a probability of 82%. Thus, the part may have a tensile strength of 1.003 lbs, but we may only know that its tensile strength will fall between 1.000 and 1.010 with a probability of 82%.

Random variables come in two basic flavors: continuous and discrete. Continuous random variables take on values that can be described by decimal numbers, such as tensile strength, length, electrical current, etc. Discrete random variables take on integer values.

All random variables, continuous or discrete, are associated with a function that describes any probability one might like to know. Unfortunately, the exact form of this function, called the cumulative distribution function (CDF), is almost always unknown. The job of statistics is to make a guess about the form of the CDF, using observations, or data. The process of making guesses using data is called estimation and inference. More on that later.

S. A. Pardo, *Statistical Methods and Analyses for Medical Devices*,
https://doi.org/10.1007/978-3-031-26139-8_1

Often we put some structure on the form of the CDF for some random variables. Structure means a model. The model is a mathematical representation of the CDF, with constants, or parameters, whose values can be estimated using data. Such functions will be referred to as "parametric distributions." The parametric distributions form families, whose mathematical forms are the same but whose parameter values vary. Parametric distributions are often described by mathematical functions that are actually the first derivative of the CDF. These functions are called probability density functions. For discrete random variables, the density function is sometimes referred to as the probability mass function.

In the case of continuous variables, the probability that a variable X is exactly equal to a particular value is always zero. This rather curious fact is based on a set of mathematical ideas called measure theory. Intuitively, the probability of finding an individual with exactly some specific characteristic (say, a weight of 2.073192648 kg, or a resistor with a resistance of 10.0372189 Ω) is, well, zero. This is not to say that once you find such an individual, you must be hallucinating. The notion of zero probability (and in fact any probability) relates to "a priori" determination, i.e., before any observation. Once an observation is made, the probability of observing whatever it is you observed is in fact 1, or 100%.

In general, capital letters, like X, will refer to a random variable, whereas lowercase letters, like x, will refer to a specific value of the random variable, X. Often, in order to avoid confusing discrete and continuous variables, the symbol $f_X(x)$ will refer to the density function for variable X, evaluated at the value x, and $p_X(x_k)$ to a probability mass function for a discrete variable X evaluated at the value x_k. The notation $\Pr\{\}$ will refer to the probability that whatever is inside the curly brackets will happen, or be observed. If the symbol "dx" means a very small range of values for X, and x_k represents a particular value of a discrete random variable, then:

$$f_X(x)dx = \Pr\{x - dx \leq X \leq x + dx\}$$

and:

$$p_X(x_k) = \Pr\{X = x_k\}$$

There is a particularly important function called the cumulative distribution function (CDF) that is the probability $\Pr\{X \leq x\}$, which is usually defined in terms of density or mass functions, namely:

$$F_X(x) = \int_{-\infty}^{x} f_X(\xi)d\xi$$

for continuous variables, and:

$$F_X(x) = \sum_{x_k \leq x} p_X(x_k)$$

for discrete variables.

Table 1.1 Some probability density and mass functions

Name	Parameters	Density or mass function	Expected value	Range of values
Normal	μ, σ	$\frac{1}{\sqrt{2\pi\sigma}}\exp\left(-\frac{1}{2}\left(\frac{x-\mu}{\sigma}\right)^2\right)$	μ	$-\infty < x < +\infty$
Gamma	ν, λ	$\frac{\lambda^\nu}{\Gamma(\nu)}x^{\nu-1}\exp(-\lambda x)$	$\nu\lambda$	$x > 0$
Chi-squared	ν	$\frac{(1/2)^{\frac{\nu}{2}}}{\Gamma\left(\frac{\nu}{2}\right)}x^{\frac{\nu}{2}-1}\exp\left(-\frac{1}{2}x\right)$	$\frac{\nu}{2}$	$x > 0$
Student's t	ν	$\frac{\Gamma\left(\frac{1}{2}(\nu+1)\right)}{\sqrt{\pi\nu}\Gamma\left(\frac{1}{2}\nu\right)}\left[1+\frac{x^2}{\nu}\right]^{-\frac{(\nu+1)}{2}}$	0	$-\infty < x < +\infty$
F	ν_1, ν_2	$\frac{\Gamma\left(\frac{\nu_1+\nu_2}{2}\right)}{\Gamma\left(\frac{1}{2}\nu_1\right)\Gamma\left(\frac{1}{2}\nu_2\right)}\frac{x^{\left(\frac{\nu}{2}\right)-1}}{(1+x)^{(\nu_1+\nu_2)/2}}$	$\frac{\nu_2}{\nu_2-2}$	$x > 0$
Poisson	λ	$\frac{\lambda^k e^{-\lambda}}{k!}$	$\frac{1}{\lambda}$	$k = 0, 1, 2, \ldots$
Binomial	n, p	$\binom{n}{k}p^k(1-p)^{n-k}$	np	$k = 0, 1, 2, 3, \ldots, n$
Beta	α, β	$\frac{\Gamma(\alpha+\beta)}{\Gamma(\alpha)\Gamma(\beta)}p^{\alpha-1}(1-p)^{\beta-1}$	$\frac{\alpha}{\alpha+\beta}$	$0 < p \leq 1$
Truncated normal	μ, σ, a, b ($\alpha = (a-\mu)/\sigma$), ($\beta = (b-\mu)/\sigma$)	$\frac{\phi(z)}{Z}$ $z = (x-\mu)/\sigma$ $Z = \Phi(\beta) - \Phi(\alpha)$ $\phi(z) =$ *std normal density*	$\mu + \frac{\phi(\alpha)-\phi(\beta)}{Z}\sigma$	$a \leq x \leq b$

As mentioned earlier, the functions $f_X(.)$ and $p_X(.)$ generally have parameters, or constants, that dictate something about the particular nature of the shape of the density curve. Table 1.1 shows the parameter lists, and density or mass functions for several common distributions.

In the case of the binomial and beta distributions, the symbol "p" was used to denote a parameter (binomial), or as a value of a random variable (beta), and not the mass function itself. The function $\Gamma(x)$ is called the gamma function (oddly enough) and has a definition in terms of an integral:

$$\Gamma(x) = \int_0^\infty \xi^{x-1}e^{-x}d\xi$$

Aside from the CDF, there are some other important functions of $f_X(x)$ and $p_X(x_k)$. In particular, there is the expected value, or mean:

$$E[X] = \mu = \begin{cases} \sum_k x_k p_X(x_k) \\ \int_{-\infty}^{+\infty} \xi f_X(\xi)d\xi \end{cases}$$

and the variance:

$$V[X] = E\left[(X-\mu)^2\right] = \sigma^2 = \begin{cases} \sum_k (x_k-\mu)^2 p_X(x_k) \\ \int_{-\infty}^{+\infty} (\xi-\mu)^2 f_X(\xi)d\xi \end{cases}$$

An important property of expectation and variance is the effect of multiplying a random variable by a constant:

$$E[aX] = aE[X]$$
$$V[aX] = a^2V[X]$$

The expected value is also called the first moment about 0, and the variance is called the second moment about the mean. In general, the *k*th moment about zero is simply:

$$E\left[X^k\right] = \mu^{(k)} = \begin{cases} \sum_i x_i^k p_X(x_i) \\ \int_{-\infty}^{+\infty} \xi^k f_X(\xi)d\xi \end{cases}$$

The *k*th moment about the mean is:

$$E[X-\mu]^k = \sigma^{(k)} = \begin{cases} \sum_i (x_i-\mu)^k p_X(x_i) \\ \int_{-\infty}^{+\infty} (\xi-\mu)^k f_X(\xi)d\xi \end{cases}$$

Commonly the Greek letter μ is used to symbolize the expected value, and σ^2 is used to represent the variance. The variance is never negative (a sum of squared values).

The square root of the variance is called the standard deviation and has its most important role in random variables having a normal distribution. The expected value has units that are the same as individual measurements or observations. The variance has squared units, so that the standard deviation has the same units as the measurements.

Another parameter that is a measure of variability is the coefficient of variation, or cv. The cv is sometimes referred to as relative standard deviation (rsd). It is the ratio of the standard deviation to the expected value:

$$c = \frac{\sigma}{\mu}$$

Inasmuch as this is a unitless quantity, it is sometimes expressed as a percentage.

Often we must deal with more than one random variable simultaneously. The density or mass function of one variable might depend on the value of some other variable. Such dependency is referred to as "conditioning." We symbolize the conditional density of X, given another variable, say Y, is equal to a particular value, say y, using the notation:

$$f_{X|Y}(x|Y=y)$$

Typically, the fact that $Y=y$ will affect the particular values of parameters. Also, we will usually drop the subscript $X|Y$, since the conditional nature of the density is made obvious by the "|" notation.

It is possible that the value of one random variable, say Y, has no effect on the probability distribution of another, X. It turns out that any two random variables have what is called a joint density function. The joint density of X and Y could be defined as:

$$f_{XY}(x,y)dxdy = \Pr\{x - dx \le X \le x + dx, \text{ AND } y - dy \le Y \le y + dy\}$$

The joint density quantifies the probability that random variable X falls in a given range and at the same time random variable Y falls in some other given range.

It turns out that this joint density can be expressed in terms of conditional densities:

$$f_{XY}(x,y) = f_{X|Y}(x|y)f_Y(y)$$

The marginal density of one variable (say X) is the density of X without the effect of Y, and is computed as:

$$f_X(x) = \int_{-\infty}^{+\infty} f_{XY}(x,y)dy$$

When X and Y are independent of each other, then:

$$f_{X|Y}(x|y) = f_X(x)$$

So that:

$$f_{XY}(x,y) = f_X(x)f_Y(y)$$

In other words, when X and Y are independent, their joint density is the product of their marginal densities.

Two random variables, X and Y, may not be independent, and their joint density may not be factorable into two functions, one of X and the other of Y. There is another function called covariance that quantifies the degree to which the two variables depend upon each other:

$$\mathrm{Cov}[X, Y] = E[(X - \mu_X)(Y - \mu_Y)]$$

Often, the covariance is "normalized," so that its values vary between -1 and $+1$. The normalized covariance is called correlation:

$$\mathrm{Corr}(X, Y) = \frac{\mathrm{Cov}(X, Y)}{\sqrt{V_X[X] V_Y[Y\}}}$$

The values μ_X and μ_Y are referred to as the marginal means of X and Y, respectively, and Var(X) and Var(Y) are the marginal variances. "Marginal" means with respect to the marginal distributions, e.g.:

$$E_X[X] = \mu_X = \int_{-\infty}^{+\infty} \xi f_X(\xi) d\xi$$

and:

$$V_X[X] = \sigma_X^2 = \int_{-\infty}^{+\infty} (\xi - \mu)^2 f_X(\xi) d\xi$$

The correlation of any random variable with itself is always $+1$.

In addition to joint distributions, the expected values and variances of sums and differences of random variables find themselves in many applications. So, if X and Y are two random variables:

$$E[aX \pm bY] = aE[X] \pm bE[Y]$$

If X and Y are independent, then:

$$V[aX \pm bY] = a^2 V[X] + b^2 V[Y]$$

where a and b are constants.

While the sign of the operator ($\pm$) follows along with the expected values, the variance of the difference is the sum of the variances.

Another set of facts we will use relating to conditional densities or mass functions is based on something called Bayes' theorem. Briefly, Bayes' theorem states that if X is a random variable with density f, and Y is a random variable with density g, then:

$$g(x|Y=y) = \frac{g(y)f(x|Y=y)}{\int_{-\infty}^{+\infty} f(x|Y=\xi)g(\xi)d\xi}$$

As long as Y is continuous, this particular formula holds even if X is discrete and f is the mass function of X. If, however, Y is discrete, and g is its mass function, then the integral is replaced with a summation:

$$g(x|Y=y) = \frac{g(y)f(x|Y=y)}{\sum_k f(x|Y=\xi_k)g(\xi_k)}$$

It should be noted that it is possible for a random variable to not actually have a density function associated with it. However, that situation probably never exists in nature, so we will assume the density always exists.

1.2 Summary

The heart of probability is the random variable. Random variables are actually functions that map outcomes of experiments to numbers. The common feature that all random variables have is the probability distribution, which in turn is a function that describes the probability of observing particular values of the random variable. Probability distributions, also referred to as distribution functions, are often described by a mathematical formula which has a set of parameters. The parameters allow for a whole family of distributions.

Reference

Meyer, P. L. (1970). *Introduction to probability and statistical applications* (2nd ed.). Addison-Wesley.

Chapter 2
Some Fundamentals: Estimation and Inference

Abstract The concept of inferential reasoning is described. Some principles of estimation such as maximum likelihood are covered. Bayesian methods are introduced.

2.1 Estimation

Statistical problems can be classified into three categories:

1. Estimation
2. Inference and decision-making
3. Model building: discrimination and prediction

Estimation is the process of using data to guess at the value of parameters or some feature of the probability distribution assumed to be governing the data-generating process. Probably the most common is estimating the expected value of a distribution. The expected value of the random variable's distribution is:

$$E[X] = \mu = \begin{cases} \sum_k x_k p_X(x_k) \\ \int_{-\infty}^{+\infty} \xi f_X(\xi) d\xi \end{cases}$$

One of the useful mathematical properties of expected value is that it is a linear operator, namely:

$$E[X_1 + X_2 + \cdots + X_n] = E[X_1] + E[X_2] + \cdots + E[X_n]$$

and:

$$E[aX] = aE[X] \text{ when } a \text{ is a nonrandom constant}$$

S. A. Pardo, *Statistical Methods and Analyses for Medical Devices*,
https://doi.org/10.1007/978-3-031-26139-8_2

An estimate based on a sample of observations from the data-generating process is:

$$\widehat{\mu} = \frac{1}{n}\sum_{i=1}^{n} x_i$$

We use the notation $\widehat{\mu}$ instead of a perhaps more well-recognized symbol $\overline{x}$, to emphasize the fact that we are using the data to estimate the expected value. There are many such estimation formulae (called estimators), and many are used in different contexts for different reasons. The main point is that data can be used to estimate parameters or other features of probability distributions. The other point is that, since estimators use data, they themselves are random variables. Thus, if two researchers studying the same population of finches each make independent observations on either two sets (samples) of birds or even on the same sample, but at two different times, and each researcher calculates an average, the two averages most likely won't be the same exactly. Similarly, if two chemists measure the heat of a reaction after a fixed amount of time on each of several samples of reactants, and each chemist computes the average heat, they will probably not obtain exactly the same numerical result.

There are different methods used to derive estimator formulas for various parameters. Perhaps the best known is called the method of maximum likelihood. The idea is that if you have a random sample of measurements (*X*), you can find values of parameters that maximize something called the likelihood function, which generally depends on assuming the form of the distribution for the data-generating process. Suppose that the values $x_1, x_2, \ldots, x_n$ represent n values sampled from a normally distributed data-generating process, with unknown expected value and variance, the density function evaluated at x_i, say, would be given by:

$$f(x_i) = \frac{1}{\sigma\sqrt{2\pi}} e^{-\frac{1}{2}\left(\frac{x_i-\mu}{\sigma}\right)^2}$$

The likelihood function for the sample would be the product of all the valuations of the density function:

$$L(x_1, x_2, \ldots, x_n) = \prod_{i=1}^{n} f(x_i)$$

Of course, this likelihood function cannot be computed without knowing μ and σ. The idea of maximum likelihood is to find values $\widehat{\mu}$ and $\widehat{\sigma}$ that maximize L. Usually the log of the likelihood function is taken before attempting to solve the maximization. Maximizing the log of L is equivalent to maximizing L, since the log is a monotonic increasing function. The log of a product is the sum of the logs of the factors:

$$\log L = \sum_{i=1}^{n} \log(f(x_i))$$

Maximizing the sum is easier mathematically than maximizing the product.

What is important to note is that first we had to pick a parametric form for the density function of the random variable from which we were sampling, the parameter values are unknown, and our guess for the parameter values is based on a criterion that gives us the "best guess." It turns out that for the normal model, the maximum likelihood estimators for μ and σ^2 are:

$$\widehat{\mu} = \frac{1}{n}\sum_{i=1}^{n} x_i$$

and:

$$\widehat{\sigma^2} = \frac{1}{n}\sum_{i=1}^{n} (x_i - \widehat{\mu})^2$$

In general, the kth sample moment about the mean is given by:

$$\widehat{\sigma^{(k)}} = \frac{1}{n}\sum_{i=1}^{n} (x_i - \widehat{\mu})^k$$

Some may notice that the maximum likelihood estimator for σ^2 differs from the formula used in most elementary texts, in that it divides by n and not $n - 1$. Dividing the sum by $n - 1$ to estimate σ^2 gives the formula a property known as unbiasedness. While this is important, in the case of this estimator, the effects are fairly small. Another estimation method is called least squares. Rather than maximizing a likelihood function, least squares choose estimators that minimize an "error" function. A common context for least squares estimation is linear regression. More will be said about least squares. For now, just recognize it as a method for estimating parameters.

Statistical estimates, since they are based on a finite sample of observations or measurements made on individuals taken from some population or data-generating process, have themselves a random variation component. Inasmuch as a statistical estimate is attempting to help make a guess about a parameter, it would be good to know that the formula used to compute the estimate has a reasonable chance of getting "close" to the actual value of the parameter. One such property has already been described, namely, maximum likelihood. Another property that is desirable is called unbiasedness, which was also mentioned earlier. An estimation formula is said to be unbiased if its expected value is equal to the parameter to be estimated. For example, assuming a random sample, $x_1, x_2, \ldots, x_n$, then the expected value of each x_i is the population mean, μ, and:

$$E[\widehat{\mu}] = \frac{1}{n}\sum_{i=1}^{n} E[x_i] = \frac{1}{n}\sum_{i=1}^{n} \mu = \mu$$

Thus, our arithmetic mean estimator for μ is in fact unbiased. Conversely, the maximum likelihood estimate of σ^2 is not unbiased (or, in other words, biased). It turns out that:

$$E\left[\widehat{\sigma^2}\right] = \frac{1}{n}\sum_{i=1}^{n} E\left[(x_i - \widehat{\mu})^2\right] = \frac{n-1}{n}\sigma^2$$

Thus, the maximum likelihood estimator of σ^2 slightly underestimates the variance. The point of the discussion about unbiasedness is that estimation formulae are themselves random variables, and as such we will need to consider their probabilistic characteristics.

2.2 Inference

Inference is about making "a priori" guesses about parameter values and then using data to decide if you were correct. Suppose, for example, you guessed that the average duration of a courtship display was 30 s. How would you decide whether to believe your guess, or not? First you would gather data, by timing courtship displays of several individuals, say *n*. Then you would probably compute the maximum likelihood estimates of mean and variance. Suppose the estimate of the mean was 31.5 s, and the standard deviation (square root of variance) estimate was 3 s. OK, so it wasn't 30. Were you wrong? The question becomes one of how much variation there might be if the experiment were repeated. The idea of statistical inference is to make a decision about what to believe and not what actually is the truth. Our decision has risk associated with it, namely, the risk (or probability) of saying our guess is wrong when in fact it is correct and the risk of saying our guess is correct when in fact it is not. There is a formalism for expressing the notions of inference. There are two competing "hypotheses," or guesses about the parameter or parameters of interest. One is called the "null" hypothesis, symbolized as "H_0." The logical negation of the null hypothesis is called, not surprisingly, the alternate hypothesis and is often symbolized as "H_1." So, in the example of the courting display question, we might have:

H_0: $\mu = 30$
H_1: $\mu \neq 30$

The process of formulating hypotheses about parameters and then using data to decide which of the two (mutually exclusive) hypotheses to believe is referred to as "hypothesis testing" (Hoel, 1971).

The error of deciding that H_0 is false when in fact it was true is called Type I error. The error of believing H_0 is true when it is not is called Type II error. The next thing required is a rule, based on data, that lets the decision-maker decide whether to believe H_0 or H_1. Since data are subject to variation, the rule is necessarily probabilistic. It turns out that, conveniently, the calculation:

$$t = \frac{\widehat{\mu} - 30}{\widehat{\sigma}/\sqrt{n}}$$

has a known probability distribution, the familiar Student's t, provided that the null hypothesis is actually correct (i.e., that $\mu = 30$). This formula is known as a test statistic, because it is the quantity we will use to decide whether to believe (accept) the null hypothesis or disbelieve it (reject). The use of this statistic to test hypotheses is referred to as a t-test. It is important to note that the t statistic has Student's t distribution under the assumption that the data were sampled from a population having a normal distribution. However, it is true that as the sample size gets "large," sample means will have a normal distribution with expected value:

$$E[\widehat{\mu}] = \mu$$

and variance:

$$V[\widehat{\mu}] = \frac{\sigma^2}{n}$$

The convergence of the sample mean's distribution to normal is called the "central limit theorem" (Hoel, 1971). In fact, it is one of several forms of the central limit theorem but may be the most recognizable. Using these facts allows for finding closed-form probabilistic limits on the test statistic t.

A common feature of all inference is determining the distribution of the test statistic if H_0 were actually true. The probability of making Type I error is symbolized with the letter α. The probability of Type II error is traditionally symbolized with the letter β. We can find a range of values that t would fall in between with probability $1 - \alpha$, given that H_0 is true, even before we gathered any data. In fact, the range of possible values only depends on the sample size, n, and the desired probability content of the range. If, for example, the sample size was $n = 10$, and we wanted the probability content to be $100(1 - \alpha)\% = 95\%$, then the range of values for t we would expect if the null hypothesis was correct would be approximately ± 2.228. The range ($t \leq -2.228$, $t \geq +2.228$) is called the *critical region* of size α. If the value of the test statistic falls in the critical region, we say that the test statistics is significant, and we REJECT the null hypothesis is favor of the alternative. The particular region for this example is partially based on the presumption that we computed the maximum likelihood estimate for standard deviation. If after getting data, we computed the value of t using the formula above, and its value fell within the range ± 2.228, then we would continue to believe the null hypothesis, because there is a fairly "high" (95%) chance of t falling inside this range if H_0 is correct. Conversely, there is a relatively "low" chance that t would fall outside the "critical" range if H_0 was correct. Unfortunately, we cannot make the same statement about the alternative hypothesis, H_1, since there are an infinite number of possible values (anything other than 30) that would make it correct. Thus, it is easier to fix the chance of making the mistake of deciding that H_0 is false when in fact it is true. Once this risk is decided upon, the decision rule for either believing H_0 or not believing it is fairly easy to compute, provided we know something about the distribution of the test statistic, given that the null hypothesis is true.

Another way of determining a rule for rejecting or accepting the null hypothesis is to *compute a probability* of observing the data you got *if* the null hypothesis was actually correct. This probability is usually referred to as a "p-value." Thus, in our example, if in fact $\mu = 30$, then the test statistic:

$$t = \frac{\widehat{\mu} - 30}{\widehat{\sigma}/\sqrt{n}}$$

has a Student's t distribution with degrees-of-freedom parameter equal to n (since we used the maximum likelihood estimate of σ). Suppose we had data that yielded a sample estimate of μ, say:

$$\widehat{\mu} = \frac{1}{n}\sum_{i=1}^{n} x_i = 31.5$$

and an estimate of σ^2:

$$\widehat{\sigma}^2 = \frac{1}{n}\sum_{i=1}^{n} (x_i - \widehat{\mu})^2 = 9$$

If $n = 25$, then the sample test statistic would be:

$$t = \frac{31.5 - 30}{3/\sqrt{25}} \approx 2.50$$

Since the alternative hypothesis is $\mu \neq 30$, we compute the probability that the test statistic would be outside the range $(-2.50, +2.50)$. To compute this, we can use the R function *pt*():

$$pt(q = -2.5, df = 25, \text{lower.tail} = \text{TRUE}) \approx 0.01934$$

and

$$pt(q = +2.5, df = 25, \text{lower.tail} = \text{FALSE}) \approx 0.01934$$

The "two-sided" p-value is $0.01934 + 0.01934 = 0.03868$.

Since Student's t distribution is symmetric about 0, the probability for the "lower tail" of $-t$ is equal to the probability for the "upper tail" of $+t$.

If our threshold of p-values is $\alpha = 0.05$, then since 0.03868 is less than 0.05, we will no longer believe that the null hypothesis is correct, and reject it. With a sample size of $n = 25$, and $1 - \alpha = 0.95$, then the critical region is $t \leq -2.06$, $t \geq +2.06$. Since $t = 2.50 > +2.06$, we would reject the null hypothesis. Regardless of whether you determine a critical region of size α, or choose α to be a threshold for p-values, the conclusions would be identical.

Another methodology that is somewhere between estimation and inference is called confidence interval building. The confidence interval again employs that risk level, α, but in a slightly different manner. Until now, estimates were only a single number, such as $\widehat{\mu}$. Such an estimate is referred to as a "point" estimate. Suppose we wanted to know the values of the parameters that would correspond to the limits of the critical range for the test statistic. Using the previous example, let:

$$t_{\text{low}} = -2.06 = \frac{\widehat{\mu} - \mu_{\text{low}}}{\widehat{\sigma}/\sqrt{n}}$$

and:

$$t_{\text{high}} = -2.06 = \frac{\widehat{\mu} - \mu_{\text{high}}}{\widehat{\sigma}/\sqrt{n}}$$

Solving for μ_{low} and μ_{high} gives:

$$\mu_{\text{low}} = \widehat{\mu} - 2.06\frac{\widehat{\sigma}}{\sqrt{n}}$$

and:

$$\mu_{\text{high}} = \widehat{\mu} + 2.06\frac{\widehat{\sigma}}{\sqrt{n}}$$

The range of values $(\mu_{\text{low}}, \mu_{\text{high}})$ is called the $100(1 - \alpha)\%$ "confidence interval" for parameter μ. It can be thought of as a feasible range for the unknown values of μ. That is, we are not certain about the actual value of μ, but we are nearly certain ($100(1 - \alpha)\%$ certain) that it lies somewhere in the interval $(\mu_{\text{low}}, \mu_{\text{high}})$. Technically, we say that the interval $(\mu_{\text{low}}, \mu_{\text{high}})$ has a $100(1 - \alpha)\%$ chance of covering, or containing, the parameter value. In any case, we are nearly certain that this interval contains the actual population mean. So, in our example with point estimates $\widehat{\mu} = 31.5$, $\widehat{\sigma} = 3$, and $n = 25$, the 95% confidence interval would be:

$$\mu_{\text{low}} = 31.5 - 2.06\frac{3}{\sqrt{25}} \approx 30.26$$

$$\mu_{\text{high}} = 31.5 + 2.06\frac{3}{\sqrt{25}} \approx 32.74$$

Since the hypothetical value for μ, namely, 30, is not contained in the confidence interval (30.26, 32.74), we do not believe that 30 is a feasible value for μ.

When the null hypothesis is rejected, we say that the difference between our estimate of the parameter and the null value is *statistically significant at the 100α% level*. Another way of stating the same thing is that if we reject the null hypothesis, we would believe that the results of our analyses are repeatable.

2.3 Model Building

Model building is a special application of estimation, but it usually has some inference associated with it. The idea is to postulate some mathematical relationship between some variables, some random and some without any random component. Then we estimate the values of model parameters. Finally, we test to see if we should believe that the form of the model we postulated was reasonable. Models can be predictive or discriminatory/classificatory. A simple example of a predictive model would be a simple linear regression. Suppose there is a continuously valued random variable, Y, and another continuously values non-random variable, X. Y could be things such as response time, elapsed time, distance traveled, or other random variables that can be expressed as a decimal number. In this simple case, we are assuming the X variable is not random. In other words, X is something whose value would have no random variation and whose value is known perfectly without error. Y is referred to as the response variable, and X is the predictor or regressor variable. The general form of the linear model is:

$$Y = \beta_0 + \beta_1 X + \varepsilon$$

The coefficients β_0 and β_1 are unknown and need to be estimated. The variable ε represents random "noise," indicating that the value of Y is on the average a linear function of X, but the actual observed values may have some perturbations, or noise, or sometimes called error associated with them.

Once the values of the parameters are estimated, a predicted value of Y can be computed for a given value of X. We would not usually consider X to have random noise associated with it. That is, when we get a value for X, we are (mostly) certain that the value would not vary if we measured or observed it a second time under exactly the same conditions.

Rather, we suppose that given the value of X, we can predict on the average what Y would be, with the understanding that Y might vary from this average.

Another closely related type of model is also linear but is classificatory or discriminatory. The "X" variables are not continuous but are discrete categories. The goal is to determine if particular groupings of individuals actually discriminate between individuals. In other words, we want to know if individuals in different groups actually differ from each other with respect to Y. Perhaps the simplest example is the one-way analysis of variance (ANOVA). In this case, the single X variable is a set of discrete categories, and Y is the continuous random variable response. The question is not to find a "prediction" of Y for a given value of X, per se. Rather, the question is to estimate the difference in the average Y between the different categories. In the case of ANOVA, often the inferential part of modeling is of greater interest, namely, whether the difference in average values of Y between the different groups of the X categories is in fact repeatable.

There are certainly more types of both predictive and classificatory modeling. The key notion here is that data can be used to create these sorts of models, through a combination of estimation and inference.

This is the classical parametric methodology for statistical inference. There is another set of methods, sometimes called nonparametric or distribution-free, neither of which term is strictly true. The idea is that the distribution of test statistics should not depend on the distribution of the data-generating process. The basic idea is still the same; you formulate a test statistic, you determine the "critical range" or "critical value" based on α, you get some data, and then you compute the test statistic to decide if you should accept or reject the null hypothesis.

A special set of nonparametric techniques is sometimes referred to as resampling methods. This book will in fact emphasize resampling methods where appropriate. The resampling techniques will generally fall into the bootstrap estimation process or the permutation hypothesis testing process. Both of these methods are computer-based, but given modern computing software such as R, they are fairly easy to perform.

Bayesian statistics is an alternate view of parameters, not as particular values to estimate or about which to make a guess about their true values, but treating them as if they themselves are random variables. Like the classic "frequentist" approach, Bayesian methods employ a likelihood function. However, these methods incorporate prior information about the parameters of interest. "Prior" to making observations, the analyst posits a distribution of the parameters of interest. The "prior" distribution expresses the knowledge about the parameter prior to performing the "next" experiment. So, for example, perhaps the mean response time to a stimulus is guessed to be most likely 10 s, but it could be as fast as 5 s and as delayed as 15 s. Rather than simply hypothesizing that the mean is exactly 10 s, the Bayesian method is to postulate a distribution that expresses the current level of knowledge and uncertainty in the parameter. Then, once data are gathered, Bayes' theorem is used to combine the prior distribution with the likelihood function, to update the prior knowledge. The updated distribution for the parameter is called the posterior distribution. So, if $f_{\text{old}}(\tilde{\mu})$ represents the prior density function for the parameter $\tilde{\mu}$, and $L[x_1, x_2, \ldots, x_n|\tilde{\mu}]$ the likelihood function for the sample, given a particular value of $\tilde{\mu}$, then the updated density function (called the posterior density) is:

$$f_{\text{new}}(\tilde{\mu}|x_1, x_2, \ldots, x_n) = \frac{f_{\text{old}}(\tilde{\mu})L[x_1, x_2, \ldots, x_n|\tilde{\mu}]}{\int_{-\infty}^{+\infty} f_{\text{old}}(\xi)L[x_1, x_2, \ldots, x_n|\xi]d\xi}$$

2.4 Summary

The three main divisions of statistical analysis can be thought of as estimation, inference, and model building. Estimation is all about how to use sample data to make educated guesses about the values of some population parameters, such as the expected value and standard deviation of some random variable over the entire (possibly infinite) population. Inference is all about trying to decide whether or not some a priori guess about population parameters should be believed or not, based on sample data. Finally, model building is a sort of hybrid between estimation and inference. Models are often used to make predictions about the values of random variables based on sample data and also to decide whether certain conditions or treatments have an effect on the values of those random variables (e.g., does the color of your shirt affect the price of a stock?).

Reference

Hoel, P. (1971). *Introduction to mathematical statistics*. Wiley.

Chapter 3
Confidence

Abstract Emphasis is placed on interpretation of confidence. In particular, there is a discussion of what confidence is and what it is not. Some confidence interval formulas are presented.

3.1 What Does Confidence Mean?

Confidence is an English word meaning, among other things, *trust*, *reliance*, or *self-assurance*, *boldness* (Merriam-Webster, 2005). Well, the word "confidence" is used in the science of statistics to mean something very technical. It means the probability that an interval[1] (a range of numerical values), constructed from empirical observations, contains the actual, but unknown, value of some particular parameter or set of parameters governing the population's distribution of values from which those empirical observations are sampled.

This is the definition (more or less) formulated by Jerzy Neyman, back in the 1920s (Armitage, 1971). Other definitions contended for such (and related) intervals, most notably the "fiducial limits" of Sir Ronald A. Fisher, Sc.D., FRS. However, Professor Neyman's definition won the prize.

Wow. So much for Merriam-Webster. This statistical definition requires some unpacking.

First, what we mean by "values" must be understood. "Values" refer to (quantified) observations or measurements of random variables. Some examples of random variables are the "fasting" (say 9 h after eating last) blood glucose (FBG) of adults with type 1 diabetes, the shoe size of children, the color of a person's hair (say, 1 = blonde, 2 = brown, 3 = red), the fuel consumption rate of the Abrams M-1 Main Battle Tank, or the maximum speed of the Boeing 767 aircraft.

The next part is the notion of "population" which is the entire collection of all individuals upon which we could make the measurement of some particular random

[1] We speak about intervals in one dimension, but confidence regions can be constructed in a multidimensional parameter space situation. However, for now we are restricting ourselves to an interval for a single "scalar" parameter and not a parameter vector.

S. A. Pardo, *Statistical Methods and Analyses for Medical Devices*,
https://doi.org/10.1007/978-3-031-26139-8_3

variable. So, for the fasting blood glucose example, our population might be all adults who have type 1 diabetes, whether they are alive now, were alive at one time, or have yet to be born and grow into adults. You might say, "But, you cannot measure the fasting blood glucose of people who are no longer living or of those people yet to be..." While this is true, the point is that those people either did or will have a fasting blood glucose, assuming that there was at least one 9-h period in which they consumed (or will consume) no food.

For the children's shoe size example, similarly we could talk about all children who ever existed or who may exist in the future.

The point is that a population is a complete collection (possibly of infinite number) of individuals having some things in common, where those things are the factors that we use to define the population.

The idea of cumulative distribution functions for random variables was described earlier. The CDF gives the probability that the random variable can have values falling within various intervals. The range of values for a random variable could be infinite, half-infinite, or completely bounded. So, for example, fasting blood glucose cannot be a negative number, but there is no defined upper boundary. Thus, the distribution of fasting blood glucose is half-infinite, since fasting blood glucose can never be less than zero. Now, you might argue that the distribution of fasting blood glucose is really bounded, because no living human (or human who was once living, or human who will come into existence and grow to adulthood) could have a fasting blood glucose value of, say, 10,000 mg/dL (555.06 mmol/L). While that may be true, since no one could define the exact upper bound, above which it is impossible to find any value, we will treat fasting blood glucose as half-infinite. What we can do is to say that the interval of values from, say, 2000 mg/dL to positive infinity ($+\infty$) has an extremely low probability.

While we are making definitions, let us loosely define probability to mean the "frequency of occurrence."

So, populations can have random variables defined as some measurement or quantified observations having values for each individual in the population, and these values in turn have a distribution.

The next part of the confidence definition we will discuss is the word "parameter." Again, as explained earlier, the CDF and associate density or mass function can often be described by a mathematical expression that has some set of fixed "parameters" or numbers that describe the shape of the curve. In fact, even if we do not know the form of this mathematical function, we can postulate that it has certain constants, or parameters, that would govern its shape. We may be more interested in the values of those parameters and not so much interested in the general shape of that distribution function. One such parameter we are commonly interested in is the mean, or average, value.

The statistical problem is that we do not know the values of those parameters, but we wish we did. So, for example, consider the frequency curve for average fasting blood glucose. Suppose we sampled $n = 100$ adults with type 1 diabetes (never mind about how the sample is drawn, for the time being), and for each of those individuals, we measured a fasting blood glucose (once per individual). We could make a histogram of the values we observed. Figure 3.1 shows such a histogram.

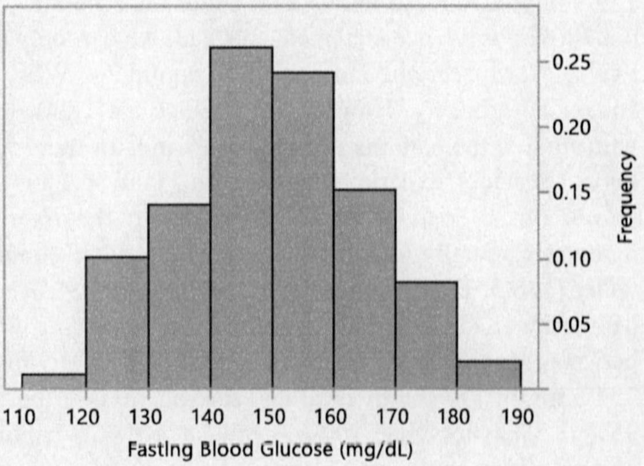

Fig. 3.1 Histogram of $n = 100$ fasting blood glucose (FBG) measurements

We see a frequency plot, where we computed the proportion (frequency) of values observed between 110 and 120 mg/dL, 120–130 mg/dL, ..., and 180–190 mg/dL. This histogram was generated using measurements from only $n = 100$ individuals sampled from a population of effectively an infinite number of individuals. Suppose we wanted to know the average FBG of this population. The average value is a parameter of the population's FBG distribution. Well, since we have only $n = 100$ values, we cannot actually know the average FBG for the whole population. Rather, all we can do is estimate that parameter. It turns out that the "best" way to estimate the population mean is the sample's arithmetic average, which in this case is 150.4 mg/dL. So, is 150.4 mg/dL the average FBG of *all* adults with type 1 diabetes? If we got a different sample of people, or even if we used the same 100 people, but had them repeat the experiment a month later, would we get the same sample average?

What we would like to do is find a "feasible" range of values in which we are fairly certain (confident) that the population's mean FBG falls into. With a sample size of $n = 0$ (!!!!), we can say with absolute certainty (100% confidence) that the population's average FBG falls somewhere between 0 and $+\infty$ mg/dL. That is 100% confidence! How much better can you get? Well, you might say, "excuse me, but could you narrow down that range just a bit?" Well, we can, but there are two prices to pay:

1. We will have to sample more than $n = 0$;
2. We will have to give up on having 100% confidence.

Instead of 100% confidence, we will only be able to say we are "fairly certain." We will define quantitatively what we mean by "fairly certain" in a little bit. The narrowness of our "feasible range" for the population parameter depends mostly on the sample size, n. The bigger the sample, the narrower the range. In this case, we can use some mathematical formulae (not specified here) to find that our feasible range for being fairly certain is 147.5–153.3 mg/dL. That is, based on these data, we are fairly certain that the population's average FBG is somewhere between 147.5 and

153.3 mg/dL. OK, so that is a lot better than the interval (0, +∞). But, we gave up the 100% confidence statement. Instead, we can only say "fairly certain." Quantitatively, "fairly certain" means 95% confident. Why 95%? Well, it is ultimately somewhat arbitrary. However, ever since the 1880s (thanks to Sir Gilbert Walker, who studied the patterns of Monsoon winds in India for the British Royal Meteorological Service), experimenters and data analysts have conventionally accepted a 5% (1 out of 20) risk of being wrong about the interval in which the population parameter actually lies. If we wanted, say, 99% confidence, our interval would be wider (146.5, 154.3 mg/dL). If we wanted only 90% confidence, our interval would be narrower (147.9, 152.8 mg/dL). So, choosing 95% confidence is a tradeoff between specificity (a narrowness of 147.5, 153.3 mg/dL is narrower than 146.6, 154.3 mg/dL) of the interval and risk of missing the true value (a risk of 100% − 95% = 5% is less than 100% − 90% = 10%). Sometimes we are only interested in one side of a confidence interval. For example, we might only want to know the lower confidence bound on the probability of "success"; in that case, the upper bound is not interesting, because we would be perfectly happy if the probability of success was 100%, but we may want to know the worst it could "feasibly" be, based on the data we have gathered. Regardless of whether you want a two-sided interval or a single bound or limit, use any confidence level you want as long as it is 95%. OK, there are some special circumstances under which 90% or 99% confidence levels should be used, but we will not discuss those circumstances in this document.

We must understand that confidence intervals are constructed from empirical observations, or data, from a finite sample of individuals. The sample is a subset of the population. If we took a different sample, we would most likely get a slightly different interval. The point is that generally we cannot observe every single member of a population, nor can we continually sample and construct many intervals. We must get as much information as we can from our single sample. The confidence interval gives us such information and quantifies it with a confidence level.

The simplest way in which values are selected from a population is called random sampling. That means every individual in the population of interest is equally likely to be selected to be part of the sample. Sampling paradigms can be very complex and affect the formulas used to compute confidence intervals. We, however, are not going to discuss any of that here. In fact, at this point, we are only assuming that the sample was obtained in some legitimate fashion, so it is possible to compute a confidence interval without invalidating any necessary assumptions.

3.2 What Confidence Does *Not* Mean

The term "confidence" can *only* be applied to an interval (whether it is bounded or unbounded on one or both sides). It makes sense, for example, to say: "We are 95% confident that the probability a blood glucose (BG) meter has ±15% error is between 98.5% and 99.5%." It would also make sense to say, "We are 95% confident that there is *at least* a 95% probability of meter errors falling within ±15% of lab results." Note that the phrase "at least" implies an interval, namely, 95% or 100%.

It does not make sense to say, "We are 95% confident that the average glucose meter error is 8.5%." In this statement, the value 8.5% is not an interval. It also does not make sense to ask the question, "What sample size will give us 95% confidence to pass the test?" You can ask "What sample size will make the width of a confidence interval for the probability of success to be no more than ±1%?" The width of the confidence interval is the difference between the lower bound to the upper bound of the interval. More commonly we refer to the half-width (e.g., ±1%, and not the full width of 2%). The width of the confidence interval is mostly dependent on the sample size. The confidence level (e.g., 90%, 95%, 99%) also affects the width, but it also affects the interpretation (i.e., the risk of missing the value of the population parameter).

Likewise, it does not make sense to ask, "What confidence do we have of passing the test if there is a 90% probability of success on each attempt?" Confidence is a probability that is applied after data have been gathered, and confidence can only apply to an interval. We could make confidence statements about a hypothetical situation. For example, we could say, "If we sample $n = 100$ people, and find that 70 of them are right-handed, then we would be 95% confident that the percent of right-handed people in the population is somewhere between 60% and 79%." You can ask the statistician to find a sample size that would yield a particular interval width. Or you can ask the statistician to tell you, for a given sample size, the smallest value of a sample statistic (e.g., the sample arithmetic average, or the percent of individuals satisfying a particular criterion) required in order for the lower/upper confidence limit (lower/upper bound on a confidence interval) to have some particular value.

The probability of passing or failing a statistical test of some given hypothetical condition or value of an unknown parameter is called "power." It is often mistaken for confidence. Don't make that mistake. Power is the probability of concluding from data that some hypothesis about a parameter is incorrect. Confidence is the degree of feasibility that an interval has for containing the actual value of the parameter. We can determine a sample size before performing an experiment (a priori) that would give us a particular power to reject a hypothesis given it is "false," but we cannot compute a sample size to give us some particular level of confidence about a parameter's value. We can, however, determine a sample size that would yield a confidence interval with some specific confidence level and a specified width. It is true that there is a relationship between power and confidence, but they are not the same, and their relationship is a topic for another day.

In summary, confidence intervals are *feasible ranges* for the actual value of a population parameter. Confidence intervals have an associated confidence level, which is a probability that indicates the degree to which we can be certain the interval actually contains the true (but unknown) value of the parameter. The confidence level ought to be 95%, except in some special circumstances. Confidence intervals are constructed after data are collected (a posteriori). The sample size governs the specificity (width) of the interval (although confidence level also has an effect).

3.3 Some Consequences of Confidence Limits

When we compute some value, such as a mean, standard deviation, or proportion, from sample data, the results of those computations are known as "point" estimates of a corresponding population parameter. So, for example, suppose that in a sample of $n = 100$ results, 95 of them are considered to be "good" or "successful" and 5 of them not so much. Our point estimate for the probability of obtaining a "good" result on a new "try" is:

$$0.95 = \frac{95}{100} = 95\%$$

We know that this value, 0.95, may not remain if we sample yet another 100 "tries." The confidence interval will be a range of values, starting with something below 0.95 (lower limit) and bounded above by some value greater than 0.95. In the example of $n = 100$, with 95 "good" results, the 95% confidence interval for the probability of observing a "good" result is about (0.8998, 0.9834). Clearly, although the point estimate for this probability is 0.95, the lower limit is 0.8998, which is less than 0.95. Similarly, the upper limit is 0.9834, which is greater than 0.95. Thus, if one desires the lower limit of a confidence interval to be above some value, then the point estimate must be greater than that desired lower bound. In this case, if we desired the lower confidence limit to be greater than 0.95, then the smallest point estimate we would need with $n = 100$ is 0.99. In other words, to have the lower confidence limit for a 95% confidence interval to be greater than 0.95 with $n = 100$, we would have to observe at least 99 "good" results. The larger the sample size, the smaller the range of the confidence interval. So, if instead of $n = 100$, we had $n = 1000$, then we would need to observe at least 963 (96.30%) "good" results in order to have the lower confidence limit to be greater than 0.95. The 95% confidence interval would be 0.9505–0.9734. Clearly, 0.963 falls within this interval.

Throughout the above discussion, we referred to confidence intervals have a finite lower and upper limit. It is possible to have confidence intervals where one of the limits is either $+\infty$ or $-\infty$. Such intervals are referred to as "one sided." In the case of percentages (or proportions), the "infinite" limit would be truncated at either 0 or at 1. So, for example, suppose we observed $n = 100$ "tries," which resulted in 95 "good" values. The lower one-sided 95% confidence interval for the probability of obtaining another "good" result would be 0.9099, 1.000. In other words, the lower 95% confidence limit is 0.9099, or 90.99%. We would usually refer to the result 90.99% as simply the one-sided 95% lower confidence limit.

3.4 Some Confidence Limit Formulas

3.4.1 Proportions

If X is a binomially distributed random variable with parameters n and p, then there are several possible approximation formulas for a confidence limit on the probability, p. One of the more popular formulas is by Clopper and Pearson (1934):

$$P_{\mathrm{L}} = \frac{X * F_{\alpha/2}^{-1}(2X, 2(n-X+1))}{n-X+1+X * F_{\alpha/2}^{-1}(2X, 2(n-X+1))}$$

and:

$$P_{\mathrm{U}} = \frac{X * F_{1-\alpha/2}^{-1}(2X, 2(n-X+1))}{n-X+1+X * F_{1-\alpha/2}^{-1}(2X, 2(n-X+1))}$$

P_{L} = lower limit, P_{U} = upper limit, X = number of events or results of interest (e.g., test is positive or test is negative), $F_q^{-1}(\text{num}, \text{denom}) = 100q$th percentile of an F distribution with numerator degrees of freedom num and denominator degrees of freedom denom.

Another approximation was developed by Wilson (1927).

The (1–2α)100% confidence interval (L.L., U.L.) is:

$$\text{L.L.} = \frac{\widehat{p} + \frac{z_\alpha^2}{2n} + z_\alpha \sqrt{\frac{\widehat{p}(1-\widehat{p})}{n} + \frac{z_\alpha^2}{4n^2}}}{1 + \frac{z_\alpha^2}{n}}$$

$$\text{U.L.} = \frac{\widehat{p} + \frac{z_{1-\alpha}^2}{2n} + z_{1-\alpha} \sqrt{\frac{\widehat{p}(1-\widehat{p})}{n} + \frac{z_{1-\alpha}^2}{4n^2}}}{1 + \frac{z_{1-\alpha}^2}{n}}$$

$$\widehat{p} = \frac{X}{n}$$

3.4.2 Means

The 100(1 − α)% confidence interval for the mean is:

$$\overline{X} \pm t_{1-\alpha/2}(n-1) \frac{S}{\sqrt{n}}$$

$t_{1-\alpha/2}(n-1)=$ the 100(1 – α/2)th percentile of a Student's t distribution with $n - 1$ degrees of freedom, $\overline{X}=$ sample mean, $S =$ sample standard deviation.

3.4.3 *Standard Deviations*

Often only an upper confidence limit is computed for standard deviations and other measures of variability, inasmuch as lower variability tends to be more desirable, so an upper confidence limit gives a sort of "worst case." The 100(1 – α)% limit formula is:

$$\widehat{\sigma}_{1-\alpha}=S\sqrt{\frac{n-1}{\chi_{\alpha}^{2}}}$$

where χ_{α}^{2} is the 100α percentile of a Chi-Squared distribution with $n - 1$ degrees of freedom.

3.5 Summary

Confidence intervals provide a range of feasible values for population parameters. The level of "feasibility," or confidence, is stated as a probability. This probability is chosen a priori, to reflect the experimenter's tolerance of risk and uncertainty. Confidence level can be anything from 0% to 100%. However, in order to have 100% confidence, a sample size of $n = 0$ must be drawn, and the interval will be the entire range of possible values for the parameter (e.g., $\pm\infty$, or maybe $(0, +\infty)$, or (0%, 100%)). Generally, a sample size of 0 does not provide very helpful information. As a result, most are willing to obtain some sample data and sacrifice the ability to say "100% confident." The other point is that as you increase your desired level of confidence, you also increase the width of the confidence interval. Thus, a more or less happy medium has traditionally been a confidence level of 95%.

References

Armitage, P. (1971). *Statistical methods in medical research* (4th ed.). Blackwell Scientific Publications.

Clopper, C. J., & Pearson, E. S. (1934). The use of confidence or fiducial limits illustrated in the case of the binomial. *Biometrika, 26*, 404–413.

Merriam-Webster. (2005). *The Merriam-Webster dictionary* (New ed.). Merriam-Webster.

Wilson, E. B. (1927). Probable inference, the law of success, and statistical inference. *Journal of the American Statistical Association*, 22(158), 209.

Chapter 4
Power and Hypothesis Testing

Abstract More details about hypothesis formulation and testing are introduced. Power curves are discussed, and bootstrap methods are described.

4.1 Inductive Statistical Reasoning

What is the probability that a "fair" coin will land with the "heads" side up in a "fair" toss? Is it 50%? Yes. Do you need to flip the coin 1,000,000 times and count up the number of flips that resulted in a "heads" to know this? No. Why? Well, for a fair coin with a fair flip, there are exactly two possible outcomes, or results, after a fair flip: "tails" or "heads." The probability of any outcome is 1 divided by the total number of possible outcomes. Hence, the probability of a "heads" on a fair toss is ½. An experiment is a process defined in terms of the particular sorts of outcomes that may be observed after the execution of the process. An event is a collection of outcomes that satisfy some criterion. The probability of an event is the number of outcomes that satisfy the definition of the event divided by the number of possible events that could occur for the given experiment. In the coin-flipping example, the experiment is flipping the coin and observing whether the coin lands with heads or tails facing up. In other words, the defining event is whether the coin lands with heads up, or not.

Suppose we symbolized a "heads" outcome with the letter H and a "tails" outcome with T. Suppose we did a slightly different "experiment," where we flipped the coin five times, and we wanted to know the probability of the event that exactly one flip out of the five would result in H. How many different ways are there of getting such a result? They are:

HTTTT
THTTT
TTHTT
TTTHT
TTTTH

S. A. Pardo, *Statistical Methods and Analyses for Medical Devices*,
https://doi.org/10.1007/978-3-031-26139-8_4

So, there are five possible ways, or outcomes, for getting exactly one H out of five flips. How many possible results of five flips are there? There are a lot. In fact, there are:

$$\sum_{k=0}^{5}\binom{5}{k} = \sum_{k=0}^{5}\frac{5!}{k!(5-k)!} = 2^5 = 32$$

This is the number of possible results of five flips. So, the probability of getting exactly one H out of five flips is:

$$\frac{5}{32} = 15.625\%$$

Do we need to perform 1,000,000 times the experiment of flipping the coin five times, and counting up the number of the five flip experiment that resulted in exactly one H? No, we surely do not. We know the total number of possible results. We know how many ways the event of interest can occur, so calculating the probability that the event of interest would occur is relatively straightforward.

What is the probability that a new Blood Glucose Monitoring System (BGMS) will yield a result that would be within $\pm 10\%$ of a corresponding laboratory instrument result (LAB) for the same sample of blood (or at least the same person from which the blood was drawn)? How many different ways can the event (LAB $= X$, BGMS $= Y$) occur? Oh, that would be pretty close to infinity. In fact, the number of ways that the two numbers, X and Y, be chosen so that:

$$-10\% \leq \frac{Y-X}{X} * 100\% \leq +10\%$$

is also pretty much infinite (we are not saying these numbers are actually infinite, since both meters and laboratory instruments have finite numerical precision. So, for example, the numbers 120.4321 and 120.3987 would both be represented as 120 on the meter screen).

Can we ever actually know the probability that the BGMS would have an error within $\pm 10\%$? No, we cannot. Rather, we must use data to induce, rather than deduce, the truth. Hence, we can only do one of two things, both of which require inductive reasoning: estimation and Inference.

4.2 Estimation

We did not need empirical evidence for the coin flipping example, but in the case of the BGMS, we do. Consider the experiment where we get $n = 100$ pairs of glucose results, say from one person, where the pairs are LAB $= X$, BGMS $= Y$. For each pair, we calculate the percent relative difference:

$$\text{RD} = \frac{Y - X}{X} * 100\%$$

Then we count up the number of times RD is within ±10%. Suppose K represents the number of RD values within ±10%. Then the proportion of BMGS values within ±10% of corresponding LAB values is:

$$p = \frac{K}{100}$$

The value of p is a point estimate for the probability that the BMGS would yield a result within ±10% of a corresponding LAB result.

So, p is a point estimate, but it is not the actual probability. That probability remains unknown. We can, however, get some idea of where the truth lies. We can compute a feasible range for the truth. Such a feasible range is called a confidence interval. While the technical definition of confidence interval is, well, technical, we will treat it as a feasible range for the truth. How feasible? That is based on the level of confidence we choose. Confidence is a value between 0% and 100%. The width of the confidence interval tells you something about how useful the information you have might be. The width of the confidence interval gets smaller as the sample size gets larger. There are only two special cases: 0% and 100%. We are 0% confident that the true probability is exactly equal to any one number, no matter how large the sample size. This presumes that the probability can be ANY number between 0% and 100%. Conversely, we are 100% confident that the true probability is *somewhere* between 0% and 100%, even with a sample size of $n = 0$. Generally, we fix the confidence level to 95%, a long-standing traditional compromise between confidence and interval width. The confidence interval is only computed *after* data are gathered; it does not make sense to say we have 95% confidence of anything before the data are observed (we will not discuss the notions of Bayesian inference, which actually dispenses entirely with the concept of confidence, but I digress).

OK, so suppose that out of our $n = 100$ pairs of glucose measurements, 97 out of 100, or 97%, have *RD* within ±10%. Then 97% is our "point" estimate for the probability. The 95% confidence interval is approximately 91.48–99.38%. Any percentage between these two limits is a feasible value for the truth. You might then say, "Which is it? 99.38% is good, but 91.48%, not so much." OK, increasing the confidence level won't help. It would only make the range wider. The only reasonable alternative, if this interval is too wide to make a reasonable decision, is to increase the sample size. Instead of $n = 100$, suppose we had obtained $n = 300$ pairs and had observed 291 pairs with *RD* within ±10%. Then the 95% confidence interval for the probability that *RD* would be within ±10% is 94.38–98.62%. The range has narrowed, but the confidence level has not changed. Suppose we were hoping that the lower limit of the confidence interval would be at least 95%. Well, if we had a sample size of $n = 500$, and had observed 485 pairs with *RD* within ±10%, then the 95% confidence interval would have been 95.01–98.31%. Again, the interval width is narrower, with the limits getting closer and closer to the point estimate as the sample size increases. The confidence level, however, remains unchanged, at 95%. The confidence level quantifies the risk of being right about the interval actually containing the truth.

4.3 Inference and Hypothesis Tests

Suppose we do the experiment again, and we actually use $n = 100$ pairs of measurements again. Do you expect that we would get exactly 97 pairs with RD within ±10%? Probably not. We might get 96. We might get 98. There is no telling. So, what do we do? Suppose we knew that we actually *hoped* the probability of RD falling within ±10% was at least 95%. Furthermore, suppose all we could afford is $n = 100$ pairs of tests (we are still talking about the same, unfortunate individual we keep sticking and bleeding; we will address that issue later).

If we knew that we would be using $n = 100$ values of RD, and that we hoped the probability RD would be within ±10% is at least 95%, can we determine the smallest number we expect of measurement pairs yielding an RD within ±10%? Would it be 95? Define an event to be getting exactly 95 "good" results out of $n = 100$ pairs of glucose measurements. We can actually calculate the probability of getting 95 "good" results out of $n = 100$, *if* the actual probability of getting an RD within ±10% was exactly 95%. That probability is given by the formula:

$$\Pr\{\text{exactly 95 good results}|n = 100, p = 0.95\} = \binom{100}{95}(0.95)^{95}(0.05)^{5} \approx 0.1800$$

In other words, there is only about an 18% chance of getting exactly 95 good results out of $n = 100$. But, you say, we are really interested in the event that *at least* 95 good results out of 100. In that case, the probability is:

$$\Pr\{\text{at least } 95|n = 100, p = 0.95\} = \sum_{k=95}^{n}\binom{n}{k}(p)^{k}(1-p)^{n-k} \approx 0.43598$$

So, if the actual probability of RD falling within ±10% is 95%, then there is approximately a 43.6% chance of obtaining at least 95 good results out of $n = 100$.

It seems less than reasonable to expect at least 95 good results out of $n = 100$ if we are hoping that the probability of RD falling within ±10% could be as low as 95% and still be acceptable. So, with $n = 100$, there are two questions we might ask:

1. If $p = 95\%$ is in fact acceptable, then what would be the lowest number of good results we would expect to have out of $n = 100$, with, say, about a 95% chance (instead of 43.6%)?
2. If we really want to have at least 95 good results out of $n = 100$, how high must the probability that RD fall within ±10% be in order to have about a 95% chance of getting those 95 good results?

The answer to #1 is 92. There is about a 93.69% chance of getting at least 92 good results out of $n = 100$, if the probability of a "good" results (RD within ±10%) is 95%. Why wouldn't we choose, say, 91, as our acceptance criterion? The probability of getting at least 91 good results out of 100 is about 97.18%. The process by which we set expectations for results given beliefs or hopes about the true state of nature makes an assessment of the risk, or probability of meeting those expectations, and then making a decision after the experiment is completed and the results are assessed is called "inference."

The minimum number of good results we require is called the "critical number" or "critical value." It is in fact our acceptance number, and we symbolize it as X_c, that is, the whole idea is to make a test concerning the probability of obtaining a good result. If out of n tries, we observe at least the critical number, X_c, of good results, then we *pass* the test, and we can believe that the probability of obtaining a good result is sufficiently high. So we need to be a little conservative. After all, if we set our acceptance number $X_c = 2$ good results out of $n = 100$, and the probability of a good result was 95%, then we would have virtually a 100% chance of passing. The problem is that we would pass with virtually a 100% chance the criterion of getting at least two good results out of $n = 100$ *even if* the actual probability of obtaining a good result was only 19%. So, we find the largest number of good results such that the probability of obtaining that number out of n tries is no greater than 95%.

The answer to #2 is that the actual probability of getting a good result would need to be at least 97.35% in order to have approximately a 95% chance (actually 94.96%) of obtaining at least 95 good results out of $n = 100$. This 95% chance is referred to as power. Unlike confidence intervals, the critical number to provide a given power is determined prior to actually obtaining any data. Power is actually a continuum of risk; the further the truth is from our a priori hope, or the hypothetical value for the truth, the lower the chance of *pass*ing the test. Figure 4.1 shows the power "curve" for the test with $n = 100$ and a critical number of $X_c = 95$ "good" results.

In both #1 and #2, the probability of obtaining a good result is strictly hypothetical. Either we specify the hypothetical value first and then determine the critical value for a given sample size, or we specify a critical value for a particular sample size first and then determine a hypothetical value for the probability of obtaining a good result. In both cases, we seek to have the probability of *pass*ing the test to be as close to 95% as possible without exceeding 95%. This 95% is referred to as power. Of course, even if we *pass* the test, we still do not actually know the probability of obtaining a good result; there is some risk that we would *pass* even if the probability of obtaining a good result is something lower than what we want. Inductive reasoning is *always* accompanied by the risk of making the wrong decision.

Setting a hypothetical probability for obtaining a good result, and determining a critical number of good results out of n tries, is called a *hypothesis test*. It is the fundamental tool of inference. The reason we call it a hypothesis test is that we describe the question in terms of two "hypotheses" that are logical negations of each other, for example:

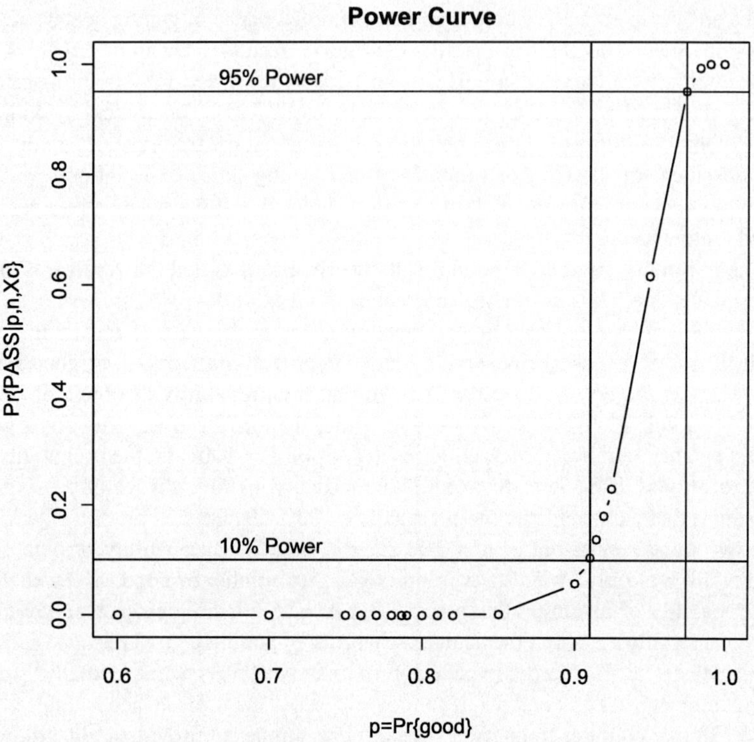

Fig. 4.1 A power curve, $n = 100$, $X_c = 95$

$$H_0 : \Pr\{\text{good result}\} < 0.95$$
$$H_1 : \Pr\{\text{good result}\} \geq 0.95$$

We use the data to decide whether we should believe H_0, called the null hypothesis, or H_1, called the alternate hypothesis. In other words, if the number of good results is at least our critical value, then we *reject* H_0, in favor of believing the alternate, H_1. This is just a formalism used to describe in symbols what we already know: if the number of good results is at least the critical number, we *pass* the test.

4.4 More Hypothesis Tests

Hypotheses about virtually any parameter of any probability distribution can be formulated. Perhaps the most familiar are those concerning means and standard deviations. Using the assumption that data are sampled from a normal population, the distribution of the sample variance is based on:

$$\frac{(n-1)S^2}{\sigma^2} \sim \chi^2_{n-1}$$

that is, under the null hypothesis:

$$H_0 : V[X] = \sigma_0^2$$

the statistic:

$$\widehat{\chi}^2 = \frac{(n-1)S^2}{\sigma_0^2}$$

has a Chi-squared distribution with $n-1$ degrees of freedom. Thus, if $\widehat{\chi}^2 > \chi^2_{1-\alpha}$, we would reject H_0 in favor of the alternative:

$$H_1 : V[X] > \sigma_0^2$$

It is possibly more common to have a one-sided hypothesis test for standard deviations (or variances), inasmuch as usually we are only concerned about variability being too large.

There are cases where the distribution of the sample statistic is either unknown or intractable. An example might be the shape or scale parameters of a gamma distribution. The mean of a gamma variable, Y, is the product of the shape (ν) and scale (λ) parameters:

$$E[Y] = \nu\,\lambda$$

The variance is:

$$V[Y] = \nu\,\lambda^2$$

One way to estimate these parameters is via the method of moments, or mom (NIST, 2022):

Set:

$$\bar{y} = \widehat{\nu}\widehat{\lambda}$$

and:

$$S^2 = \widehat{\nu}\widehat{\lambda}^2$$

Solving for the two parameter estimates:

```
setwd("<your path here>")

#
set.seed(26)
gamma.sample <- c()
actual.sample <- c()
nu.boot <- c()
lamda.boot <- c()
N <- 10000 #number of transitions to get to steady state
size <- 50
nu <- 2 #shape parameter
lamda <- 5 # scale parameter
actual.sample <- rgamma(n=size,shape=5,scale=2)
mean.actual <- mean(actual.sample)
sd.actual <- sd(actual.sample)
mean.sq.actual <- mean.actual**2
var.actual <- sd.actual**2
nu.mom <- mean.sq.actual / var.actual
lamda.mom <- var.actual / mean.actual
for (i in 1:N) {
 gamma.sample <- rgamma(n=size,shape=nu,scale=lamda)
 gamma.mean <- mean(gamma.sample)
 gamma.sd <- sd(gamma.sample)
 var.boot <- gamma.sd**2
 mean.sq <- gamma.mean**2
 nu.boot[i] <- mean.sq / var.boot
 lamda.boot[i] <- var.boot / gamma.mean
 }
F.nu <- ecdf(nu.boot)
F.lamda <- ecdf(lamda.boot)
p.low.nu <- F.nu(nu.mom)
p.low.lamda <- F.lamda(lamda.mom)
p.low.nu
p.low.lamda
#
```

Fig. 4.2 Script for simulating gamma variables and method of moments estimators

```
hist.nu <- hist(nu.boot,plot=FALSE)
hist.nu$density <- 100*hist.nu$counts / sum(hist.nu$counts)
plot(hist.nu,freq=FALSE,col="light gray",main="Histogram of Bootstrap Estimates:
Shape Parameter",xlab="Shape",ylab="Percentage",xaxt="n")
axis(side=1,at=seq(from=0,to=15,by=1))
abline(v=nu.mom,lwd=3)
dev.new()
hist.lamda <- hist(lamda.boot,plot=FALSE)
hist.lamda$density <- 100*hist.lamda$counts / sum(hist.lamda$counts)
plot(hist.lamda,freq=FALSE,col="light gray",main="Histogram of Bootstrap Estimates:
Scale Parameter",xlab="Scale",ylab="Percentage",xaxt="n")
axis(side=1,at=seq(from=1,to=15,by=1))
abline(v=lamda.mom,lwd=3)
#####
```

Fig. 4.2 (continued)

$$\widehat{\nu} = \frac{\overline{y}^2}{S^2}$$

$$\widehat{\lambda} = \frac{S^2}{\overline{y}}$$

The method of moments works by setting sample moments equal to theoretical population moments and solving for the parameters.

It would be very difficult to determine the distribution of either of these sample estimates. One method for testing an hypothesis about either of these parameters is via a simulation approach, that is, suppose the null hypothesis is $\nu = 2$ and $\lambda = 5$. For N repetitions, simulate n observations from a gamma distribution with parameters $\nu = 2$ and $\lambda = 5$. For each simulated sample, compute the sample statistics:

$$\widehat{\nu} = \frac{\overline{y}^2}{S^2}$$

and:

$$\widehat{\lambda} = \frac{S^2}{\overline{y}}$$

Make histograms of the N values of these simulated sample statistics. Then find the percentiles for the parameter values $\nu = 2$ and $\lambda = 5$. If those values fall beyond the 100α or $100(1 - \alpha)$ percentiles of one of the bootstrap statistic distributions, then the null hypothesis is rejected for the relevant parameter.

As an example, suppose that despite hypotheses, the actual shape parameter is 5 and the actual scale is 2. Figure 4.2 shows a script for generating a sample of

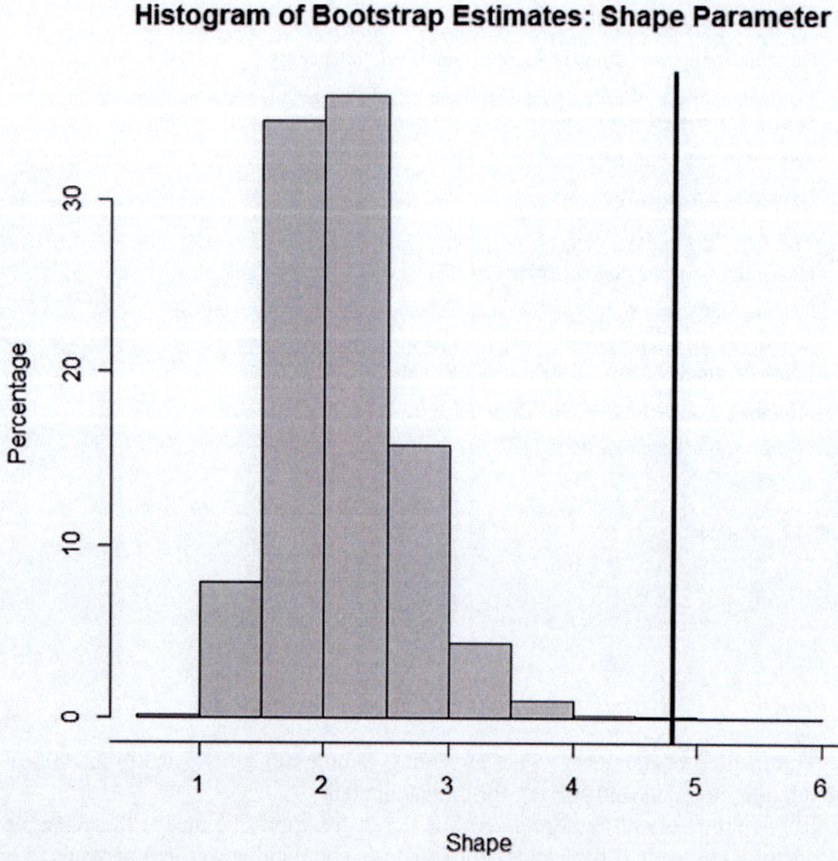

Fig. 4.3 Histogram of shape parameter estimates from simulation

"actual" data ($n = 50$) and the simulated "null" bootstrap distributions. Then the script computes the percentile of the method of moments (mom) estimators for shape and scale. Figures 4.3 and 4.4 show histograms of the estimate results of the simulated data for ν and λ, respectively. The bold vertical lines on the histograms show where the mom estimates for the parameters lie relative to the simulated null distribution estimates.

The mom estimates from the "actual" sample data were $\widehat{\nu} = 4.796$ and $\widehat{\lambda} = 2.083$. The estimated percentiles for the mom estimates were 99.98% for shape and 00.00% for scale. If the null hypothesis had been true, we would have expected the sample mom estimates to fall somewhere near the center of the simulated distributions. However both estimates of shape and scale were in the extreme tails. Thus, we would reject the null hypotheses that $\nu = 2$ and $\lambda = 5$.

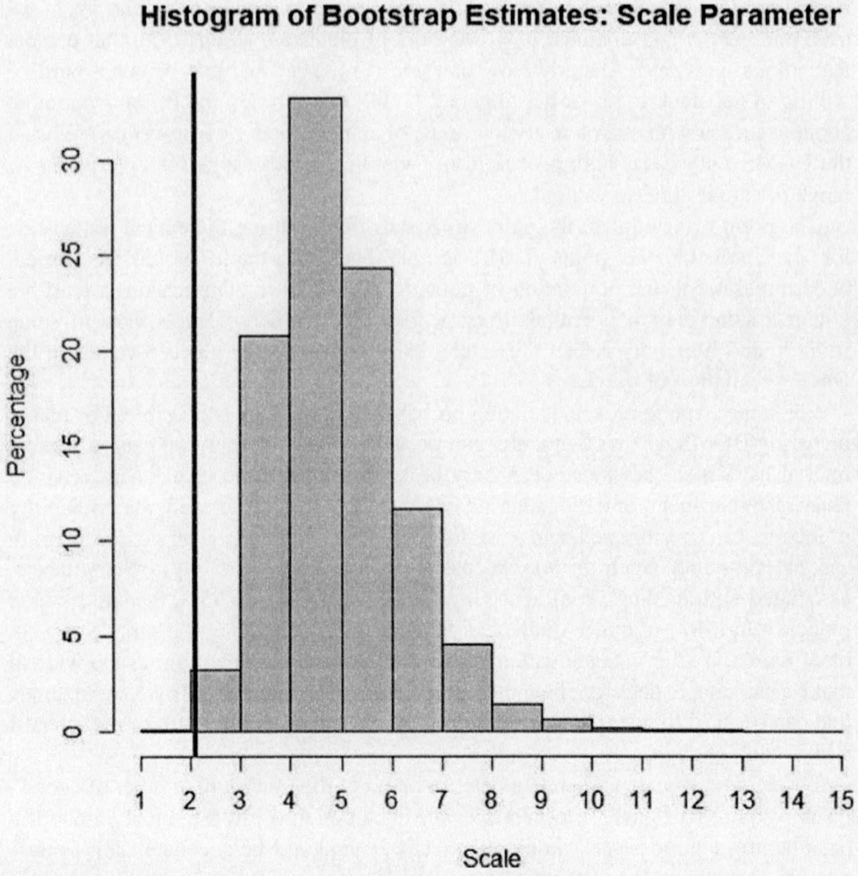

Fig. 4.4 Histogram of scale parameter estimates from simulation

4.5 Summary

In answering both questions #1 and #2, we have specified or calculated a hypothetical probability of obtaining a good result, where that hypothetical probability represents what we hope to be true. It can be thought of as a system specification, or requirement. Then, we decided to do an experiment by obtaining $n = 100$ pairs of glucose measurements. We figured out the minimum number of "good" results we would need out of $n = 100$ tries, in order to have about a 95% chance of obtaining that minimum number *if* our hypothetical probability were the actual truth. In case #1, our hypothetical probability was 95%, and our minimum, or "critical" number of good, result was 92 out of $n = 100$. In case #2, our hypothetical probability was 97.35%, and our critical number was 95 out of $n = 100$.

You may recall that all our talk so far has been about an experiment where we used one person and obtained $n = 100$ pairs of glucose readings from that person. Ridiculous, you say? Absurd? How can you (a) expect that one person would be willing to get stuck in his or her fingers $2 \times 100 = 200$ times and (b) know enough about accuracy with such a restricted range of actual blood glucose values? What if the BGMS only has a high probability of yielding good results for a very narrow range of actual glucose values?

The point (a) is ethical; the point (b) is statistical. Putting the ethical issue aside for the moment, the point in (b) is that such a sample would be biased, underrepresenting the population of potential users. So for this reason at least we obtain a sample of n potential users, whose blood glucose levels vary in some fashion, and hopefully reflect the actual distribution of such glucose values in the whole population of users.

For some experiments, calculating probabilities of events is possible. For many, including BGMS accuracy, it is just not possible to know. Thus, we can only make inferences, which themselves can only be known with some level of uncertainty. Once an experiment, or data-gathering exercise, has been performed, the probability of interest can be estimated, and a feasible range, or confidence interval, for the truth can be computed. Such an interval has a fixed level of certainty, or confidence, associated with it. While confidence level is generally kept to 95%, the width of the range, from lower to upper limits, can be adjusted by changing the sample size. In other words, if after data are gathered, the width of the feasible range is too wide to make a reasonable decision, then the sample size was inadequate. There are methods that can be used to guess at the minimum sample size required to make the interval width small enough.

Before any data are gathered, a determination of the minimum number of "good" results required to "pass" some test depends in part on how low the actual probability of obtaining a good result can hypothetically be and still be adequate. This critical number also depends on desired power, or risk (typically ~95%), and the sample size. In some cases, the process of constructing the hypothesis test is reverse-engineered, beginning with a sample size and a critical value and then determining what hypothetical value for the probability of obtaining a good result would yield approximately a 95% chance of *pass*ing by observing the critical number of good results out of n tries.

Of course, even if after the experiment is completed, and observing the number of good results, the actual probability of obtaining a good result remains unknown. However, a confidence interval can be constructed, to at least provide a feasible range for the truth.

Reference

National Institute of Standards and Technology (NIST). (2022). *Engineering Statistics Handbook*. NIST.

Chapter 5
Least Squares: Regression and ANOVA

Abstract Some fundamental concepts relating to linear models are introduced. Least squares estimation is discussed as a method for computing estimates of linear model parameters. ANOVA is described in a least squares context as it relates to linear regression.

5.1 Motivation for Least Squares: Simple Linear Regression

Suppose there are two variables, call them Y and X, both of which can be measured or observed on the same experimental unit. For example, suppose there are two methods for measuring the same analyte on a sample of blood. Let X represent the standard method and Y a new, experimental method. The experimenters would like to know whether the new method yields the "same" results as the standard, or comparator, method. Furthermore, suppose that the evaluation method has some variability associated with it, so that even if the same sample, having the same concentration of analyte, was measured several times, the same value would not be obtained each time. Furthermore, suppose that the comparator method might also have some variability, but much less than that of the valuation method. Thus, whatever value of X was obtained for a given sample, we can presume that this value would be obtained no matter how many times the sample was measured/ assayed. Now suppose that samples were collected from n individuals and each sample was assayed with both the comparator and evaluation methods. It is expected that the samples will vary in their analyte concentrations. Figure 5.1 shows a scatter diagram of the pairs of assay results (x_i, y_i) for $i = 1, n$.

One might imagine that somehow Y is related to X, at least on the average, in a linear fashion. In fact, one might postulate that Y is a linear function of X with some "error" associated. Mathematically this relationship is expressed as:

$$Y = \beta_0 + \beta_1 X + \varepsilon$$

S. A. Pardo, *Statistical Methods and Analyses for Medical Devices*,
https://doi.org/10.1007/978-3-031-26139-8_5

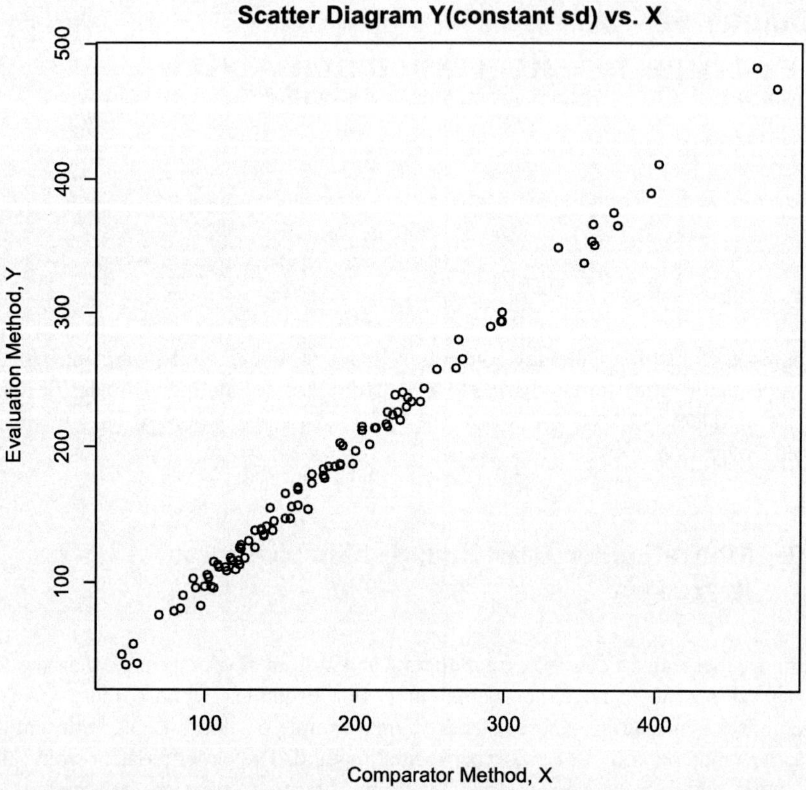

Fig. 5.1 Assay results for n individuals (x_i, y_i)

So β_0 is the intercept, β_1 is the slope of the line, and ε represents the "error" between the line and the actual value of Y.

The error term can be expressed for each sample result as:

$$\varepsilon_i = Y_i - \beta_0 - \beta_1 X_i$$

Since there must be exactly one intercept and one slope, the "errors" must vary in some fashion. So we might want to find out what values of β_0 and β_1 would minimize the average value of the errors, ε_i. Averaging the errors directly would yield a number close to 0, regardless of the values of β_0 and β_1, so either average the absolute values of the errors or square the errors. It turns out that it is easier to minimize the average of the squared errors:

$$\frac{1}{n}\sum_{i=1}^{n}\varepsilon_i^2 = \frac{1}{n}\sum_{i=1}^{n}(Y_i - \beta_0 - \beta_1 X_i)^2$$

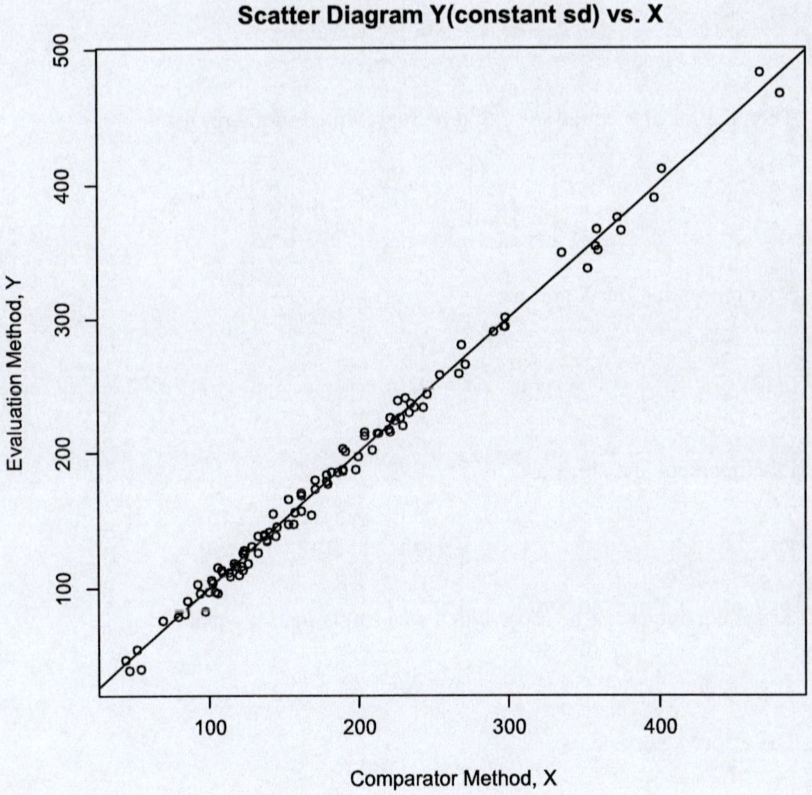

Fig. 5.2 Scatter diagram with least squares regression line

The solutions to minimizing the average squared errors are referred to as the least squares solutions. The process of finding these "optimal" values for intercept and slope is also called regression, or linear regression, or least squares regression. The R function *lm()* will perform this regression. Figure 5.2 shows the data together with the least squares regression line.

The least squares problem can be expressed in terms of matrices and vectors. Suppose the Y values are arrayed in a column vector:

$$\mathbf{y} = \begin{bmatrix} y_1 \\ \vdots \\ y_n \end{bmatrix}$$

The X values can also be represented in the same way. However, consider that each y_i is postulated to be a linear function of the corresponding x_i, plus noise, i.e.:

$$y_i = \beta_0 + \beta_1 x_i + \varepsilon_i$$

If the errors, ε_i, are represented by the vector:

$$\boldsymbol{\varepsilon} = \begin{bmatrix} \varepsilon_1 \\ \vdots \\ \varepsilon_n \end{bmatrix}$$

Then the set of n equations could be represented in matrix form:

$$\begin{bmatrix} y_1 \\ \vdots \\ y_n \end{bmatrix} = \begin{bmatrix} 1 & x_1 \\ \vdots & \vdots \\ 1 & x_n \end{bmatrix} \begin{bmatrix} \beta_0 \\ \beta_1 \end{bmatrix} + \begin{bmatrix} \varepsilon_1 \\ \vdots \\ \varepsilon_n \end{bmatrix}$$

So representing the X part as:

$$\boldsymbol{X} = \begin{bmatrix} 1 & x_1 \\ \vdots & \vdots \\ 1 & x_n \end{bmatrix}$$

and the intercept and slope as:

$$\boldsymbol{\beta} = \begin{bmatrix} \beta_0 \\ \beta_1 \end{bmatrix}$$

the set of equations can be represented as a single matrix equation:

$$\boldsymbol{y} = \boldsymbol{X\beta} + \boldsymbol{\varepsilon}$$

The error vector is thus:

$$\boldsymbol{\varepsilon} = \boldsymbol{y} - \boldsymbol{X\beta}$$

The squared errors are obtained by multiplying the vectors by their transposes:

$$\boldsymbol{\varepsilon}^T \boldsymbol{\varepsilon} = (\boldsymbol{y} - \boldsymbol{X\beta})^T (\boldsymbol{y} - \boldsymbol{X\beta})$$

Simplifying, taking derivatives with respect to the parameters β_0 and β_1, setting them equal to 0, and solving give the least squares solutions for the intercept and slope:

$$\widehat{\boldsymbol{\beta}} = [\boldsymbol{X}^T \boldsymbol{X}]^{-1} \boldsymbol{X}^T \boldsymbol{y}$$

The "hat" over the parameter vector indicates that this solution is a sample estimate, based on the data (x_i, y_i). For those readers who may not be very familiar with matrix and vector notation, see the Appendix A.

5.2 Multiple Regression and ANOVA

There are cases where the "Y" variable (called the "response") is associated simultaneously with multiple "X" variables (called the "regressors"). Suppose there are only two regressors, $X1$ and $X2$. Then perhaps the "model," or equation that describes the relationship between Y, $X1$, and $X2$, is:

$$Y = \beta_0 + \beta_1 X1 + \beta_2 X2 + \varepsilon$$

The only thing that changes to the least squares solution is the dimensionality of the $\boldsymbol{X}$ matrix and the $\boldsymbol{\beta}$ vector, that is, the X matrix has an additional column and the $\boldsymbol{\beta}$ vector another row:

$$\boldsymbol{X} = \begin{bmatrix} 1 & x1_1 & x2_1 \\ \vdots & \vdots & \vdots \\ 1 & x1_n & x2_n \end{bmatrix}$$

and:

$$\boldsymbol{\beta} = \begin{bmatrix} \beta_0 \\ \beta_1 \\ \beta_2 \end{bmatrix}$$

Analysis of variance (ANOVA) has least squares at its heart. ANOVA is a method for determining whether there is any effect of multiple conditions/groups on the response simultaneously. The ANOVA effects are really coefficients in a linear model, just as in the case of multiple regression. The trick is that the regressors in ANOVA are coded variables used to indicate the groups or conditions. For example, suppose an experiment is performed to determine whether there is a difference in the tensile strength of cables made by three different suppliers, called A, B, and C. The supplier could be coded using two X variables, which is a fashion indicated by Table 5.1.

Thus, for each cable tested, the tensile strength, y_i, would be associated with two X variables whose values would indicate the supplier. One might ask why not simply code the supplier with a single X variable having three numerical values, say 1, 2, and 3?

$$Y = \beta_0 + \beta_1 X1 + \beta_2 X2 + \varepsilon$$

Table 5.1 Coding X variables for three suppliers

Supplier	*X1*	*X2*
A	1	0
B	0	1
C	0	0

With the two-regressor coding for suppliers A, B, and C, the predicted values of Y would be:

$$Y_{\mathrm{A}} = \beta_0 + \beta_1$$
$$Y_{\mathrm{B}} = \beta_0 + \beta_2$$
$$Y_{\mathrm{C}} = \beta_0$$

It turns out that the least squares estimates of these β would yield predicted values for Y of:

$$\widehat{Y}_{\mathrm{A}} = \overline{Y}_{\mathrm{A}}$$
$$\widehat{Y}_{\mathrm{B}} = \overline{Y}_{\mathrm{B}}$$
$$\widehat{Y}_{\mathrm{C}} = \overline{Y}_{\mathrm{C}}$$

In other words, the two-regressor coding resulted in the predicted values simply being the average for each supplier. This would not be the case had we used a single regressor with three values. The two-regressor coding also allows us to make the following comparisons:

$$\mathrm{A–C} = \beta_0 + \beta_1 – \beta_0 = \beta_1$$
$$\mathrm{B–C} = \beta_0 + \beta_2 – \beta_0 = \beta_2$$
$$\mathrm{A–B} = \beta_0 + \beta_1 – \beta_0 – \beta_2 = \beta_1 – \beta_2$$

Had we used a single regressor, these comparisons would not have been so easy to perform. Furthermore, the comparisons are all made under that same level of noise, i.e., a single standard deviation, rather than, say, three separate *t*-tests. This provides a sort of even playing field, so that each comparison has the same level of noise to overcome in order to determine that the difference is significant. Also, it helps to avoid the problem of inflating the chances of type I error, rejecting the null hypothesis of no difference when in fact there was no difference in the groups or conditions. Keep in mind that in the case of ANOVA, we are more interested in making comparisons than predictions.

The point of this discussion is not to instruct the reader in the many aspects of regression, ANOVA, and least squares. Rather, it is to provide some insight into the underpinnings of those techniques and hopefully to demystify them.

There are many texts on regression and ANOVA. Two excellent examples are *Applied Regression Analysis*, by Draper and Smith (1981), and the *Design and Analysis of Experiments* (2001), by Montgomery. Another superb book is *Statistics for Experimenters*, by Box, Hunter, and Hunter (1978). Although these texts make use of matrix/vector notation, these are not overly mathematical (no theorem-proof format).

5.3 Multiple Comparisons

Carlo Emilio Bonferroni (1892–1960) was a mathematician who generalized a theorem of George Boole (1815–1864). He did not intend for his work to be applied to the problems of making many hypothesis tests simultaneously (i.e., using the same set of data). In particular, Bonferroni's inequality (his theorem) is most often applied to making multiple confidence intervals for pairwise differences between treatments, or factor levels. So, the idea is the following.

Suppose that an experiment consists of comparing k groups or k treatments or k levels of some factor that is hypothesized to alter the average value of some response variable. Then if A_{ij} represents the confidence interval for the difference in average response between group/treatment/level i and j, then:

$$\Pr\left\{\bigcup_{i \neq j} A_{ij}\right\} \leq \sum_{i \neq j} \Pr\{A_{ij}\}$$

In other words, the whole confidence is less than or equal to the sum of the parts. So, if you wanted the "whole" confidence to be at least 95%, each of the parts would need to have more than a 95% confidence level. For example, suppose that in an experiment there are four treatment groups, I, II, II, and IV. Suppose further that confidence intervals for the difference in average response for all pairs of treatment groups are to be computed. That would give:

$$\binom{4}{2} = 6$$

pairs of confidence intervals. If we wanted overall to have 95% confidence for the six intervals, then each interval would require 1–0.05/6 ≈ 0.9917, or about 99.17% confidence. Of course, as confidence level increases, the width (imprecision) of the interval increases. Just as confidence levels for individual intervals can be "adjusted" to insure, based on the inequality of Bonferroni, that the overall confidence level is at least 95% (say), so too significance levels of individual tests can be adjusted. For example, if the experimenter desired an overall significance level of 0.05, then for the case with the four treatments, each comparison must have a significance level of 0.05/6 ≈ 0.0083. That means any p-value for a given pair of treatments must be no more than 0.0083 in order to be considered statistically significant.

There are potentially several issues associated with applying Professor Bonferroni's inequality to making sure the whole confidence level is 95%. Three of them are:

1. When the number of confidence intervals required gets large, the precision of each individual interval gets, well, imprecise.
2. Especially when applying the adjustment to significance levels, there is a tacit assumption that the null hypothesis of no difference applies to all individual comparisons.

3. Bonferroni's theorem is an inequality; thus, applying it to multiple confidence intervals is at best conservative.

There are in fact many such "adjustments" to confidence or significance levels, most of which are based on Bonferroni's original inequality. For example, the Šidák adjustment for significance level with k comparisons is:

$$\alpha_{ij} = 1 - (1 - \alpha)^{1/k}$$

So, again using the four-treatment example with six possible comparisons, and assuming a desired overall significance level of $\alpha = 0.05$, then each of the comparisons would have a significance level of:

$$\alpha_{ij} = 1 - (1 - 0.05)^{1/6} \approx 0.0085 > 0.0083$$

The Šidák adjustment appears to be a little less stringent than the pure Bonferroni adjustment.

One of the controversies concerning multiple comparison adjustments is exactly when one should apply an adjustment. There are two basic schools of thought:

1. Whenever a post-hoc test is performed, after a general test has concluded that a factor or treatment grouping is significant. Then a pairwise test can be performed to decide on which of the groups or factor levels differ from each other. Furthermore, adjustments should be made only within levels of a given factor or grouping.
2. Whenever more than one test is performed, or more than one confidence interval is constructed, using data from one experiment.

We have already indicated that one problem with multiple comparison adjustments is that when there are many simultaneous comparisons to be made, the significance level gets too close to 0, or the confidence level too close to 1, which is very meaningful. Benjamini and Hochberg (1995), and later Benjamini and Yekutieli (2001), developed a method of multiple comparison adjustment that is well-adapted to making very large numbers of comparisons or confidence intervals. The method will be referred to here as the Benjamini-Hochberg-Yekutieli false detection rate (BHY-FDR) adjustment. The BHY-FDR may be applied to a set of p-values in the following manner to determine the significance level.

Suppose you have N p-values:

1. Pick a proportion, q (a decimal number between 0 and 1, not including 0 or 1).
2. Sort the p-values from smallest to largest: $p(1), p(2), \ldots, p(N)$.
3. Calculate the values: $\frac{i*q}{N}$ for each $i = 1$ to N.
4. Find the largest value of i, call it k, where $p(i) < \frac{i*q}{N}$.
5. The value $p(k)$ is the critical value. If $p \leq p(k)$, then reject the null hypothesis.

As an example, consider:

Table 5.2 List of sorted *P*-values

i	$p(i)$	$i * q/N = i * 0.2/10$
1	0.03	0.02
2	0.04	0.04
3	**0.04**	**0.06**
4	0.10	0.08
5	0.11	0.10
6	0.13	0.12
7	0.14	0.14
8	0.17	0.16
9	0.20	0.18
10	0.25	0.20

The bold values are the ones that satisfy the FDR criterion

$N = 10$ p-values, as shown in Table 5.2 (sorted from smallest to largest)
Let $q = 0.20$, and compute the $\frac{i*q}{N}$:

$$1^*0.20/10, 2^*0.2/10, 3^*0.2/10, 4^*0.2/10, 5^*0.2/10, 6^*0.20/10, 7^*0.2/10, \\ 8^*0.2/10, 9^*0.2/10, \\ 10^*0.2/10$$

The largest p-value that is smaller than the corresponding value of i^*q/N is $p(3) = 0.04$. Therefore, the critical value for p-values is 0.04. Reject any null hypothesis in the group with a p-value less than or equal to 0.04. To apply this for constructing confidence intervals, each interval should have confidence level $1 - \frac{k*q}{N} = 1 - \frac{3*0.20}{10} = 0.94$.

The R function *p.adjust*() will take as input a vector of p-values and return a vector of adjusted p-values. Among the adjustment methods are Bonferroni and the BHY-FDR methods. To adjust the p-values by the Bonferroni method, simply multiply each p-value by N, the number of p-values. The adjusted p-value is the lesser of this product or 1. To adjust via the Benjamini-Hochberg-Yekutieli method, compute:

$$q(i) = \frac{p(i) * N}{i}, i = 1, N$$

The adjusted p-value, $p_{\text{adj}}(i)$ is the smallest value of $q(j), j = i, N$.

Note that the choice of q is arbitrary and subjective. Then again, choosing 0.05 as the critical value is also arbitrary and subjective.

As an example, consider again the list of p-values in Table 5.2. Figure 5.3 shows some R code for using the *p.adjust*() function, together with a computation of the BHY-FDR adjusted p-values. Table 5.3 shows the output.

Hsu (1996) is an excellent reference for most types of multiple comparisons and adjustments.

5.4 Collinearity and Correlation of Regressors

There are some techniques that can be useful when there are either more regressors than observations or if some of the regressors are correlated with each other (also referred to as collinearity).

```
setwd("your path here")
#
#
df1 <- read.csv("20180529 p-values.csv")
#
# index
# p.k
#
q.k <- c()
p.sort <- c()
qstar <- c()
q.list <- c()
attach(df1)
N <- length(p.k)
p.Bon <- p.adjust(p=p.k,method="bonferroni")
p.BH <- p.adjust(p=p.k,method="BH")
#
# This code produces a modified version of the BH method:
#
for (m in 1:length(p.k)){
 q.k[m] <- p.k[m]*length(p.k)/index[m]
 }
#
for (i in 1:N){
 limit <- N - i + 1
 for (j in 1:limit){
```

Fig. 5.3 *P*-value adjustments in R

```
    jind <- i + j - 1
    q.list[j] <- q.k[jind]
    }
  qstar[i] <- min(q.list) #this is the vector of BHY adjusted p-values
  q.list <- c()
  }
p.sort <- sort(p.k)
thresh20 <- index*.2/N

df2 <- cbind(index,p.k,p.Bon,p.BH,qstar,p.sort,thresh20)
write.csv(df2,"20180530 Adjusted P-Values.csv")
#####
```

Fig. 5.3 (continued)

Table 5.3 Adjusted *P*-values

Index	p.k	p.Bon	p.BH	qstar	p.sort	thresh20
1	0.03	0.30	0.13	0.13	0.03	0.02
2	0.04	0.40	0.13	0.13	0.04	0.04
3	0.04	0.40	0.13	0.13	0.04	0.06
4	0.10	1.00	0.20	0.20	0.10	0.08
5	0.11	1.00	0.20	0.20	0.11	0.10
6	0.13	1.00	0.20	0.20	0.13	0.12
7	0.14	1.00	0.20	0.20	0.14	0.14
8	0.17	1.00	0.21	0.21	0.17	0.16
9	0.20	1.00	0.22	0.22	0.20	0.18
10	0.25	1.00	0.25	0.25	0.25	0.20

p.k original *p*-values, *p.Bon* Bonferroni adjusted, *p.BH* BHY adjusted from p.adjust(), *qstar* BHY adjusted from direct computation, *p.sort* *p.k* sorted from smallest to largest, *thresh20* index * 0.20/*N*

5.4.1 Partial Least Squares

Suppose we want to get a predictive equation using the usual linear model:

$$\mathbf{y} = \mathbf{X}\boldsymbol{\beta} + \boldsymbol{\varepsilon}$$

Of course, if the fundamental requirement for OLS that regressors are uncorrelated, then we are out of luck. However, there are several potential solutions; one is called partial least squares (PLS). In PLS, we attempt to find a "weighting" matrix, **W**, and then compute a new "regressor" matrix:

$$\mathbf{T} = \mathbf{X}\mathbf{W}$$

The **W** matrix is computed (we will discuss how) so that the model:

$$\mathbf{y} = \mathbf{T}\boldsymbol{q} + \boldsymbol{\varepsilon}$$

can be fit using ordinary least squares. The parameter estimates, $\boldsymbol{q}$, can be related to the original regression parameters:

$$\boldsymbol{\beta} = \mathbf{W}\boldsymbol{q}$$

There are multiple ways of getting the **W** matrix. One way is called NIPALS (nonlinear iterative partial least squares). It works like this:

Let $\mathbf{M}_1 = \mathbf{X}^\mathbf{T}\mathbf{X}$, $\mathbf{A}_1 = \mathbf{X}^\mathbf{T}\boldsymbol{y}$, $\mathbf{C}_1 = \mathbf{I}_k$
For $j = 1, k$

Compute $q_j = \mathbf{A}_j^\mathbf{T}\mathbf{A}_{j.}$
Compute $\boldsymbol{w}_j = \mathbf{C}_j\mathbf{A}_j q_j$; normalize $\boldsymbol{w}_j$ by dividing by its length, and replace it with the normalized version.
Store normalized $\boldsymbol{wj}$ as the jth column of a $k \times k$ matrix, **W**.
Compute $c_j = \boldsymbol{w}_j^\mathbf{T}\mathbf{M}_j\boldsymbol{w}_j$, $\boldsymbol{p}_j = \mathbf{M}_j\boldsymbol{w}_j/c_j$, $\boldsymbol{q}_j = \mathbf{A}_j^\mathbf{T}\boldsymbol{w}_j/c_j$
Store qj as the jth row in the $k \times 1$ vector $\boldsymbol{q}$
Store $\mathbf{A}_j = \mathbf{A}_j - c_j\boldsymbol{p}_j\boldsymbol{q}_j^\mathbf{T}$
Store $\mathbf{M}_j = \mathbf{M}_j - c_j\boldsymbol{p}_j\boldsymbol{q}_j^\mathbf{T}$
Store $\mathbf{C}_j = \mathbf{C}_j - \boldsymbol{w}_j\boldsymbol{p}_j^\mathbf{T}$
Loop (next j)

In the end, you will have a weight matrix, **W**, that is $k \times k$, where k is the number of parameters (regressors). The PLS coefficients (parameter estimates) for the regressors would then be:

$$\widehat{\boldsymbol{\beta}} = \mathbf{W}\boldsymbol{q}$$

There are other algorithms for computing the PLS estimates, but there is no reason to believe one is "better" than any other in terms of their statistical properties. Note that PLS can be easily extended to include multivariate regression, where there are multiple response variables all being predicted by the same regressors. Another method useful for regression in the face of collinearity is called principal component regression (PCR). It is very similar in concept to PLS.

PLS and PCR are particularly useful when there are more regressors than observations ($k > n$).

5.4.2 Ridge Regression

Ridge regression was invented to compensate for regressors that are correlated with each other. As mentioned earlier, another term for correlation between regressors is "collinearity." One of the consequences of having more regressors than observations is that correlation between regressors is induced. Thus, ridge regression and its related methods can be used to fit models when $k > n$.

In OLS, the idea is to find values of the parameters that minimize the squared "error":

$$\epsilon^2 = (\boldsymbol{y} - \mathbf{X}\boldsymbol{\beta})^{\mathrm{T}}(\boldsymbol{y} - \mathbf{X}\boldsymbol{\beta})$$

As we have noted earlier, this problem cannot be solved when $k > n$. In ridge regression, they add another condition to the minimization problem. Instead of finding the parameter values that minimize the squared error, ridge regression attempts to minimize:

$$\eta^2 = (\boldsymbol{y} - \mathbf{X}\boldsymbol{\beta})^{\mathrm{T}}(\boldsymbol{y} - \mathbf{X}\boldsymbol{\beta})$$

subject to:

$$\|\boldsymbol{\beta}\|_2 \leq t$$

where t is some predetermined maximum allowable value for the "norm" $\|\boldsymbol{\beta}\|_2$.

The symbol $\|\boldsymbol{x}\|_2$ is the Euclidean length of vector x or, as it is also called, the L_2 norm of vector $\boldsymbol{x}$. So, if $\boldsymbol{\beta}$ is the vector of parameters (regression coefficients):

$$\boldsymbol{\beta} = \begin{bmatrix} \beta_1 \\ \vdots \\ \beta_k \end{bmatrix}$$

then:

$$\|\boldsymbol{\beta}\|_2 = \left(|\beta_1|^2 + |\beta_2|^2 + \ldots + |\beta_k|^2\right)^{\frac{1}{2}}$$

It turns out that trying to minimize η^2 when some of the regressors are correlated with each other is possible. It often gives estimates that are very close to 0 for some of the parameters.

In general, the L_p norm of $\boldsymbol{\beta}$ is:

$$\|\boldsymbol{\beta}\|_p = (|\beta_1|^p + |\beta_2|^p + \ldots + |\beta_k|^p)^{\frac{1}{p}}$$

5.4.3 Least Absolute Shrinkage and Selection Operator (LASSO)

The name of LASSO makes it sound very special, but it isn't. It is just like ridge regression, with a slight difference:

LASSO attempts to minimize:

$$\eta^2 = (\mathbf{y} - \mathbf{X}\boldsymbol{\beta})^{\mathrm{T}}(\mathbf{y} - \mathbf{X}\boldsymbol{\beta})$$

subject to:

$$\|\boldsymbol{\beta}\|_1 \leq t$$

That is to say, the constraint condition on LASSO is the L_1 norm of the parameter vector, instead of the L_2 norm. There is actually no reason to believe that LASSO is somehow better than ridge regression.

The so-called Bayesian LASSO is just LASSO under the assumption that each of the regression parameters, β_i, has a Laplace prior distribution. In other words, Bayesian LASSO assumes that the parameters have Laplace prior distributions with mean 0 and that they are all independent of each other (which had no relation to the independence or dependence of regressors from each other). The Laplace distribution (also called "double exponential") has the density function:

$$f(\beta) = \frac{1}{2\theta} \exp\left(-\frac{|\beta - \mu|}{\theta}\right)$$

In the case of the Bayesian LASSO, μ is assumed to be 0, and θ is assumed to be the same for all the regression parameters.

If you were to use Bayes' theorem to compute a posterior mean for the regression parameters with the Laplace prior assumption, you would get the same result as LASSO.

Note that in software, you might see as output the value of a parameter called lamda; this "parameter" is a product of the method by which the ridge regression or LASSO minimization proceeds. It is called Lagrange's method of undetermined multipliers. The value of lamda is the undetermined multiplier and has no meaning at all. It is just used to convert the minimization problem subject to a constraint into an easier-to-solve problem. It is of no modeling consequence.

Package *pls* has a partial least squares function, *plsr*(). Package *MASS* has a ridge regression function, *lm.ridge*(), and package *glmmLasso* has a LASSO function, *glmmLasso*(). These functions are illustrated in the R code contained in Fig. 5.4. Figure 5.5 shows a plot of predicted values from four models plotted against the observed response.

```
setwd("<your path here>")
df1 <- read.csv("20191125 Example 6.1 Collinearity.csv")
#library(car) #needed to use function Anova() - not the same as anova()
# Variables
#
# X.01
# X.02
# X.03
# X.04
# X.05
# X.06
# X.07
# X.08
# X.09
# X.10
# Response
attach(df1)
regressors <- as.matrix(cbind(const=1,X.01,X.02,X.03,X.04,X.05,X.06,X.07,X.08,X.09,X.10))


library(pls) #has function plsr
library(MASS) #has function lm.ridge
library(glmmLasso) #has function glmmLasso
model.pls <- plsr(Response ~.,data=df1,ncomp=8,method="simpls")
model.pcr <- pcr(Response ~
X.01+X.02+X.03+X.04+X.05+X.06+X.07+X.08+X.09+X.10,data=df1,ncomp=8,method="svdp
c")
model.ridge <- lm.ridge(Response ~
X.01+X.02+X.03+X.04+X.05+X.06+X.07+X.08+X.09+X.10,data=df1)
model.Lasso <-
glmmLasso(fix=Response~X.01+X.02+X.03+X.04+X.05+X.06+X.07+X.08+X.09+X.10,rnd=N
ULL,lambda=1,data=df1)
plot(model.pls,ncomp=7,main="PLS,PCR,RIDGE,LASSO: Actual vs. Predicted",xlab="Actual
Response",pch=1,ylim=c(50,750))
pred.pcr <- predict(model.pcr,ncomp=7)
pred.ridge <- regressors%*%coef(model.ridge)
pred.Lasso <- predict(model.Lasso)
```

Fig. 5.4 R code for regression methods useful for collinear data

```
points(x=Response,y=pred.pcr,pch=2)
points(x=Response,y=pred.ridge,pch=3)
points(x=Response,y=pred.Lasso,pch=4)
legend(x=200,y=600,legend=c("PLS","PCR","RIDGE","LASSO"),pch=c(1,2,3,4))
```

Fig. 5.4 (continued)

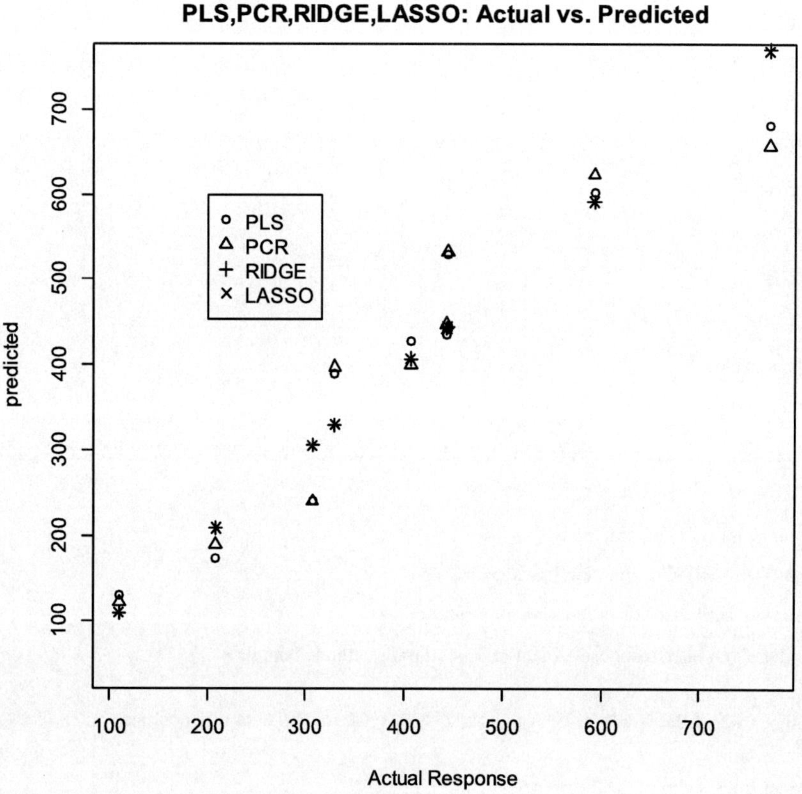

Fig. 5.5 Predicted vs. actual response: four methods

5.5 Summary

Least squares is a method for picking a mathematical expression or geometric construct that "best" reflects the process that generated the observed data. There are tow common applications, regression, and analysis of variance (ANOVA). Regression is useful for finding that mathematical expression for predicting future values of random variables, given some non-random variables that can be observed. For example, suppose one wanted to predict the energy of activation required given certain proportions of the reactants. Regression may be a useful tool for developing an equation. ANOVA uses least squares not to find a predictive equation but to decide whether or not some discrete variables have any effect on some random variable. So, for example, one might want to know if using one of several materials to make wire cable has any effect on tensile strength of wires with some particular diameter.

ANOVA would by itself only provide an answer to the question, "Does the factor have any effect on the response?" Often the following question is, "Which value or level of the factor is best?" or "Do all the factor levels have the same effect?" To answer these questions, post-hoc multiple comparison tests can be used. One particular test is Tukey's honestly significant difference (HSD) test. In general, to avoid the inflation of type 1 error when multiple comparisons are made, several methods have been devised to mitigate the problem of finding "significant" differences that are not truly significant. The methods of Bonferroni, Sidak, and Benjamini-Hochberg-Yekutieli are described.

References

Benjamini, Y., & Hochberg, Y. (1995). Controlling the false discovery rate: A powerful approach to multiple testing. *Journal of the Royal Statistical Society, Series B, 57*, 289–300.

Benjamini, Y., & Yekutieli, D. (2001). The control of the false discovery rate in multiple testing under dependency. *Annals of Statistics, 29*, 1169–1188.

Box, G. E. P., Hunter, W. G., & Hunter, J. S. (1978). *Statistics for experimenters*. John Wiley and Sons.

Draper, N. R., & Smith, H. (1981). *Applied regression analysis* (2nd ed.). John Wiley and Sons.

Hsu, J. (1996). *Multiple comparisons: Theory and methods*. Chapman and Hall/CRC.

Montgomery, D. C. (2001). *Design and analysis of experiments*. John Wiley and Sons, Inc..

Chapter 6
Product Design: Factorial Experiments

Abstract The notion of factorial experimentation is discussed as a method for making design decisions. In particular, two-level fractional factorials and second-order designs such as central composite and Box-Behnken designs are presented. Taguchi methods are also discussed.

6.1 General

This chapter will be largely concerned with making plans, called "factorial experiments," for deciding which regressors are actually related to the response and how to construct a mathematical equation, or model, that approximates those relationships. The objective is to either simply decide whether a regressor actually has a relationship to the response or to decide on "optimal" values for the regressors to provide the best response.

The regressors will be referred to as "factors," a term steeped in the history of factorial experimentation.

The idea is that every product has features, and the values or characteristics of these features are chosen by design. The questions are:

1. Which features (factors) actually affect important performance variables (responses)?
2. What are the best values to choose for the important factors, in order to optimize the critical response variables?

The process of answering those two questions involves performing a sequence of experiments. Each experiment will consist of a set of runs, or conditions under which the response variables will be measured/observed.

S. A. Pardo, *Statistical Methods and Analyses for Medical Devices*,
https://doi.org/10.1007/978-3-031-26139-8_6

6.2 Eliminating the "Unimportant" Factors: First-Order Model

The initial screening of factors, to potentially eliminate some of them from further experimentation, involves a linear (first order) approximation to the response function. For each input variable, the question to be answered is whether changing the variable increases or decreases the average value of the response or if changing the input variable induces no change on the average response. Admittedly, this assessment is crude. However, if the experimenter is faced with more input variables that can be simultaneously optimized in an economically feasible fashion, it may be the best assessment that can be made.

In order to fit a kth order polynomial of the form:

$$y = \beta_0 + \beta_1 x + \beta_2 x^2 + \ldots + \beta_k x^k$$

At least $k + 1$ different values of x would be required, that is, we would require $k + 1$ pairs of observations (x_i, y_i), $i = 1, k + 1$.

Suppose, for example, there were five input variables that were of potential interest. In order to fit a fourth-order polynomial in each input variable simultaneously, the experimenter would require $5^5 = 3125$ points. Imagine if the experimenter wanted to duplicate results at each point. The experimenter would require $N = 6250$ measurements of the response variable. This is probably infeasible. If, however, the experimenter desires to decide whether any of the input variables could be disregarded initially, he should consider using a linear approximation to the response function.

In order to fit a linear function to data, two points are minimally required for each input variable. The potential values to be included in the experiment for a given factor are called "levels." In the case of five input variables, the number points would be $2^5 = 32$, corresponding to all possible combinations of 2 levels per factor, with $r = 5$ factors. This is considerably fewer than 3125.

A first-order model may have the form:

$$y = \beta_0 + \sum_{i=1}^{r} \beta_i x_i + \sum_{i \neq j} \sum \gamma_{ij} x_i x_j + \varepsilon$$

The symbol ε (error) represents the random "noise" that is associated with response values observed or measured under "identical" conditions, at least in terms of the input variables. The experimenter could include higher-order cross-product terms in the first-order model, e.g.:

$$x_i x_j x_k (3 - \text{way})$$
$$x_i x_j x_k x_l (4 - \text{way})$$
$$\vdots$$
$$x_1 x_2 x_3 \ldots x_r (r - \text{way})$$

Generally, for a first-order model used primarily to decide which inputs are truly important, anything beyond the two-way cross-products is overkill.

In order to obtain statistical estimates of the coefficients β_i and γ_{ij}, some minimum number of points must be selected for data gathering. Suppose there were k input variables, and the experimenter wanted to estimate coefficients for a "complete" linear model, i.e., including all possible cross-product terms. Then, as mentioned earlier, the minimum number of points would be 2^k. So, if $k = 5$, $2^5 = 32$ points would be required. With duplicate response values at each point, a total of $n = 64$ measurements would be required.

6.3 Assessing the Effect of Each Factor

In the case of a single input variable, or factor, to assess the linear effect of the factor, x, on a response variable, y, we require observations of y at two different values, or levels, of x. The levels chosen should cover the range of interest, that is, the lower of the two levels should be the lowest value for which x could feasibly be set, and the highest level should be the highest value at which x could feasibly be set. Denote the low and high levels of x by x^- (for low) and x^+ (for high). Let y^- represent the value of response y^+ observed at x^- and y^+ the value of the response at x^+. The effect of x is estimated by:

$$E = y^+ - y^-$$

If n observations or measurements of the response variable are obtained at each condition, then the average value of the response under each condition is used to compute the effect:

$$E = \overline{y}^+ - \overline{y}^-$$

Notice that the slope of the line segment joining the points, (x^-, y^-) and (x^+, y^+) is given by:

$$b = \frac{\overline{y}^+ - \overline{y}^-}{x^+ - x^-} = \frac{E}{x^+ - x^-}$$

Now suppose we computed the midpoint between x^- and x^+:

$$m = \frac{x^- + x^+}{2}$$

Then transform, or code, the x variable into a new variable whose range is (−1, +1) with midpoint 0:

$$H = \frac{x - m}{\frac{1}{2}(x^+ - x^-)}$$

When $x = x^-$, $H = -1$; when $x = x^+$, $H = +1$; when $x = m$, $H = 0$. Now the new slope, call it b', with respect to the coded input variable, is given by:

$$b' = \frac{(\bar{y}^+ - \bar{y}^-)}{2} = \frac{E}{2}$$

This coding transformation is called Helmert coding, named after Professor Dr. Friedrich Robert Helmert. Unless otherwise specified, we will assume that all input variables are expressed as Helmert-coded. Thus −1 will represent the lowest level, and +1 the highest level, of each factor in the experimental design.

We would like to gather data under various conditions determined by multiple input factors in such a way as to allow the independent estimation of all coefficients in the first-order model. Each unique condition will be called a "run." The collection of conditions used to gather the data for fitting the model (i.e., estimating the coefficients) is called an experiment. The collection of conditions is also called an experimental design, or simply "design."

6.4 Assessing the Cross-Product or Interaction Effects

Suppose there is more than one input factor that may have some effect on the response. It is possible that the first-order approximation model should include at least the two-way cross-product terms. Cross-product terms are also called *interaction effects* in that the levels of all factors included in the cross-product "interact" in their joint effect on the response. Just as in the main effect of each factor individually, an interaction effect can be computed. The two-way interaction effect for any two factors would be the difference between the effect of the first factor when the second is at its "low" (−1) level and the effect of the first factor when the second is at its "high" (+1) level. The choice of which factor is "first" and which is "second" is arbitrary. If there were exactly two input factors, x_1 and x_2, coded to (−1, +1), the experimental runs could be symbolized as in Table 6.1.

If the symbols $y(-, -)$, $y(-, +)$, $y(+, -)$, and $y(+, +)$ represent the average responses in each run, then the effect of x_1 at each level of x_2 can be computed as:

Table 6.1 A two-factor, two-level experiment

Run	x_1	x_2	y
1	−1	−1	y(−, −)
2	−1	+1	y(−, +)
3	+1	−1	y(+, −)
4	+1	+1	y(+, +)

$$E_{x2=-1} = y(+,-) - y(-,-)$$

$$E_{x2=+1} = y(+,+) - y(-,+)$$

The interaction effect is computed as:

$$E_{x1,x2} = \frac{E_{x2=+1} - E_{x2=-1}}{2}$$

The coefficient corresponding to this cross-product term is:

$$b_{12} = \frac{E_{x1,x2}}{2}$$

It turns out that the coefficients computed in the fashion described are actually the least squares estimates.

6.5 A Three-Factor Example

Consider the following example. Hypodermic needles have several features that can affect the force required to penetrate skin, beyond that of the diameter of the needle (usually measured in gauge units). In particular, a needle may have a "bevel," or angled cut, both on its face and on the sides of the edge of the point. In addition, the point length can be varied. In this experiment, the face bevel is to be varied at 10° and 15°, as are the side bevels (assume that both side bevels will have the same angle). The point length will be set at 0.75 mm or 1.25 mm. The response variable will be the force (P.Force), measured in Newtons, at the time in which a needle penetrates simulated skin on a test fixture. The levels and their coded values for all the runs are shown in Table 6.2.

There are $2^3 = 8$ runs, representing all possible combinations of levels for the three factors. The model to be fit is:

$$y = \beta_0 + \beta_1 H_1 + \beta_2 H_2 + \beta_3 H_3 + \gamma_{12} H_1 H_2 + \gamma_{23} H_2 H_3 + \gamma_{13} H_1 H_3 + \delta_{123} H_1 H_2 H_3 + \varepsilon$$

The three-way cross-product term is probably not necessary, but since all possible combinations of factor levels are included in the design, it is possible to estimate the

Table 6.2 Experimental design: hypodermic needle

	Natural units			Helmert coded		
Run	Face bevel	Side bevel	Point length	H.F	H.S	H.L
1	10	10	0.75	−1	−1	−1
2	10	10	1.25	−1	−1	1
3	10	15	0.75	−1	1	−1
4	10	15	1.25	−1	1	1
5	15	10	0.75	1	−1	−1
6	15	10	1.25	1	−1	1
7	15	15	0.75	1	1	−1
8	15	15	1.25	1	1	1

Table 6.3 Hypodermic needle experiment

Run	Rep	Face.Bevel	Side.Bevel	Point.Length	H.F	H.S	H.L	P.Force
1	1	10	10	0.75	−1	−1	−1	4.87
2	1	10	10	1.25	−1	−1	1	6.22
3	1	10	15	0.75	−1	1	−1	5.04
4	1	10	15	1.25	−1	1	1	5.83
5	1	15	10	0.75	1	−1	−1	2.86
6	1	15	10	1.25	1	−1	1	5.29
7	1	15	15	0.75	1	1	−1	2.56
8	1	15	15	1.25	1	1	1	5.10
1	2	10	10	0.75	−1	−1	−1	5.22
2	2	10	10	1.25	−1	−1	1	5.45
3	2	10	15	0.75	−1	1	−1	5.10
4	2	10	15	1.25	−1	1	1	5.50
5	2	15	10	0.75	1	−1	−1	3.26
6	2	15	10	1.25	1	−1	1	4.88
7	2	15	15	0.75	1	1	−1	3.21
8	2	15	15	1.25	1	1	1	4.78

coefficient for this term. Suppose further that $n = 2$ duplicate values are obtained for each run. The data are given in Table 6.3.

There are two very important properties of this experimental design that are made apparent by the use of Helmert coding. The first is balance; for any input factor, there are equal numbers of observations made when the input factor is set to its low (-1) and high $(+1)$ values. If you sum all the -1's and $+1$'s in any column, the result is zero. The second is orthogonality; if each of the columns in Table 6.3 is thought of as column vectors, the dot product of any two columns is zero. Balance and orthogonality greatly simplify the calculations of least squares.

Once the data are gathered, the analysis begins with a least squares fit, or estimation of the coefficients in the first-order model.

With two-level experiments, where each input variable is Helmert-coded, there is a simple way to compute the effects and coefficients for each term in the first-order model. Suppose h_{ij} represents the jth coded value for the ith input variable, and y_j represents the jth value of the response variable. If there are $m = 2^k$ unique runs in the experiment, with n replicate values for each run, then the estimate of the effect, E_i, and the least squares estimate of the coefficient, β_i, are given by:

$$\widehat{E}_i = \frac{1}{\frac{1}{2}nm} \sum_{j=1}^{nm} h_{ij} y_j$$

$$b_i = \frac{\widehat{E}_i}{2} = \frac{1}{nm} \sum_{j=1}^{nm} h_{ij} y_j$$

The least squares estimates of any two-way interaction coefficients are computed in a similar fashion:

$$b_{ij} = \frac{\widehat{E}_{ij}}{2} = \frac{1}{nm} \sum_{k=1}^{nm} h_{ik} h_{jk} y_{jk}$$

Table 6.4 illustrates the computations of some of the coefficient estimates for the volume yield data.

The only reason these simple formulas for the least squares estimates of the coefficients work is because this is a two-level, balanced, orthogonal design, where each input factor is Helmert-coded.

Figure 6.1 shows the R code for analyzing the penetration force data. Figure 6.2a shows box plots of the data from all eight runs. Figure 6.2b shows boxplots for point length, Fig. 6.2c shows the box plots for face bevel, and Fig. 6.2d shows the box plots for side bevel.

Now that the estimates of the first-order approximation model are obtained, the next stage is to decide which input variables are likely to have a non-zero effect on the response.

Figure 6.3 shows some output that helps indicate which terms are actually important.

In Fig. 6.3, in the Parameter Estimates section, along with the "Estimate" column are three others: Std. error, t value, and Pr(>|t|). The Std. error (SE) column is the standard error of the estimated coefficient, or parameter estimate, given by:

$$\text{SE} = \sqrt{\frac{\text{MS}_{\text{error}}}{nm}} = \sqrt{\frac{\text{MS}_{\text{error}}}{n2^k}}$$

The t value is:

Table 6.4 Penetration force data with coefficient estimates

Run	Rep	Face. Bevel	Side. Bevel	Point. Length	H. F	H.S	H. L	P. Force	H. F•Point. Length	H. S•Point. Length	H. L•Point. Length
1	1	10	10	0.75	−1	−1	−1	4.87	−4.87	−4.87	−4.87
2	1	10	10	1.25	−1	−1	1	6.22	−6.22	−6.22	6.22
3	1	10	15	0.75	−1	1	−1	5.04	−5.04	5.04	−5.04
4	1	10	15	1.25	−1	1	1	5.83	−5.83	5.83	5.83
5	1	15	10	0.75	1	−1	−1	2.86	2.86	−2.86	−2.86
6	1	15	10	1.25	1	−1	1	5.29	5.29	−5.29	5.29
7	1	15	15	0.75	1	1	−1	2.56	2.56	2.56	−2.56
8	1	15	15	1.25	1	1	1	5.10	5.1	5.1	5.1
1	2	10	10	0.75	−1	−1	−1	5.22	−5.22	−5.22	−5.22
2	2	10	10	1.25	−1	−1	1	5.45	−5.45	−5.45	5.45
3	2	10	15	0.75	−1	1	−1	5.10	−5.1	5.1	−5.1
4	2	10	15	1.25	−1	1	1	5.50	−5.5	5.5	5.5
5	2	15	10	0.75	1	−1	−1	3.26	3.26	−3.26	−3.26
6	2	15	10	1.25	1	−1	1	4.88	4.88	−4.88	4.88
7	2	15	15	0.75	1	1	−1	3.21	3.21	3.21	−3.21
8	2	15	15	1.25	1	1	1	4.78	4.78	4.78	4.78
								Sum:	−11.29	−0.93	10.93
								Coeff:	−0.706	−0.058	0.683

$$t = \frac{b}{\mathrm{SE}}$$

where b represents the estimate of the coefficient for the term in question. The formula for SE only has this simple form due to Helmert coding.

Finally, the column labeled "Pr(>|t|)" is the p-value for testing the hypothesis that the coefficient is actually 0. P-values less than some prespecified level (usually 0.05) are considered statistically significant, meaning that the actual coefficient does not appear to be 0. Any terms for which the p-value is above the predetermined threshold are candidates to be excluded from the model.

In examining the fit of this model, we notice several things. One is that all the two factors, H.L and H.F, are significant at the 0.05 level. Thus, it appears the side bevel, represented by H.S, can be ignored. Another is that the interaction H.F-H.L is significant.

Yet another thing is that the adjusted R^2 is relatively high (0.9133), indicating that possibly the first-order polynomial may be a good enough approximation.

It turns out that the least squares-based predicted value of the response (also called the least-squares mean) for any run is actually the arithmetic average response for that run, provided the experiment is balanced. This is another convenient consequence of the two-level experimental design. Many software systems provide least squares means as outputs.

```
setwd("<your path here>")

df1 <- read.csv("20210826 Example 6.1 Hypodermic Needle.csv")
library(car)
#
# Variables:
# Run
# Face.Bevel
# Side.Bevel
# Point.Length
# H.F
# H.S
# H.L
# P.Force
attach(df1)
Fac.H.F <- as.factor(H.F)
Fac.H.S <- as.factor(H.S)
Fac.H.L <- as.factor(H.L)
linear.model.fac <- lm(P.Force ~ Fac.H.F + Fac.H.S + Fac.H.L + Fac.H.F:Fac.H.L +
Fac.H.F:Fac.H.S + Fac.H.S:Fac.H.L + Fac.H.F:Fac.H.S:Fac.H.L)
#Anova(linear.model.fac,type=3)
anova(linear.model.fac)
linear.model.reg <- lm(P.Force ~ H.F + H.S + H.L + H.F:H.L + H.F:H.S + H.S:H.L +
H.F:H.S:H.L)
#Anova(linear.model.reg,type=3)
anova(linear.model.reg)
boxplot(P.Force ~ Face.Bevel + Side.Bevel + Point.Length,main="Box Plots of Data by
Factor Combinations",ylab="P.Force (N)")
dev.new()
boxplot(P.Force ~ Face.Bevel,main="Box Plots of Data by Face Bevels",xlab="Face
Bevel Angle (degrees)",ylab="P.Force (N)")
dev.new()
boxplot(P.Force ~ Side.Bevel,main="Box Plots of Data by Side Bevels",xlab="Side
Bevel Angle (degrees)",ylab="P.Force (N)")
dev.new()
boxplot(P.Force ~ Point.Length,main="Box Plots of Data by Point Lengths",xlab="Point
length (mm)",ylab="P.Force (N)")
#####
```

Fig. 6.1 R Code for hypodermic needle analysis

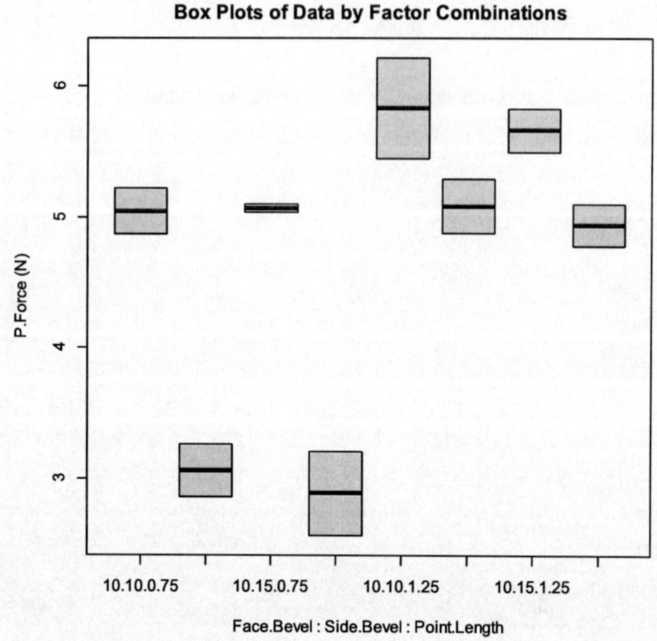

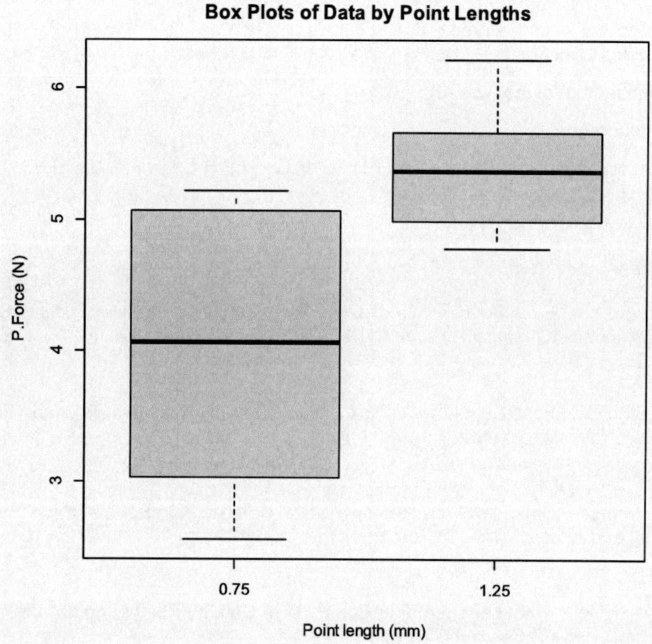

Fig. 6.2 (**a**) Boxplots: all runs. (**b**) Boxplots: point length. (**c**) Boxplots: face bevel. (**d**) Boxplots: side bevel

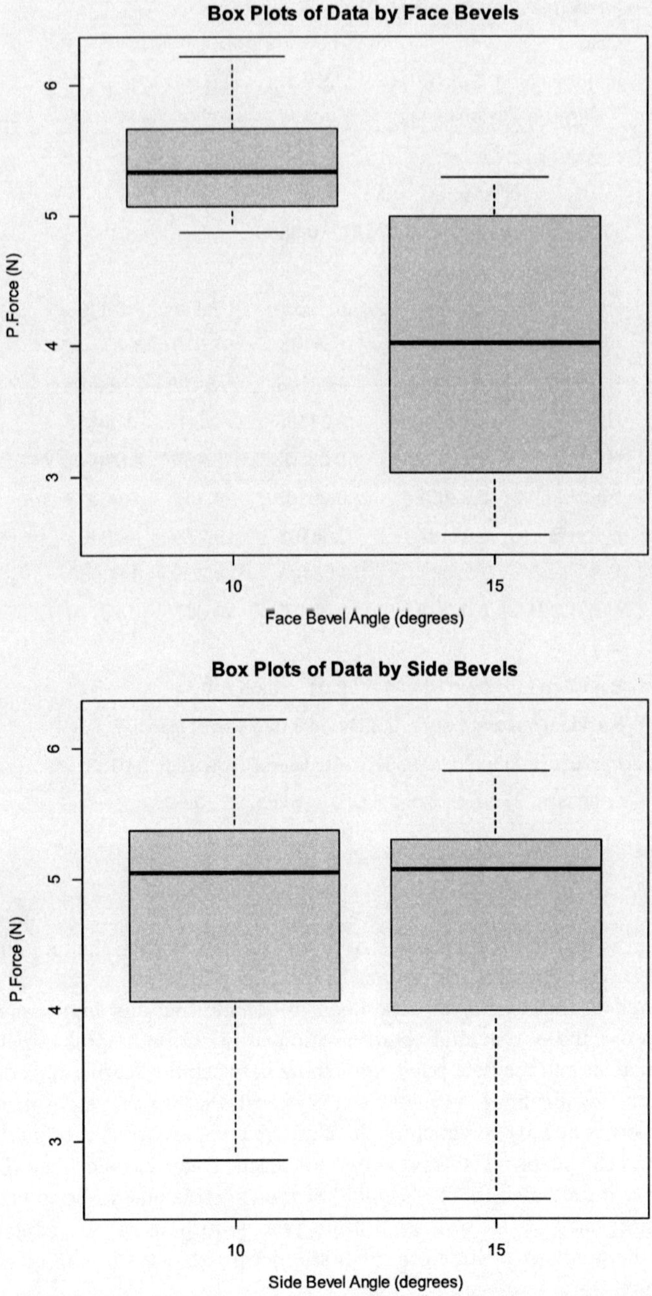

Fig. 6.2 (continued)

```
> summary(linear.model.reg)
Call:
lm(formula = P.Force ~ H.F + H.S + H.L + H.F:H.L + H.F:H.S +
   H.S:H.L + H.F:H.S:H.L)
Residuals:
   Min      1Q  Median      3Q     Max
-0.3850 -0.1812 0.0000 0.1812 0.3850
Coefficients:
             Estimate   Std. Error   t value   Pr(>|t|)
(Intercept)   4.69812    0.08103     57.979    8.70e-12 ***
H.F          -0.70563    0.08103     -8.708    2.36e-05 ***
H.S          -0.05813    0.08103     -0.717    0.49359
H.L           0.68312    0.08103      8.430    2.99e-05 ***
H.F:H.L       0.33687    0.08103      4.157    0.00318 **
H.F:H.S      -0.02187    0.08103     -0.270    0.79403
H.S:H.L      -0.02063    0.08103     -0.255    0.80550
H.F:H.S:H.L   0.02813    0.08103      0.347    0.73748
---
Signif. codes: 0 '***' 0.001 '**' 0.01 '*' 0.05 '.' 0.1 ' ' 1
Residual standard error: 0.3241 on 8 degrees of freedom
Multiple R-squared: 0.9537,  Adjusted R-squared: 0.9133
F-statistic: 23.57 on 7 and 8 DF, p-value: 9.773e-05
```

Fig. 6.3 Coefficient estimates and tests of significance

It is quite possible that the chosen design factor levels are suboptimal. The presumption is that the chosen range of input factors in which the experiment was conducted at least contains the optimal point. Secondly, there is no way using the data gathered in this two-level experiment to decide that any higher-order terms would improve the polynomial approximation. It might be helpful to perform the experiment at an intermediate point, something between the "corners" as defined by the high and low levels of the input factors. Recall that the factors in natural units were Helmert-coded by subtracting the midpoint values from the high and low values of each factor. The midpoints for each factor would then be coded (or mapped) to the value 0. The "center" of the experimental space in coded units would then be at $H_i = 0$, $i = 1, 2, 3$. If the average response at this center point is "close" to the predicted value from the first-order model, then it is more believable that no higher-order terms are necessary.

The center point conditions for the volume yield experiment are face and side bevels, 12.5°, and point length 1.00 mm. The engineer obtained two replicate values for the center point, for an average response of 5.425 N. The predicted value at the center point is the intercept, namely, 4.698 N. The standard error of residuals is 0.3241 N. With 8 degrees of freedom, we might expect the observed value of the response at the center point to be within:

Intercept $\pm$ 0.3241*t(0.975, df = 8) = 4.698 $\pm$ 0.3241*2.0306 $\approx$ (3.951 N, 5.445 N). The observed center point average was 5.425 N, which falls within this range. Now there must be a decision to either spend the resources to perform another experiment and fit a second-order model or use the current model to predict values of the response.

6.6 Fractional Factorial Designs

We have seen the economy of choosing two levels for each regressor. However, suppose that the experimenter has five or even six factors to consider. A full two-level design would then have either $2^5 = 32$ or $2^6 = 64$ distinct runs. There is a way to do only a fraction of these runs and still have a reasonable ability to fit a model. The issue is that the model will not be able to have all possible interaction effects. That is to say, due to the elimination of some possible runs, some interaction effects will become confounded with others.

Confounding is the state where two or more terms will have effects that are estimated in exactly the same way. To illustrate confounding, consider a three-factor experiment, with coded factors $H1$, $H2$, and $H3$. A full two-level experiment is illustrated in Table 6.5.

The interaction columns are computed by multiplying across the columns, as in Table 6.6.

Suppose the experimenter decided to eliminate the shaded runs. Note that the patterns of -1's and 1's for columns $H1$ and $H2$ are identical for the unshaded rows. So are the patterns for $H3$ and the three-way interaction $H1 * H2 * H3$. Thus, if we used only the first two and last two runs described in Table 6.6 to fit a model, then the estimates of the coefficients for factors $H1$ and $H2$ would be identical in the model:

Table 6.5 A three-factor design with coded factors

$H1$	$H2$	$H3$
−1	−1	−1
−1	−1	1
−1	1	−1
−1	1	1
1	−1	−1
1	−1	1
1	1	−1
1	1	1

Table 6.6 The three-factor design with the interaction columns added

H1	H2	H3	H1 * H2	H1 * H3	H2 * H3	H1 * H2 * H3
−1	−1	−1	1	1	1	−1
−1	−1	1	1	−1	−1	1
−1	1	−1	−1	1	−1	1
−1	1	1	−1	−1	1	−1
1	−1	−1	−1	−1	1	
1	−1	1	−1	1	−1	−1
1	1	−1	1	−1	−1	−1
1	1	1	1	1	1	1

$$y = \beta_1 H_1 + \beta_2 H_2 + \beta_3 H_3 + \varepsilon$$

factors *H*1 and *H*2, and *H*3 and *H*1 * *H*2 * *H*3 are said to be confounded. Sometimes confounding is referred to as aliasing, although aliasing generally refers to a slightly different but related concept. Nevertheless, confounding two main effects (*H*1 and *H*2) is problematic. There is however a way to control confounding when executing only a fraction of the total number of possible combinations of factor levels. Designs that incorporate a fraction of the possible factor level combinations are referred to as fractional factorial designs (Box et al., 1978).

To illustrate the process of generating a fractional factorial experiment, consider a $k = 5$ factor, two-level experiment. There are $2^5 = 32$ possible runs. Suppose we could only afford half, or 16, of the possible runs. We would induce some confounding, but suppose we only confounded the main effects with higher-order interactions, and the same for the two-way interactions, so that we could fit a model of the form:

$$y = \beta_0 + \sum_{i=1}^{5} \beta_i H_i + \sum_{i=1}^{k} \sum_{j=i+1}^{k} \gamma_{ij} H_i H_j + \varepsilon$$

That is to say, we would have unique coefficient estimates for all the main effect and two-way interaction terms. The effects (coefficients) would all be confounded, but only with the coefficients of higher-order interactions. To generate such a design, first write down all the runs for a $k - 1 = 4$ factor design. Then create a fifth column by computing the product of the coded levels of the first four factors. Table 6.7 shows the results. Now compute the two-way interactions, and the *H*2 * *H*3 * *H*4 three-way interaction, as illustrated in Table 6.8. The two shaded columns, for the *H*1 * *H*5 and the *H*2 * *H*3 * *H*4 interactions, are identical. These two interaction effects are confounded. The model we fit can only have one of these two terms in it. Generally, the higher order the interaction, the less likely it is to be important.

Table 6.7 A half-fraction of a 2^5 experiment

Run	$H1$	$H2$	$H3$	$H4$	$H5 = H1 * H2 * H3 * H4$
1	−1	−1	−1	−1	1
2	−1	−1	−1	1	−1
3	−1	−1	1	−1	−1
4	−1	−1	1	1	1
5	−1	1	−1	−1	−1
6	−1	1	−1	1	1
7	−1	1	1	−1	1
8	−1	1	1	1	−1
9	1	1	−1	−1	1
10	1	1	−1	1	−1
11	1	1	1	−1	−1
12	1	1	1	1	1
13	1	−1	−1	−1	−1
14	1	−1	−1	1	1
15	1	−1	1	−1	1
16	1	−1	1	1	−1

Table 6.8 Two-way interactions with the $H2 * H3 * H4$ three-way interaction

Run	H1 * H2	H1 * H3	H1 * H4	H1 * H5	H2 * H3	H2 * H4	H2 * H5	H3 * H4	H3 * H5	H4 * H5	H2 * H3 * H4
1	1	1	1	−1	1	1	−1	1	−1	−1	−1
2	1	1	1	1	1	−1	1	−1	1	−1	1
3	1	1	−1	1	−1	1	1	−1	−1	1	1
4	1	1	−1	−1	−1	−1	−1	1	1	1	−1
5	−1	−1	1	1	−1	−1	−1	1	1	1	1
6	−1	−1	1	−1	−1	1	1	−1	−1	1	−1
7	−1	−1	−1	−1	1	−1	1	−1	1	−1	−1
8	−1	−1	−1	1	1	1	−1	1	−1	−1	1
9	1	1	−1	1	−1	−1	1	1	−1	−1	1
10	1	1	−1	−1	−1	1	−1	−1	1	−1	−1
11	1	1	1	−1	1	−1	−1	−1	−1	1	−1
12	1	1	1	1	1	1	1	1	1	1	1
13	−1	−1	−1	−1	1	1	1	1	1	1	−1
14	−1	−1	−1	1	1	−1	−1	−1	−1	1	1
15	−1	−1	1	1	−1	1	−1	−1	1	−1	1
16	−1	−1	1	−1	−1	−1	1	1	−1	−1	−1

Therefore, we would choose to include the $H1 * H5$ two-way interaction in the model.

As an exercise, compute the other three-way interactions to see with which two-way interactions they are confounded.

The important thing to notice is that none of the main effects are confounded with each other or with any two-way interactions, and none of the two-way interactions are confounded with any other two-way interactions.

A half-fraction design is referred to as a 2^{k-1} fractional factorial. A quarter fraction would be 2^{k-2}. Fractional designs are categorized in terms of the level of confounding, referred to as resolution:

Resolution V Main effects are not confounded with each other or with two-way interactions; two-way interactions are confounded with three-way or higher-order interactions, but not with each other.

Resolution IV Main effects are not confounded with each other or with two-way interactions; two-way interactions are confounded with other two-way interactions as well as with higher-order interactions.

Resolution III Main effects are confounded with two-way interactions, but not with other main effects.

In general, ResV designs are useful for building predictive models. ResIII designs are useful for eliminating from further experimentation some factors. ResIV designs can be useful for building predictive models, as long as it is known or at least suspecting from previous experience or knowledge that certain two-way interactions are not meaningful.

Readers who are interested (and I encourage all of you to be interested) should read one of several excellent texts on experimental design, including Box, Hunter, and Hunter or Montgomery (2001).

6.7 Second-Order Models and Designs

While a center point run can give some indication that the response is related to at least one of the regressors in a higher order than linear, it alone cannot provide a means to construct a higher-order model. Usually, regression modeling does not progress beyond a second-order polynomial, that is, often a quadratic model provides a sufficient predictive approximation.

In order to fit such a model to data, the simple two-level designs are not enough, that is, we need some intermediate points, between the low and high levels of the regressors. Clearly including three levels of each regressor would allow such modeling, but that could get expensive rapidly. That is to say, if there are k regressors, and each one can take on three values, then the total number of combinations of regressor levels is 3^k. That gets large very fast. There are several alternatives. One is called the central composite design (CCD) and another is the Box-Behnken design (BBD) (Box & Behnken, 1960).

Table 6.9 A three-factor CCD

Run	$H1$	$H2$	$H3$
1	−1	−1	−1
2	−1	−1	1
3	−1	1	−1
4	−1	1	1
5	1	−1	−1
6	1	−1	1
7	1	1	−1
8	1	1	1
9	0	0	−1
10	0	0	1
11	0	−1	0
12	0	1	0
13	−1	0	0
14	1	0	0
15	**0**	**0**	**0**

6.7.1 *Central Composite Design (CCD)*

The CCD consists of a two-level (usually fractional) design with some added points, called face-centered or axial points. A face-centered point is one in which two of the factors are each held at a middle point, generally in the center of the low (-1) and high ($+1$) values. The other factors will be at their low or high values, depends. So, in a three-factor CCD, the first set of runs will be at they are for a full 2^3 design. Table 6.9 shows the CCD with three Helmert-coded factors $H1$, $H2$, and $H3$.

The unique advantage of the CCD is that it has a two-level fractional factorial design as a subset. Thus, if for some reason data are lost at the face centers, the experimenter will still have a two-level design for modeling and analysis. The disadvantage is that the CCD will require a few more runs than the BBD.

6.7.2 *Box-Behnken Design (BBD)*

The BBD only has "middle point" runs, unlike the CCD. While it can be used to fit second-order (quadratic) polynomials, it does not have a first-order design embedded in it. The runs are the centers of the hyper-cube defined by the low and high values of the factors, together with the face centers. Table 6.10 illustrates a three-factor BBD.

Note that the kind of models to be fit using data from either a CCD or BBD are identical. That is to say, they would have the form:

$$y = \beta_0 + \sum_{i=1}^{k} \beta_i H_i + \sum_{i=1}^{k} \delta_i H_i^2 + \sum_{i=1}^{k} \sum_{j=i+1}^{k} \gamma_{ij} H_i H_j + \varepsilon$$

Often, the only run that is replicated for either CCD or BBD experiments is the center point.

Table 6.10 A three-factor BBD

Run	*H*1	*H*2	*H*3
1	0	−1	−1
2	0	−1	1
3	0	1	−1
4	0	1	1
5	−1	0	−1
6	−1	0	1
7	1	0	−1
8	1	0	1
9	−1	−1	0
10	−1	1	0
11	1	−1	0
12	1	1	0
13	**0**	**0**	**0**

The value of having a design that allows for second-order terms is in finding the big bump or big value in a surface. The actual response function may have many peaks and troughs, but the objective of fitting a second-order function is not to completely characterize the response. Rather, it is to help the experimenter get closer to an "optimal" set of design features or operating conditions. We could potentially add enough points to fit any order polynomial, and for "smooth" response functions, polynomials provide a good approximation. Generally, however, the budget (both time and money) is limited, so experiments must be kept relatively small.

6.8 Non-continuously Valued Input Factors and Multiple Comparisons

Often factors are ordinal or even nominal valued. That is to say, a setting on a machine might have only discrete, albeit ordered, values (e.g., slow, medium, and fast speeds) or a factor's levels may have no order at all (e.g., colors, such as white or red). In these cases, Helmert coding can still be used, but the interpretation of the output is very different. Instead of fitting the model to be predictive, the model is discriminatory. That is to say, rather than interested in estimating the model coefficients, we would be more interested in estimating the effect and testing whether the effect for a given term is significantly different from 0. Most the models we have created so far were predictive polynomial approximations to the relationship between the response and the factors. The number of parameters associated with a factor or an interaction was one, and it was interpreted as a slope. When factors are discrete, the notion of slope does not really apply. Even in the case where a discrete factor has only two levels, the effect, as described in the beginning of this chapter, is a more meaningful measure than slope. In Fig. 6.3, the reader may have noticed tables with columns labeled "source" and "DF." The source was the factor or regressor variable, and DF was the degrees of freedom associated with the factor, and it was always 1. In the case of discrete factors, the degrees of freedom associated with the factor are the number of levels minus 1. The degrees of freedom for a factor are actually the number or parameters required to represent the factor's effect. In the case of two-level factors, and in fact for any continuously valued factor, only one parameter is required. If a factor is continuous, then the parameter can be interpreted as a slope. In the case of discrete factors, the parameters associated with those factors are only useful inasmuch as they are used to assess the factor's effect.

Usually, in the case of discrete factors, the engineer wants to know more specifically which combinations of levels for a factor differ significantly from each other. Tests of significance comparing specific combinations of factor levels are called multiple comparison tests (Montgomery, 2001). These tests are generally only applied if the overall effect is significant and are most useful when a factor has more than two levels. The tests are designed to control for inflating the chance of concluding that a significant difference between pairs of specific combinations of factor levels exists, when in fact it does not. One such test is the Tukey-Kramer test (Montgomery, 2001), or sometimes referred to as Tukey's honestly significant difference (HSD) test (Adler, 2010).

Figure 6.4 shows R code for an example with one discrete factor having three levels (called "first," "second," and "third"). The R output for the example, including the Tukey HSD test, is shown in Fig. 6.5. The columns labeled "lwr" and "upr" in the output for the Tukey HSD function are the lower and upper limits of simultaneous 95% confidence intervals for the differences between the average response at the

```
setwd("<your path here>")
df1 <- read.csv("20150121 Example Multiple Comparisons.csv")
attach(df1)
fF1 <- factor(F1, 1:3)
levels(fF1) <- c("first","second","third")
model123 <- aov(Y ~ fF1, na.action=na.omit)
#
# TukeyHSD requires an aov object as its first argument
#
TukeyHSD(x=model123,which="fF1",ordered=TRUE)
summary(model123)
# Note that the anova function will be the same as summary(model123)
# if model123 is an aov object
#
anova(model123)
detach(df1)
#####
```

Fig. 6.4 Example R code with Tukey HSD function

```
> setwd("<your path here>")
> df1 <- read.csv("20150121 Example Multiple Comparisons.csv")
>
> attach(df1)
>
> fF1 <- factor(F1, 1:3)
> levels(fF1) <- c("first","second","third")
>
>
> model123 <- aov(Y ~ fF1, na.action=na.omit)
> #
> # TukeyHSD requires an aov object as its first argument
> #
> TukeyHSD(x=model123,which="fF1",ordered=TRUE)
 Tukey multiple comparisons of means
  95% family-wise confidence level

Fit: aov(formula = Y ~ fF1, na.action = na.omit)

$fF1
             diff        lwr       upr     p adj
first-third     1 -0.8374262  2.837426 0.3909875
second-third    4  2.1625738  5.837426 0.0000122
second-first    3  1.1625738  4.837426 0.0007988
>
> summary(model123)

           Df  Sum Sq Mean Sq  F value  Pr(>F)
fF1         2   130.0   65.00    15.15  1.11e-05 ***
Residuals  42   180.2    4.29
---
Signif. codes:  0 '***' 0.001 '**' 0.01 '*' 0.05 '.' 0.1 ' ' 1
> #
> # Note that the anova function will be the same as summary(model123)
> # if model123 is an aov object
> #
> anova(model123)
```

Fig. 6.5 R output for Tukey HSD function

```
Analysis of Variance Table

Response: Y
          Df  Sum Sq  Mean Sq  F value  Pr(>F)
fF1        2  130.00   65.00   15.152   1.111e-05 ***
Residuals 42  180.18    4.29
---
Signif. codes:  0 '***' 0.001 '**' 0.01 '*' 0.05 '.' 0.1 ' ' 1
>
> detach(df1)
```

In this example, the factor fF1 is significant (p = 1.111e-05). While level "second" differs significantly from level "third" (p = 0.0000122) and from level "first" (p = 0.0007988), level "first" does not differ significantly from level "third" (p = 0.3909875).

Fig. 6.5 (continued)

particular levels being compared. By "simultaneous" we mean that jointly the confidence level for all the intervals is 95%. The p adj column is the *p*-value, adjusted to account for multiple comparisons.

If all the factors in the experiment, continuous or non-continuous, are restricted to two levels, then we can take advantage of all the computational efficiencies afforded by 2^k factorial designs. So, the designs and their associated characteristics are not restricted to only continuously valued factors. In fact, both discrete and continuous factors may be included in a single experimental design.

It is possible that in addition to controlling levels of discrete and continuous factors, the engineer may find that there are one or more continuously valued quantities that vary without control, but which can be measured. For example, the ambient temperature may not be completely controllable, and it may have some effect on experimental response variables. A version of ANOVA, called analysis of covariance (ANCOVA), was developed to allow the experimenter to compare the levels of discrete factors while compensating for a free-varying "covariate," which may attenuate the response signal. ANCOVA is a sort of combined ANOVA and regression. It is not the same as including a continuously varying but uncontrolled input in a regression model; rather ANCOVA is a means of adjusting the test of significance for discrete factors, to compensate for unwanted effects of the covariates. Inasmuch as this text focuses mostly on constructing predictive models, and not so much on determining the significance of discrete factors, no more will be said about ANCOVA. The Montgomery text already cited several times has an excellent coverage of this topic.

6.9 Matrix Form

Suppose **H** is the matrix of all the columns relating to the intercept and each input factor coefficient, including each cross-product terms (obtained by row-wise multiplication of the factors in the cross-product terms). Furthermore, let **Y** represent the vector of all the response results. They would look like Fig. 6.6

The first-order model can be represented by the vector equation:

$$\mathbf{Y} = \mathbf{H}\boldsymbol{\theta} + \boldsymbol{\varepsilon}$$

where:

$$\boldsymbol{\theta} = \begin{bmatrix} \beta_0 \\ \beta_1 \\ \beta_2 \\ \beta_3 \\ \gamma_{12} \\ \gamma_{13} \\ \gamma_{23} \\ \delta_{123} \end{bmatrix}$$

and $\boldsymbol{\varepsilon}$ is a nm x 1 vector of random noise variables. If $\widehat{\boldsymbol{\theta}}$ represents a vector of the ordinary least squares estimates of the model coefficients, then it can be expressed in the vector equation:

H Matrix								Y vector
Intercept	H1	H2	H3	H1H2	H1H3	H2H3	H1H2H3	y=Volume
1	-1	-1	-1	1	1	1	-1	20.8
1	-1	-1	-1	1	1	1	-1	20.8
1	-1	-1	1	1	-1	-1	1	21.6
1	-1	-1	1	1	-1	-1	1	21.3
1	-1	1	-1	-1	1	-1	1	17.8
1	-1	1	-1	-1	1	-1	1	17.7
1	-1	1	1	-1	-1	1	-1	17.9
1	-1	1	1	-1	-1	1	-1	17.7
1	1	-1	-1	-1	-1	1	1	17.6
1	1	-1	-1	-1	-1	1	1	17.8
1	1	-1	1	-1	1	-1	-1	16.5
1	1	-1	1	-1	1	-1	-1	16.7
1	1	1	-1	1	-1	-1	-1	24.1
1	1	1	-1	1	-1	-1	-1	24.4
1	1	1	1	1	1	1	1	23.3
1	1	1	1	1	1	1	1	23.0

Fig. 6.6 Matrix representation of volume yield experiment

$$\widehat{\boldsymbol{\theta}} = [\boldsymbol{H}'\boldsymbol{H}]^{-1}\boldsymbol{H}'\boldsymbol{Y}$$

That is to say, $\widehat{\boldsymbol{\theta}}$ is the solution to the least squares minimization problem (Draper & Smith, 1998), namely, to find the values of the coefficients that minimize the sum of squared errors (SSE).

Recall the formula for the standard error of predicted values:

$$\mathrm{SE}(\widehat{y}|\boldsymbol{h_0}) = s\sqrt{\boldsymbol{h}_0'[\boldsymbol{H}'\boldsymbol{H}]^{-1}\boldsymbol{h_0}}$$

where s is the root mean square error and $\boldsymbol{h_0}$ represents a particular design point (i.e., a value corresponding to each column of the $\mathbf{H}$ matrix) in the Helmert-coded input variable space. Once again, the Helmert coding simplifies the computations. The matrix $\boldsymbol{H}'\boldsymbol{H}$ has diagonal elements all equal to np, where p = number of parameters in the model = number of columns in the matrix $\boldsymbol{H}$, and all off-diagonal elements equal to 0. The inverse of such a matrix is simply the reciprocal of the diagonal elements on its diagonal and 0 everywhere else. That means:

$$\mathrm{SE}(\widehat{y}|\boldsymbol{h_0}) = s\sqrt{\boldsymbol{h}_0'[\boldsymbol{H}'\boldsymbol{H}]^{-1}\boldsymbol{h_0}} = \frac{s}{\sqrt{n}}$$

For run 7, $\boldsymbol{h}_0' = [+\mathbf{1}, +\mathbf{1}, +\mathbf{1}, -\mathbf{1}, +\mathbf{1}, -\mathbf{1}, -\mathbf{1}, -\mathbf{1}]$, and

$$SE(\widehat{y}|\boldsymbol{h_0}) = \frac{s}{\sqrt{2}}$$

With $s \approx 0.1581$, the 99% lower confidence bound on the predicted value for run 7 is:

$$\widehat{y} - t(0.99, 16-8)SE(\widehat{y}|h_0) \approx 24.25 - 2.896 * \frac{0.1581}{\sqrt{2}} \approx 23.93$$

So, now we have seen three different ways to compute the least squares estimates of the first-order polynomial coefficients using data from a balanced, 2^k factorial experiment:

1. Compute the "effect" in terms of differences in average response at the high and low levels of the input factor and then divide by 2.
2. Using the coded column for the input factor, multiply it row-wise with the corresponding response variable values, and then sum the products; divide by the total number of response values.
3. Use the matrix/vector solution to the least squares minimization problem.

All three of these methods yield the same answer. Method 2 only works for Helmert-coded, balanced, orthogonal two-level experiments.

6.10 The Taguchi Approach

Dr. Genichi Taguchi (Taguchi & Wu, 1980) described methods for using designed experiments in designing and improving products and processes. The Taguchi approach differs from the methods described previously, mostly due to their lack of fitting polynomial models to data. However, Taguchi methods can be a useful tool in developing product or process designs. We will describe some of the rudiments of his methods, together with some additional ideas. The ideas considered will be the quadratic loss function and signal-to-noise ratios, control and noise parameters, and the designs referred to as orthogonal arrays.

6.10.1 The Quadratic Loss Function

Let the variable X represent a measure of quality. Furthermore, suppose that there is a target value for X, namely, $X = \tau$.

The "loss" is the degree to which X differs from τ. For reasons that will become apparent, we will measure loss in the squared deviation:

$$L(X) = k(X - \tau)^2$$

The value of k only determines the rate at which the loss changes, so we will generally let $k = 1$.

The objective for robust design is to find values of control parameters that minimize the average loss in the face of noise parameters whose values vary in an uncontrolled fashion (except in the experimentation process).

The loss function can be applied to situations where it is desirable to minimize X or maximize X. Let $\tau = 0$. Then, the loss function is:

$$L(X) = kX^2$$

Clearly minimizing X will minimize the loss, so that the objective of minimizing loss is achieved by minimizing X. Similarly, let:

$$Y = \frac{1}{X}$$

so that the loss with respect to Y is:

$$L(Y) = kY^2 = k\frac{1}{X^2}$$

Thus, minimizing the loss is achieved by maximizing X.

One might ask why, especially in the case of minimizing or maximizing X, it is common to use the quadratic loss function. The answer lies in the expected value of $L(X)$. For our discussions, we will simply let $k = 1$. Then, if $E[X] = \mu$ is the expected value of X, and $V[X] = \sigma^2$ is the variance of X, then it can be derived that:

$$E[L(X)] = E\left[(X-\tau)^2\right] = (\mu-\tau)^2 + \sigma^2$$

Thus, minimizing the average quadratic loss simultaneously minimizes the average difference of X from target and the variability of X. When one statistic is simultaneously related to mean and variance, so that either having a mean closer to target or reducing variability gives it a more desirable value, it has been called a "concurrent" statistic (Barker, 1990).

Some people may have an aversion to minimizing, and Dr. Taguchi recognized this. So, he created what he called the "signal-to-noise" ratio (*SNR*). The *SNR* is nothing more than the loss function expressed in decibels. So, minimizing the expected value of the loss function is equivalent to maximizing the expected value of:

$$\mathrm{SNR} = -10 \log_{10} L(X)$$

Aside from an emotional tie to maximization, there is no particular reason to use the *SNR* transformation of the loss function. In fact, Taguchi considered a large number of *SNR*s, each having some intuitive appeal, but all of which are formulated so that the objective is maximization. Taguchi's original work did not concern itself with the statistical significance of parameters; he presumed that all parameters were meaningful and that their effects were already known to be repeatable. However, it may be valuable to be able to assess the significance of design parameters. As such, it will be much easier to use the loss function directly, as its statistical sample properties are easier to derive than are those of the *SNR*.

Suppose a sample of values of X is obtained, $x_1, x_2, \ldots, x_n$. Then the sample average loss function is:

$$\widehat{L}(x_1, x_2, \cdots, x_n) = \frac{1}{n}\sum_{i=1}^{n}(x_i-\tau)^2$$

If the variable X is normally distributed with expected value μ and variance σ^2, then:

$$\widehat{L} = \frac{\sigma^2}{n}\sum_{i=1}^{n}\left(\frac{x_i-\mu}{\sigma} + \frac{\mu-\tau}{\sigma}\right)^2 \sim \frac{\sigma^2}{n}\chi\prime^2\left(n, \lambda = \frac{n(\mu-\tau)^2}{\sigma^2}\right)$$

$\chi'^2\left(n, \lambda = \frac{n(\mu-\tau)^2}{\sigma^2}\right)$ stands for a non-central Chi-squared random variable with n degrees of freedom and *non-centrality parameter* λ (Johnson et al., 1995). Although both μ and σ^2 are generally unknown, an approximate 100(1 − 2α)% confidence interval is:

$$\left(\frac{\widehat{\sigma}^2}{n}\chi'^2_\alpha\left(n, \widehat{\lambda} = n\widehat{Z}_\tau^2\right), \frac{\widehat{\sigma}^2}{n}\chi'^2_{1-\alpha}\left(n, \widehat{\lambda} = n\widehat{Z}_\tau^2\right)\right)$$

where:

$$\widehat{Z}_\tau = \frac{\tau - \widehat{\mu}}{\widehat{\sigma}}$$

and:

$\chi'^2_p\left(n, \widehat{\lambda}\right) =$ the 100*p*th percentile of a non-central Chi-squared distribution with n degrees of freedom and non-centrality parameter $\widehat{\lambda}$.

The statistics $\widehat{\mu}, \widehat{\sigma}$ are sample estimates of μ and σ, respectively. Since the loss function is in squared units, it is somewhat uninterpretable as it stands. One possible remedy is to divide it by τ^2, so that the loss is expressed as a proportion (or percent) of the target value and take its square root (and the square roots of the confidence limits divided by τ^2). We will call this computation the root mean square percent loss.

Suppose we wanted to compare average loss between two different designs for a product. If the null hypothesis is that the average loss does not differ with respect to variable *X*, then if $\widehat{L}_1$ represents the average loss for a sample of n_1 observations of *X* with design *1* and $\widehat{L}_2$ the average loss for a sample of n_2 observations of *X* with design 2, then the ratio:

$$\widehat{F} = \frac{\frac{n_1}{\sigma^2}\frac{\widehat{L}_1}{n_1}}{\frac{n_2}{\sigma^2}\frac{\widehat{L}_2}{n_2}} = \frac{\widehat{L}_1}{\widehat{L}_2}$$

has a doubly non-central *F* distribution (F'') with n_1 and n_2 numerator and denominator degrees of freedom, respectively, and numerator and denominator non-centrality parameters:

$$\lambda_k = \frac{n_k(\mu-\tau)^2}{\sigma^2}, k = 1, 2$$

respectively. Now we have a means of deciding whether to believe that the average loss for one design is different from that of another. Compare the sample statistic $\widehat{F}$ to the lower and upper tails of a doubly non-central *F* with n_1 and n_2 degrees of freedom and non-centrality parameters λ_k, k = 1, 2. The only problem is that the parameters μ and σ are unknown. If we are also hypothesizing that $\mu = \tau$, then the non-centrality goes away, so we are only comparing the statistic to percentiles of the usual (central) F distribution. If, however, we are not willing to make that assumption, then one possibility is to aggregate all the data from both designs and compute the sample average and standard deviation; call them $\widehat{\mu}$ and $\widehat{\sigma}$. These can then be used to "approximate" the actual numerator and denominator non-centralities. Thus the estimated non-centrality parameters for the numerator and denominator would be:

$$\widehat{\lambda}_k = \frac{n_k(\widehat{\mu} - \tau)^2}{\widehat{\sigma}^2}, k = 1, 2$$

Inasmuch as percentiles of doubly non-central F distributions are somewhat hard to find, a reasonable approximation may be obtained with the formula described by Johnson et al. (1995) That is to say, if $F''(p, n_1, n_2, \lambda_1, \lambda_2)$ represents a percentile of a doubly non-central F, and $F(p, n_1, n_2, \lambda_1)$ the same percentile for a singly non-central F (with non-centrality λ_l in the numerator only), then:

$$F'' = \left(1 + \frac{\lambda_2}{n_2}\right)^{-1} F'$$

Figure 6.7 shows some R code for computing the approximate percentiles of a doubly non-central F.

For inputs:

nu1 = nu2 = 8, Target = 100, mu = 95, and sigma = 1.2, the outputs are:

```
> xfactor
[1] 18.36111
> Fprime_low
[1] 8.898236
> Fprime_high
[1] 51.24304
> Fdp_low
[1] 0.4846241
> Fdp_high
[1] 2.790846
```

6.10.2 *Parameter Design: Noise Parameters, Control Parameters, and Inner and Outer Arrays*

Taguchi identified the need for choosing designs that would be about to perform under the normally uncontrollable conditions of the environment in which they would be used. The optimization step is referred to as "parameter design." He divided the design parameters into two groups:

```
setwd("<your path here>")
#
#
attach(df1)

# Ref: Johnson, Kotz, Balakrishnan, (1995) Continuous Univariate Distributions
#                          Vol. 2, 2nd Ed. p. 502 John Wiley and Sons, New York


#
# inputs:
# nu1 = sample size 1
# nu2 = sample size 2
# Target
# mu = hypothetically common mean
# sigma = hypothetically common standard deviation
#
#
alpha <- 0.025
Ztau <- (Target - mu) / sigma

lamda1 <- nu1*(Ztau**2)
lamda2 <- nu2*(Ztau**2)

Fprime_low <- qf(alpha,df1=nu1,df2=nu2,ncp=lamda1) # this is the lower tail value for
#                                                      singly noncentral F
Fprime_high <- qf(1-alpha,df1=nu1,df2=nu2,ncp=lamda1) # this is the upper tail value
#                                                        for singly non-central F

xfactor <- 1 + (lamda2/nu2)

Fdp_low <- Fprime_low / xfactor; # lower tail value doubly non-central F
Fdp_high <- Fprime_high / xfactor; #upper tail value doubly non-central F

xfactor
Fprime_low
Fprime_high
Fdp_low
Fdp_high
```

Fig. 6.7 Computing percentiles of doubly non-central F distributions

Control those things whose values the researcher/experimenter can choose

Noise those things whose values that the experimenter cannot choose but must have his or her design perform well regardless of those values

For example, an engineer may choose the materials and dimensions of a cardiac lead, the cable that delivers electrotherapy from a pacemaker to the heart, but she cannot choose the impedance that the heart tissue will present. The materials and dimensions are control parameters, whereas the impedance is a noise parameter.

Dr. Taguchi suggested that two experimental designs should be performed together. The designs are referred to as the inner array (for control parameters) and the outer array (for noise parameters). The idea is that for every run in the inner array, a complete experiment in the noise parameters (outer array) should be performed. The noise array would be constructed by varying in a controlled fashion the noise parameters, in order to have a measure of loss over a wider range of noise "conditions" for each combination of the control parameters. In that way, the combination of the control parameter values that minimizes the loss (or maximizes the SNR) in face of the noise parameter variation would be the most "robust" choice for the product or process design.

The experimental designs employed by Taguchi are of a general class called "orthogonal arrays." All fractional factorials (or, for that matter, full factorials) are orthogonal arrays, but not all orthogonal arrays are fractional factorials. In order to simplify the choices of experimental design, Taguchi created a sort of catalogue of orthogonal arrays, categorized by the number of runs (rows). He used the letter "L," which stands for Latin square, together with the number of rows to designate the design. A Latin square is a k x k array of k unique items, usually symbolized with capital Latin letters, such that each item appears in each column and each row exactly once. The orthogonal array is a sort of generalization of the Latin square. Thus, the "L8" array has eight rows, or runs, and it can be used with up to seven parameters (or factors, as we have called them). The fewer the number of factors (columns) used from a given array, the greater the resolution. Figure 6.8 shows an L8 array. This is a two-level design, with the "low" level designated with a "1" and the "high" level with a "2." That is to say, Taguchi preferred to represent the levels as positive integers, rather than with Helmert coding. With $k = 7$ parameters, this is a 2^{7-4} fractional factorial design of Resolution III. If only columns *X1, X2,* and *X4* were used (i.e., only $k = 3$ parameters), this array would constitute a 2^3 full factorial design.

Fig. 6.8 The Taguchi L8 array

X1	X2	X3	X4	X5	X6	X7	Pattern
1	1	1	1	1	1	1	−−−−−−−
1	1	1	2	2	2	2	−−−++++
1	2	2	1	1	2	2	−++−−++
1	2	2	2	2	1	1	−++++−−
2	1	2	1	2	1	2	+−+−+−+
2	1	2	2	1	2	1	+−++−+−
2	2	1	1	2	2	1	++−−++−
2	2	1	2	1	1	2	++−+−−+

Fig. 6.9 The L4 array

N1	N2	Pattern
1	1	−−
1	2	−+
2	1	+−
2	2	++

X1	X2	[redacted]	X4	[redacted]	N1=1,N2=1	N1=1,N2=2	N1=2,N2=1	N1=2,N2=2
1	1		1					
1	1		2					
1	2		1					
1	2		2					
2	1		1					
2	1		2					
2	2		1					
2	2		2					

Fig. 6.10 Combined L8 and L4 arrays

The L4 array is given in Fig. 6.9. The L4 is a 2^2 full factorial design in two factors. Note that here the columns have been designated as N1 and N2, and we will use this to illustrate the noise array.

Suppose the researcher has decided that there are $k_1 = 3$ control parameters and $k_2 = 2$ noise parameters. Then he would combine the L8 and L4 arrays, so that each run in the L8 array has all four of the L4 array rows completed, represented as separate columns. Figure 6.10 illustrates the combined arrays.

The empty cells in the figure represent the quadratic loss values, given target τ. The columns not being used are blacked out. The first row represents a run where all three control parameters are set to their "low" level. The eighth row is a run where all three control parameters are set to their "high" level. For each row, observations are obtained at four different "noise" conditions.

The idea would be to average the loss over all rows where a given parameter is set to its "low" value and compare it to the loss averaged over all the rows where the parameter was set to its "high level." The parameter value to be used in the design would be the one that yielded the lowest average loss. This methodology is a considerable departure from all the methods discussed so far, in that we are not finding an approximating polynomial to predict optimal factor (or regressor, or in the Taguchi terminology, parameter) values or settings. The biggest disadvantage is that the Taguchi approach does not allow us to interpolate between levels. The advantage is that it is very simple to implement.

6.11 Summary

Factorial experiments provide a way to make decisions about product features. The analysis methods of choice are ANOVA and regression. They can be used to decide if a feature or factor actually has an effect on a key measure of quality or performance and the "best" values for those important features. An alternative methodology was developed by Dr. Genichi Taguchi, who used factorial experimentation to "optimize" those measures of quality and performance in the face of generally "uncontrollable" factors, such as raw material lots or batches.

References

Adler, J. (2010). *R in a Nutshell*, O'Reilly Media, Sebastopol.

Barker, T. B. (1990). *Engineering quality by design*. Marcel Dekker/Quality Press.

Box, G. E. P., & Behnken, D. W. (1960). Some new three level designs for the study of quantitative variables. *Technometrics*, 2, 455–475.

Box, G. E. P., Hunter, W. G., & Hunter, J. S. (1978). *Statistics for experimenters*. John Wiley and Sons.

Draper, N. R., & Smith, H. (1998). *Applied Regression Analysis* (3rd ed.). John Wiley and Sons, New York.

Johnson, N. L., Kotz, S., & Balakrishnan, N. (1995). *Continuous univariate distributions* (Vol. 2, 2nd ed.). John Wiley and Sons.

Montgomery, D. C. (2001). *Design and analysis of experiments*. John Wiley and Sons, Inc..

Taguchi, G., & Wu, Y. (1980). *Introduction to off-line quality control*. Central Japan Quality Control Association.

Chapter 7
Process Control

Abstract Statistical control is explained. Control charts are described as a method of deciding whether a process is in statistical control or not. Designed experiments and associated regression models are used to bring a process into control.

7.1 Introduction

For the purposes of this text, a process is a sequence of operations repeated generally over time, that is, ideally the operations are more or less identical. Process control is the performance of actions required to ensure that the variation in those operations is kept to an acceptable level.

The desired state of a process is referred to as "statistical control," meaning that fluctuations in observed values of response variables that measure product quality are within the range expected due to random variation. Often "statistical process control" refers to diagnostic tools used to assess the state of control, rather than the act of controlling the process. Those tools are generally known as control charts (Grant & Leavenworth, 1980). The concept of control charts was first introduced by Walter Shewhart (1931), before the use of electronic computing technology.

There is a distinction between statistical control and "in-spec." Statistical control is a state of unchanging statistical parameters; a process can be in good statistical control and have a virtually 0 probability of producing conforming items. As an extreme example, suppose that for some process with a quality variable X, the value of X is deterministically 10, but the desired value is 5, with tolerances of ± 1. The process would be in perfect statistical control but never produce an item that conformed to the specification limits ($5 - 1 = 4$, $5 + 1 = 6$).

This chapter will cover the topic of control charts first, followed by a discussion of how control chart information might be used to actually control the process.

Control charts are based on the concept of subgroups, small samples of measurements, taken in some homogenous manner, such as close in time. A statistic is generally computed for each new subgroup and plotted over time (usually time is the variable over which the process runs). Some limits, generally based on the assumed sampling distribution of the statistic, are computed. The limits are usually

S. A. Pardo, *Statistical Methods and Analyses for Medical Devices*,
https://doi.org/10.1007/978-3-031-26139-8_7

probabilistic, such as a confidence interval for the statistic of interest. Thus, if at some point in time the number of subgroups whose statistic falls outside the limits exceeds the expectations, then the process may be "out of control" and some action may be required to bring the process back into control.

7.2 The $\overline{X}$ Chart

It may be that the process engineers wish to keep the average of some continuously valued measure of quality constant. The procedure for creating the $\overline{X}$ chart is as follows:

1. Gather N values of the variable X, namely, the variable of interest.
2. Compute the mean of the N values; call it $\overline{\overline{X}}$.
3. Compute the standard deviation of the N values; call it σ''.
4. Pick a subgroup $n < N$, such that N is a multiple of n.
5. Compute limits:

$$\text{LCL} = \overline{\overline{X}} - 2.58\frac{\sigma''}{\sqrt{n}}$$

$$\text{UCL} = \overline{\overline{X}} + 2.58\frac{\sigma''}{\sqrt{n}}$$

6. Order the data in the time at which the measurements were taken, from oldest to newest.
7. Divide the data into $\frac{N}{n}$ subgroups, with each subgroup consisting of the next n measurements, in time order, and compute the mean of each subgroup.
8. Plot each subgroup mean against the average time at which the measurements were taken.
9. Plot the limits, LCL and UCL, and $\overline{\overline{X}}$ as horizontal lines, on the same plot.

This is the initial control chart. The values of ± 2.58 are the z-scores that represent approximately 99% of the hypothetical distribution of subgroup means, assuming the null hypothesis:

$$H_0 : E[X] = \overline{\overline{X}}$$

is true.

So, if more than 1% of the subgroup means fall outside the limits, we may want to reject H_0. Or, we might want to reject this null hypothesis if fewer than 99% of the subgroup means fall within the limits (LCL, UCL). One could formally test the hypothesis that the probability of finding a subgroup mean within the limits (LCL, UCL) is actually equal to 99%. That is to say, if $m = \frac{N}{n}$ is the number of subgroups, and $r =$ the number of subgroups inside, then *reject* the null hypothesis if $r < r_c$, where r_c is such that:

$$\Pr\{r \geq r_c | m, p = 0.99\} = \sum_{k=r_c}^{m} \binom{m}{k} (0.99)^k (0.01)^{m-k} = 1 - \alpha$$

The value of α is most typically chosen to be 0.05.

If this null is rejected, then it seems as though the process needs some adjustment. Methods for using these data to determine how much adjustment will be discussed alter in this chapter.

There are some other possible indicators that would suggest a process adjustment is required. If some sequences of subgroup means seem non-random, then at least an investigation is prudent. For example, suppose there is a sequence at least 5% of all subgroup means that fall on the same side of the midline, $\overline{\overline{X}}$, that would suggest a possible non-random effect. If there seems to be some sort of trend among the subgroup means (e.g., a linear function of time), then there is some non-random effect acting on the process. There are literally an infinite number of "patterns" in the subgroup means that could suggest non-randomness. So, it is somewhat incumbent upon the process experts to observe the data and decide whether or not something requires adjustment.

There is likewise no particular or universal value of N or n. Shewhart (1931) had suggested $N = 100$ and $n = 5$. That would yield $m = \frac{100}{5} = 20$ subgroups. There really is nothing magical about this number. With $m = 20$ subgroups, there is approximately a 98.31% chance of observing at least 19 out of 20 subgroups falling withing the 99% limits. In other words, with $m = 20$, and $r_c = 19$, $1 - \alpha = 0.9831$. The larger the sample size, it is possible to choose a critical value that yields $1 - \alpha$ closer to 0.9500.

Keep in mind that control charts are *not* about process "capability" (the ability of the process to produce items within specifications at a high probability). Rather, the control chart is about statistical consistency, determining whether a parameter is changing over time.

As an example, suppose there is an injection molding process with five-cavity molds. Furthermore, suppose there are three injection molding presses. The part that is being molded has a dimension that is to be about 10 mm in length with a specification range of 8.5–12.5 mm. Figure 7.1 shows an R script for computing $\overline{X}$ control charts for the three presses.

Table 7.1 shows the parameter values and the limits (LCL, UCL) for each of the three presses.

Figures 7.2, 7.3, and 7.4 show the plots of the data for $X1$, $X2$, and $X3$ $\overline{X}$ control charts.

In the case of press 1 ($X1$), all the subgroup means are within the control limits, indicating that there is a reason to believe the process mean is not changing. Press 1 is said to be in statistical control. In the case of press 2 ($X2$), all but one of the subgroup means are within control limits (one is exactly on the lower limit).

```
setwd("<your path here>")
df1 <- read.csv("20220207 Example 12 Xbar Control Charts.csv")
#
# Variables
#
# Subgroup
# Time
# X1
# X2
# X3
#
#
n.groups <- max(df1$Subgroup)
s.size <- nrow(df1) / n.groups
group <- seq(from=1,to=n.groups,by=1)
sub.size <- tapply(df1$X1,df1$Subgroup,FUN=length)
#
Tab.mean1 <- tapply(df1$X1,df1$Subgroup,FUN=mean)
Tab.sd1 <- tapply(df1$X1,df1$Subgroup,FUN=sd)
Tab.mean2 <- tapply(df1$X2,df1$Subgroup,FUN=mean)
Tab.sd2 <- tapply(df1$X2,df1$Subgroup,FUN=sd)
Tab.mean3 <- tapply(df1$X3,df1$Subgroup,FUN=mean)
Tab.sd3 <- tapply(df1$X3,df1$Subgroup,FUN=sd)
#
LL1 <- min(Tab.mean1)-1
UL1 <- max(Tab.mean1)+1
#
LL2 <- min(Tab.mean2)-1
UL2 <- max(Tab.mean2)+1
#
LL3 <- min(Tab.mean3)-1
UL3 <- max(Tab.mean3)+1

#
X.double.bar1 <- mean(df1$X1)
Sigma.prime1 <- sd(df1$X1)
#
X.double.bar2 <- mean(df1$X2)
Sigma.prime2 <- sd(df1$X2)
#
```

Fig. 7.1 R code for constructing $\overline{X}$ control charts

```
X.double.bar3 <- mean(df1$X3)
Sigma.prime3 <- sd(df1$X3)

df.subgroup <-
as.data.frame(cbind(group,sub.size,Tab.mean1,Tab.sd1,Tab.mean2,Tab.sd2,Tab.mean
3,Tab.sd3))
#
LCL1 <- X.double.bar1 - qnorm(0.995,mean=0,sd=1)*Sigma.prime1 / sqrt(s.size)
UCL1 <- X.double.bar1 + qnorm(0.995,mean=0,sd=1)*Sigma.prime1 / sqrt(s.size)
#
LCL2 <- X.double.bar2 - qnorm(0.995,mean=0,sd=1)*Sigma.prime2 / sqrt(s.size)
UCL2 <- X.double.bar2 + qnorm(0.995,mean=0,sd=1)*Sigma.prime2 / sqrt(s.size)
#
LCL3 <- X.double.bar3 - qnorm(0.995,mean=0,sd=1)*Sigma.prime3 / sqrt(s.size)
UCL3 <- X.double.bar3 + qnorm(0.995,mean=0,sd=1)*Sigma.prime3 / sqrt(s.size)
#
Real.L1 <- min(LL1,LCL1) - 1
Real.U1 <- max(UL1,UCL1) + 1
#
Real.L2 <- min(LL2,LCL2) - 1
Real.U2 <- max(UL2,UCL2) + 1
#
Real.L3 <- min(LL3,LCL3) - 1
Real.U3 <- max(UL3,UCL3) + 1
#
Real.Low <- min(Real.L1,Real.L2,Real.L3)
Real.High <- max(Real.U1,Real.U2,Real.U3)
plot(x=df.subgroup$group,y=df.subgroup$Tab.mean1,ylim=c(Real.Low,Real.High),
   main="Xbar Chart - X1",xlab="Subgroup",ylab="mean X1")
abline(h=LCL1)
abline(h=X.double.bar1)
abline(h=UCL1)
#
dev.new()
plot(x=df.subgroup$group,y=df.subgroup$Tab.mean2,ylim=c(Real.Low,Real.High),
   main="Xbar Chart - X2",xlab="Subgroup",ylab="mean X2")
abline(h=LCL2)
abline(h=X.double.bar2)
abline(h=UCL2)
```

Fig. 7.1 (continued)

```
#
dev.new()
plot(x=df.subgroup$group,y=df.subgroup$Tab.mean3,ylim=c(Real.Low,Real.High),
    main="Xbar Chart - X3",xlab="Subgroup",ylab="mean X3")
abline(h=LCL3)
abline(h=X.double.bar3)
abline(h=UCL3)

#
write.csv(df.subgroup,"20220207 Ch 12 Subgrouping Data.csv")
#####
```

Fig. 7.1 (continued)

Table 7.1 Parameter values and limits for $\overline{X}$ control charts

Variable	n	$\overline{\overline{X}}$	σ''	LCL	UCL
X1	5	10.35	1.838	8.23	12.47
X2	5	9.87	2.953	6.47	13.27
X3	5	5.30	3.264	1.54	9.06

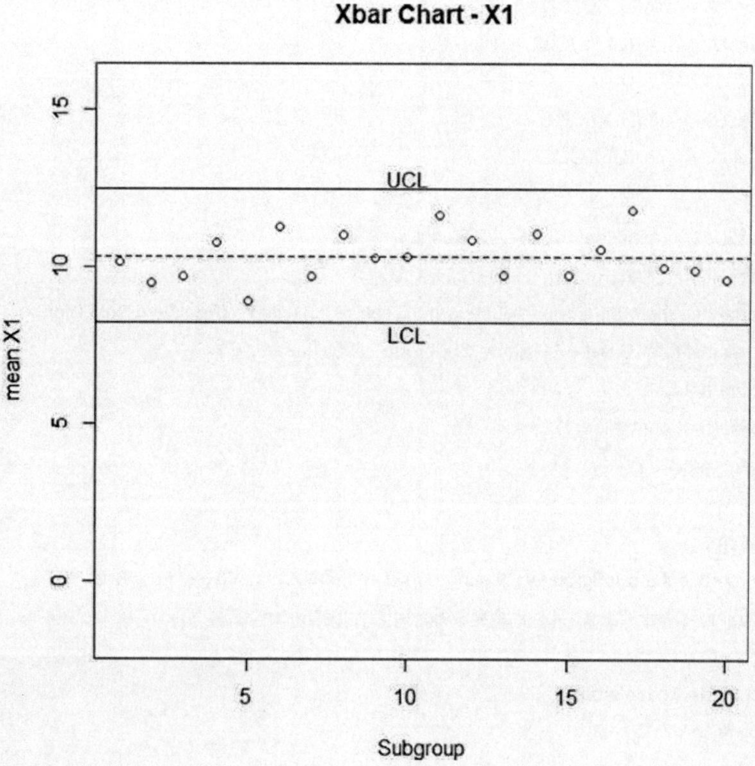

Fig. 7.2 $\overline{X}$ chart – $X1$

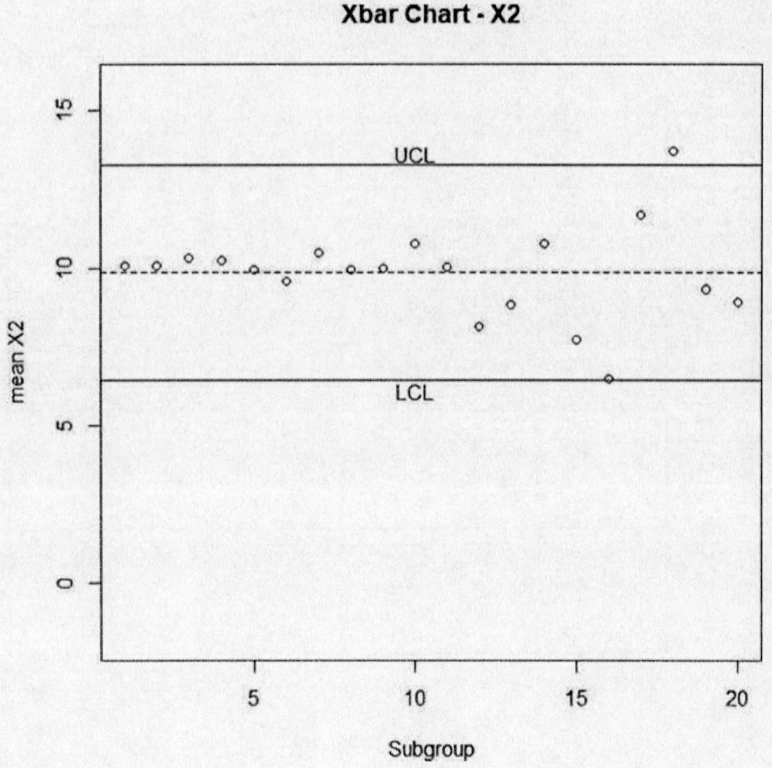

Fig. 7.3 $\overline{X}$ chart – *X*2

However, it may appear as though the subgroup means are scattering further away from the "grand" mean ($\overline{\overline{X}}$) as time (and subgroup number) increases. Thus, we may question whether the standard deviation of *X*2 is truly constant. Finally, press 3 seems to have a linear drift with time.

7.3 The S Chart

In the example of *X*2, we saw that while the process mean may be in control, other parameters may not be. In particular, the standard deviation may not be constant over time. A control chart can be constructed for the standard deviation. In the days of Walter Shewhart, simplifying computations for calculating limits was necessary to control charting accessible to workers on the factory floor. Today, we can dispense with many of those simplifications.

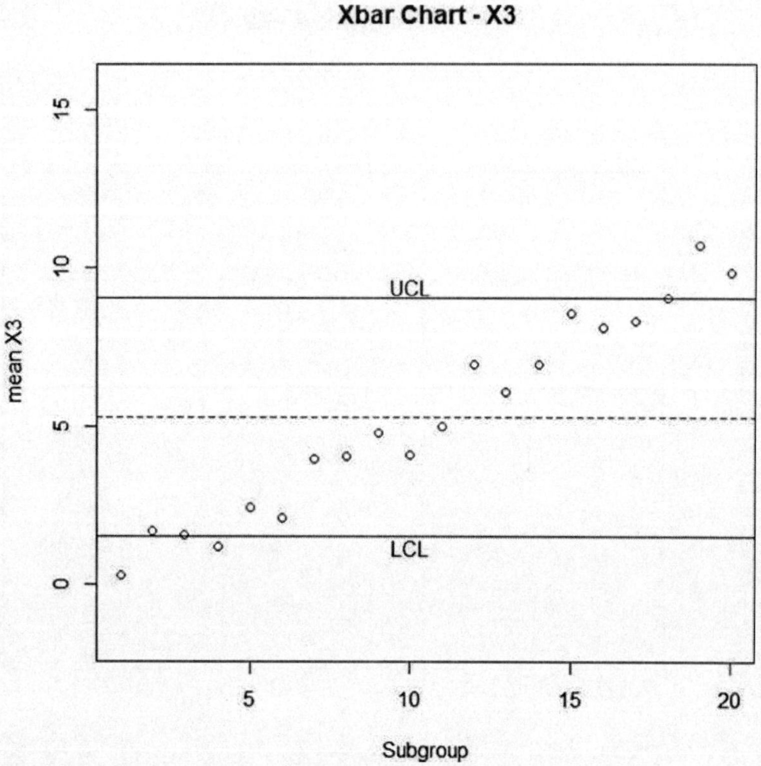

Fig. 7.4 $\overline{X}$ chart – $X3$

Inasmuch as having a standard deviation of 0 would be desirable (as unrealistic as that might be), we often only concern ourselves with an upper control limit. Using the fact that when data come from a normally distributed universe:

$$\frac{(n-1)s^2}{\sigma^2} \sim \chi^2_{n-1}$$

That is to say, the sample variance has a distribution proportional to a Chi-squared variable with $n-1$ degrees of freedom (assuming that the sample variance was computed using $n-1$ in the denominator instead of n).

This implies that the upper control limit for subgroup standard deviations ought to be:

$$\text{UCL} = \frac{\sigma''}{\sqrt{n-1}}\sqrt{\chi^2_{n-1,1-\alpha}}$$

$\chi^2_{n-1,1-\alpha}$ is the 100(1 − α)th percentile of a Chi-squared distribution with $n - 1$ degrees of freedom.

In the previous example, the values for σ'' were:

For X1: $\sigma'' \approx 1.838$
For X2: $\sigma'' \approx 2.953$
For X3: $\sigma'' \approx 3.264$

The values of the upper control limits are given in Table 7.2.

Figures 7.5, 7.6, and 7.7 show the control charts for *X*1, *X*2, and *X*3.

Figure 7.8 shows R code for constructing these charts. This code is actually appended to the code shown in Fig. 7.1

Table 7.2 Values of σ'' and UCL for three injection molding presses

Variable	n	σ''	UCL
X1	5	1.838	3.349
X2	5	2.953	5.380
X3	5	3.264	5.947

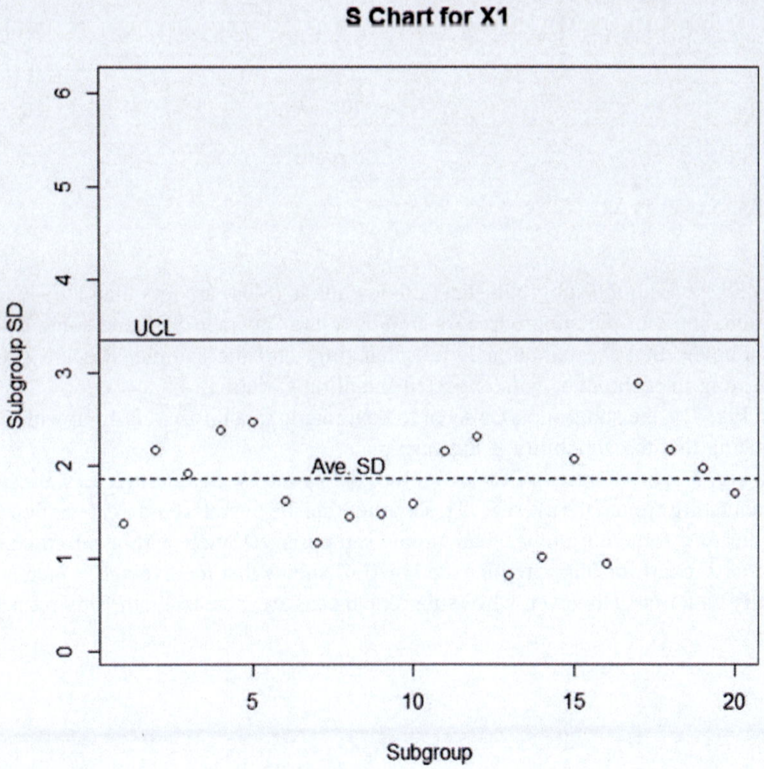

Fig. 7.5 S chart for *X*1

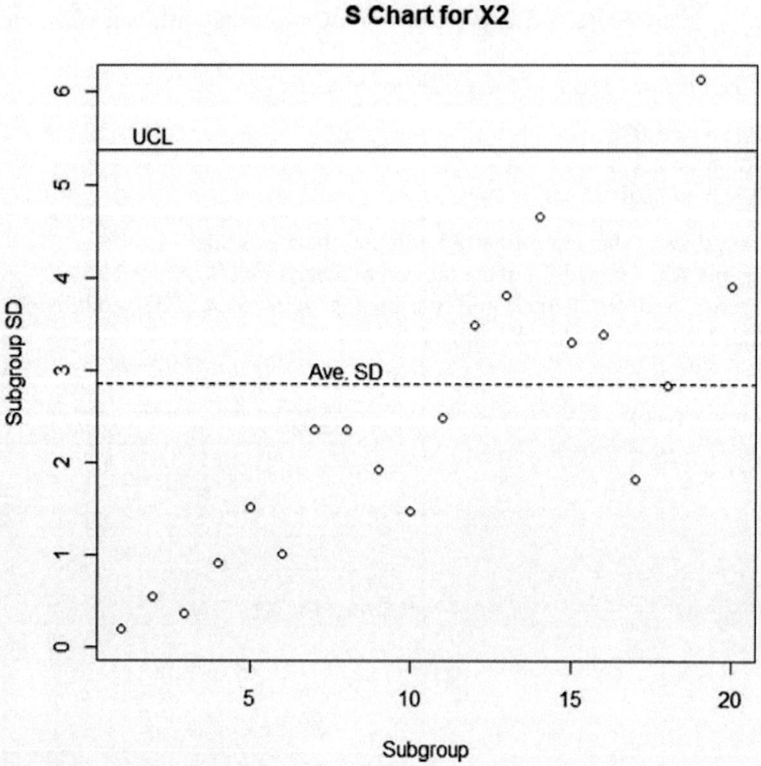

Fig. 7.6 S chart for *X*2

In Fig. 7.5, all the subgroup standard deviations (SDs) are less than the UCL. In addition, some of the subgroup SDs are above the "average SD" and some below. This implies that the variation is not changing, and the sample SDs are simply fluctuating in accordance with expected sampling variation.

In Fig. 7.6, the subgroup SDs seem to be trending in a positive fashion with time, indicating that the variability is increasing.

In Fig. 7.7, the subgroup SDs are all well below the UCL, although they appear to be fluctuating around the average. This implies that the overall standard deviation (σ'') is indicating something other than simple sampling variation within subgroups. In fact, the $\overline{X}$ chart for this variable (see Fig. 7.4) shows that the average is increasing linearly with time. However, while subgroup means are increasing, the subgroup SDs are not.

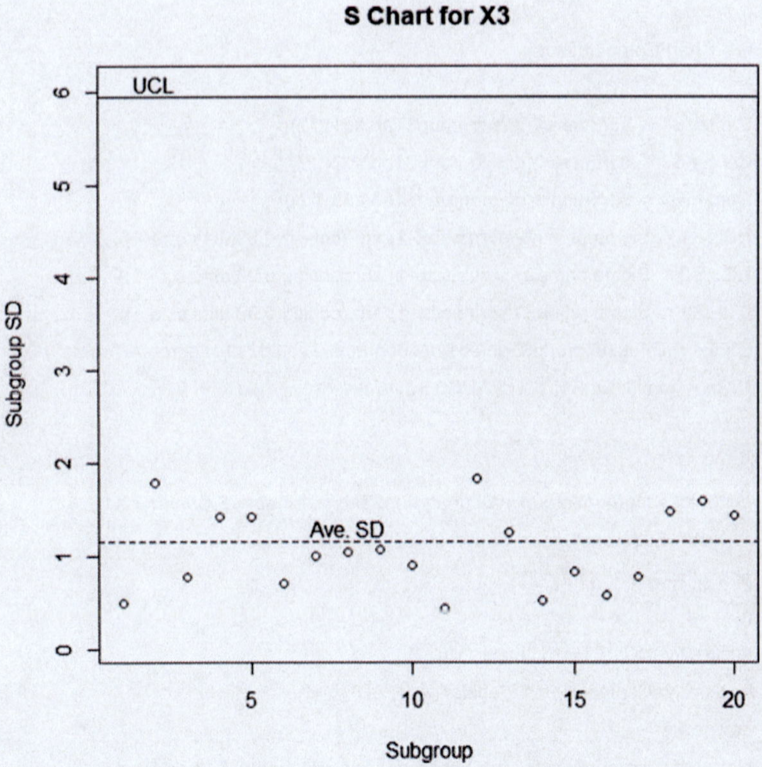

Fig. 7.7 S chart for *X*3

7.4 The CV Chart

Kang et al. (2007) approximated the sampling distribution of the sample coefficient of variation (CV) as:

$$\mathrm{CV} \sim \frac{\sqrt{n}}{T'_{n-1,\mathrm{nct}}}$$

where $T'_{n-1,\mathrm{nct}}$ is a non-central t variable with $n - 1$ degrees of freedom and non-centrality parameter:

$$\mathrm{nct} = \frac{\sqrt{n}}{c''}$$

and c'' = the population CV.

```
#
# S Chart Computations
#
Center.S1 <- sqrt(mean((df.subgroup$Tab.sd1)**2))
Center.S2 <- sqrt(mean((df.subgroup$Tab.sd2)**2))
Center.S3 <- sqrt(mean((df.subgroup$Tab.sd3)**2))
UCL.S1 <- Sigma.prime1 / sqrt(s.size-1)*sqrt(qchisq(0.99,df=s.size-1))
UCL.S2 <- Sigma.prime2 / sqrt(s.size-1)*sqrt(qchisq(0.99,df=s.size-1))
UCL.S3 <- Sigma.prime3 / sqrt(s.size-1)*sqrt(qchisq(0.99,df=s.size-1))
L.lim <- min(df.subgroup$Tab.sd1,df.subgroup$Tab.sd2,df.subgroup$Tab.sd3)-0.1
U.lim <- max(UCL.S1,UCL.S2,UCL.S3)+0.1
dev.new()

plot(x=df.subgroup$group,y=df.subgroup$Tab.sd1,main="S Chart for X1",
   xlab="Subgroup",ylab="Subgroup SD",ylim=c(L.lim,U.lim))
abline(h=Center.S1,lty=2)
abline(h=UCL.S1)
text(x=2,y=UCL.S1+.15,labels="UCL")
text(x=8,y=Center.S1+0.15,labels="Ave. SD")
dev.new()
plot(x=df.subgroup$group,y=df.subgroup$Tab.sd2,main="S Chart for X2",
   xlab="Subgroup",ylab="Subgroup SD",ylim=c(L.lim,U.lim))
abline(h=Center.S2,lty=2)
abline(h=UCL.S2)
text(x=2,y=UCL.S2+.15,labels="UCL")
text(x=8,y=Center.S2+0.15,labels="Ave. SD")
dev.new()
plot(x=df.subgroup$group,y=df.subgroup$Tab.sd3,main="S Chart for X3",
   xlab="Subgroup",ylab="Subgroup SD",ylim=c(L.lim,U.lim))
abline(h=Center.S3,lty=2)
abline(h=UCL.S3)
text(x=2,y=UCL.S3+.15,labels="UCL")
text(x=8,y=Center.S3+0.15,labels="Ave. SD")
#####
```

Fig. 7.8 R code for constructing S charts

Table 7.3 Parameter values

Variable	N	$\overline{\overline{X}}$	σ''	c''
X1	5	10.35	1.838	0.178
X2	5	9.87	2.953	0.299
X3	5	5.30	3.264	0.616

Table 7.3 shows the "parameter" values of the three variables $X1$, $X2$, and $X3$ (based on the entire sample and not just subgroups) for mean, standard deviation, and coefficient of variation.

Figure 7.9 shows the R code for computing the upper control limits and constructing the charts.

The upper control limits are computed as:

$$\mathrm{UCL} = \frac{\sqrt{n}}{T'_{n-1,\mathrm{nct}}(\alpha)}$$

$T'_{n-1,\mathrm{nct}}(\alpha)$ is the $100\alpha^{\mathrm{th}}$ percentile of a non-central t distribution with $n - 1$ degrees of freedom and non-centrality parameter nct.

As in the case of the SD charts, there is generally no concern that the CV would be too low, so only an upper limit will be computed here.

Figures 7.10, 7.11, and 7.12 show the charts for the three variables, representing the three molding presses. In these cases, $\alpha = 0.01$.

As evidenced by the chart in Fig. 7.10, the CV for $X1$ seems to be in good control. All subgroup CVs are below the UCL, and they seem to be equally likely to be above or below the center line (c''). In Fig. 7.11, there appears to be some sort of positive trend. Keep in mind Figs. 7.3 ($\overline{X}$ chart for $X2$) and 7.6 (S chart for $X2$) While Fig. 7.3 shows averages that may be increasing over time, Fig. 7.6 shows a more definite increasing trend in the SDs. Thus, the CVs also seem to be increasing. As to $X3$, Fig. 7.12 seems to indicate a decreasing trend in CV. The $\overline{X}$ chart in Fig. 7.4 together with the S chart in Fig. 7.7 provides an explanation. The subgroup means for $X3$ appear to be increasing, whereas the SDs seem to be staying fairly constant over time. Thus, the denominators of the CVs are increasing while the numerators are not.

7.5 The P Chart

So far the control charts described are for continuously valued variables. However, there are discrete variables, sometimes referred to as "attributes" (Schilling, 1982) that can also indicate whether a process is in a state of statistical control. There are several attribute-type control charts. The only attribute chart discussed in this chapter is for proportion of nonconforming items; it is called the P-chart. P is the proportion of nonconforming items produced. Generally, the proportion is small; therefore, if the subgroup size is small enough, it is likely that no nonconforming items will be

Fig. 7.9 R code for CV chart

```
# CV Chart
nct1 <- sqrt(s.size)/CV.prime1
nct2 <- sqrt(s.size)/CV.prime2
nct3 <- sqrt(s.size)/CV.prime3
CV.lim1 <- sqrt(s.size) / qt(0.01,df=s.size-1,ncp=nct1)
CV.lim2 <- sqrt(s.size) / qt(0.01,df=s.size-1,ncp=nct2)
CV.lim3 <- sqrt(s.size) / qt(0.01,df=s.size-1,ncp=nct3)
#
CV.range <- max(CV.lim1,CV.lim2,CV.lim3)
dev.new()
plot(x=df.subgroup$group,y=cv1,main="CV Chart for X1",
   xlab="Subgroup",ylim=c(0,CV.range))
abline(h=CV.lim1)
abline(h=CV.prime1,lty=2)
text(x=2,y=CV.lim1+0.05,labels="UCL")
dev.new()
plot(x=df.subgroup$group,y=cv2,main="CV Chart for X2",
   xlab="Subgroup",ylim=c(0,CV.range))
abline(h=CV.lim2)
abline(h=CV.prime2,lty=2)
text(x=2,y=CV.lim2+0.05,labels="UCL")
dev.new()
plot(x=df.subgroup$group,y=cv3,main="CV Chart for X3",
   xlab="Subgroup",ylim=c(0,CV.range))
abline(h=CV.lim3)
abline(h=CV.prime3,lty=2)
text(x=2,y=CV.lim3-0.05,labels="UCL")
#####
```

observed. For example, suppose it was true (although not necessarily known) that a process produces 0.5% nonconforming product. If the subgroup size is $n = 20$, then the probability of observing at least one nonconforming item in a subgroup is:

$$\Pr\{X \geq 1 | n = 20.p = 0.005\} = \sum_{k=1}^{n} \binom{n}{k} p^k (1-p)^{n-k} \approx 0.09539$$

Thus, with $n = 20$ and $p = 0.005$, there is about a 90% chance of having 0 nonconforming items in a subgroup. In other words, nine out of ten subgroups

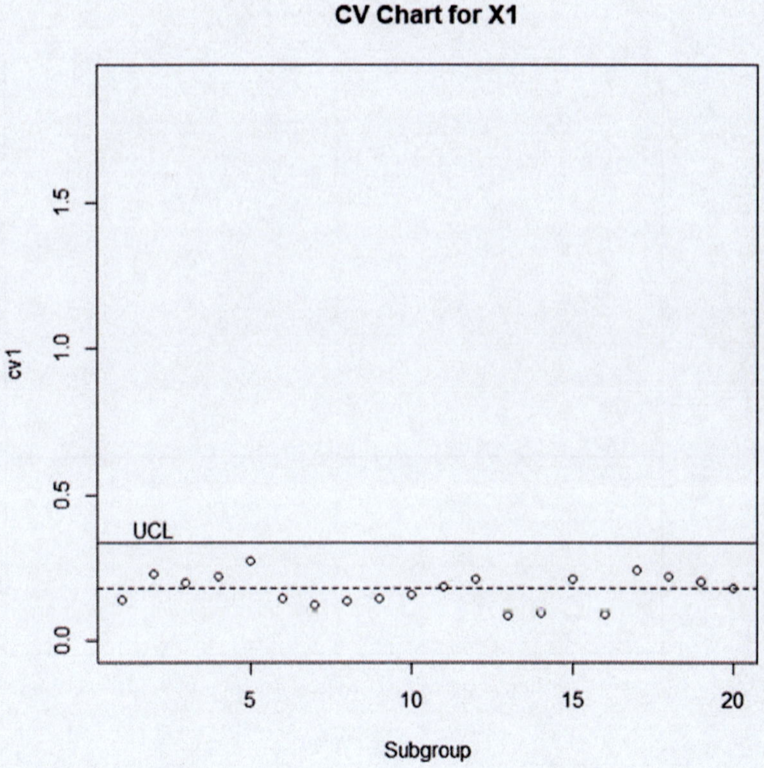

Fig. 7.10 CV chart for *X*1

would be expected to have zero nonconforming items. For discrete variables in particular, care must be used in choosing a subgroup size.

Suppose it is desired to have no more that 0.5% nonconforming items. Suppose the operators choose a subgroup size of $n = 30$, and they intend to collect 15 subgroups, for a total of $15 * 30 = 450$ items. With a subgroup size of $n = 30$, and a population percentage of nonconforming items equal to 0.5%, there is approximately a 14% chance that a subgroup would contain at least one nonconforming item.

As examples of P-charts, consider three quality "attributes," or discrete variables, *X*1, *X*2, and *X*3. Table 7.4 shows the proportions of nonconforming items for these three attributes.

Figures 7.13, 7.14, and 7.15 show the P-charts for these three attributes.

The limits for all three of these charts are based on the center line proportion. They are the 0.5 and 99.5 percentiles of binomial distributions, with $n = 30$ (subgroup size) and $p =$ center line (the total proportion of non-conforming items, aggregated over the entire sample).

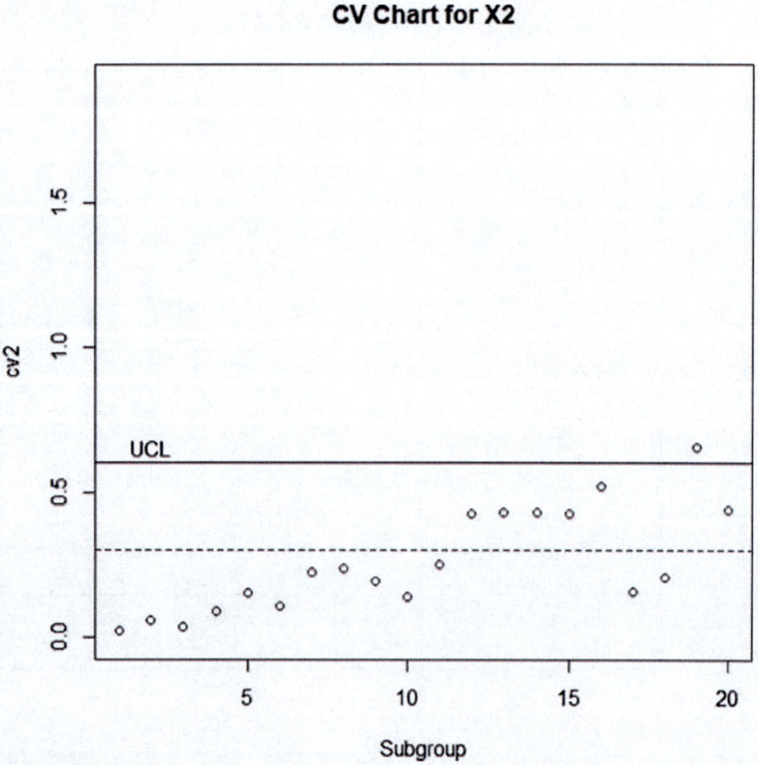

Fig. 7.11 CV chart for *X2*

The code for constructing the charts is given in Fig. 7.16.

The only chart for which any subgroup proportion was outside the control limits was for *X*3. The expected probability of a subgroup proportion within the limits is 99%. For *X*3, 2 out of 15 subgroups, or 13.33%, had proportions outside the limits. Thus, we suspect that the process is not in control for *X*3. Despite the fact that *X*2 had in the aggregate 4% nonconforming, it is not clear that the process is out of statistical control for *X*2. Recall that control charts and statistical control are not about having a low percentage of nonconforming items; rather, they are about whether that percentage is changing over time. However, generally it is desired to have processes that produce items within specifications the vast majority of time.

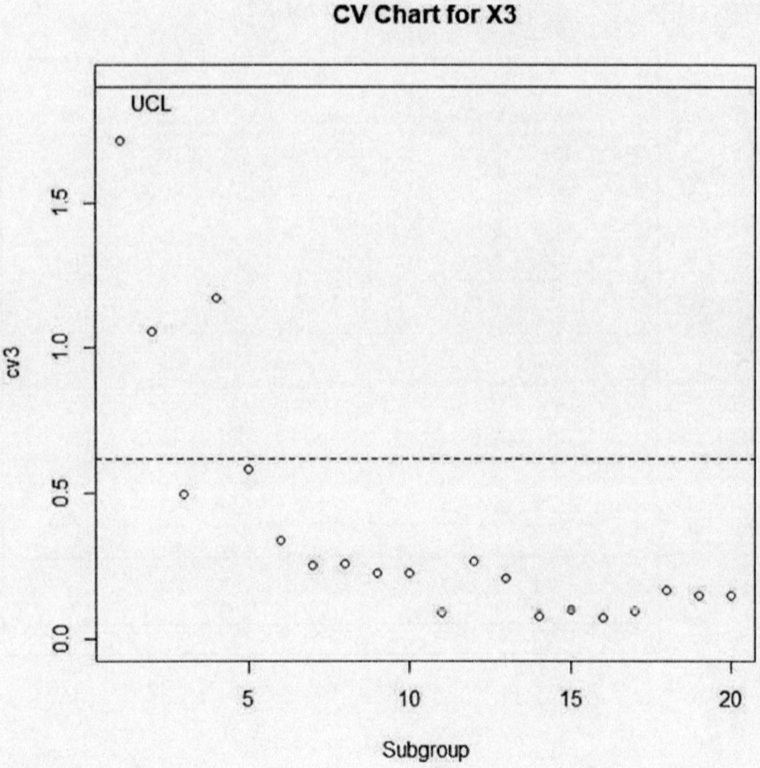

Fig. 7.12 CV chart for *X3*

Table 7.4 Proportions of nonconforming items

Attribute	*n*	*P*
X1	30	0.0067
X2	30	0.0400
X3	30	0.0760

7.6 Process Control

Although the construction of control charts is referred to as "statistical control," "quality control," or even "process control," the charts by themselves do nothing to control quality or process. Rather, they are indicators of the state of the process. If the process is found to be out of statistical control, some actions may be required by the process operators to bring the process into control.

In order to change the process state, there must be some relationship between the output measures of quality, i.e., those variables used to construct control charts, i.e., the "input" variables that are used by operators (either human or machine) to alter the process (e.g., speed, pressure, temperature, etc.). In other words, if $\boldsymbol{X}$ represents a

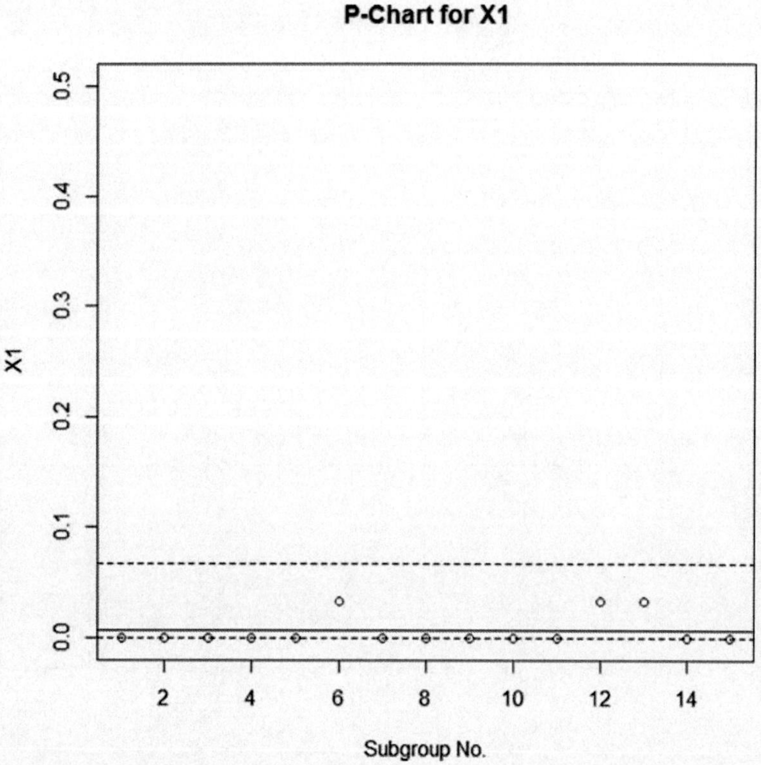

Fig. 7.13 P-chart for $X1$

vector of input variables, and $\boldsymbol{Y}$ the vector of output "response" variables, a mapping or function, $\boldsymbol{f}$, must be obtained such that:

$$\boldsymbol{Y} = \boldsymbol{f}(\boldsymbol{X})$$

The first problem is to identify the function $\boldsymbol{f}$. The methods of Chap. 6 might be used to estimate this function. The second problem is one of optimization. For example, suppose $\boldsymbol{k}$ is the vector or desired values of the response vector, $\boldsymbol{Y}$. Then the problem might be expressed as:

$$\underset{X}{\text{Minimize}}\ |\boldsymbol{f}(\boldsymbol{X}) - \boldsymbol{k}|$$

Subject to:

$$\boldsymbol{h}(\boldsymbol{X}) = \boldsymbol{0}$$

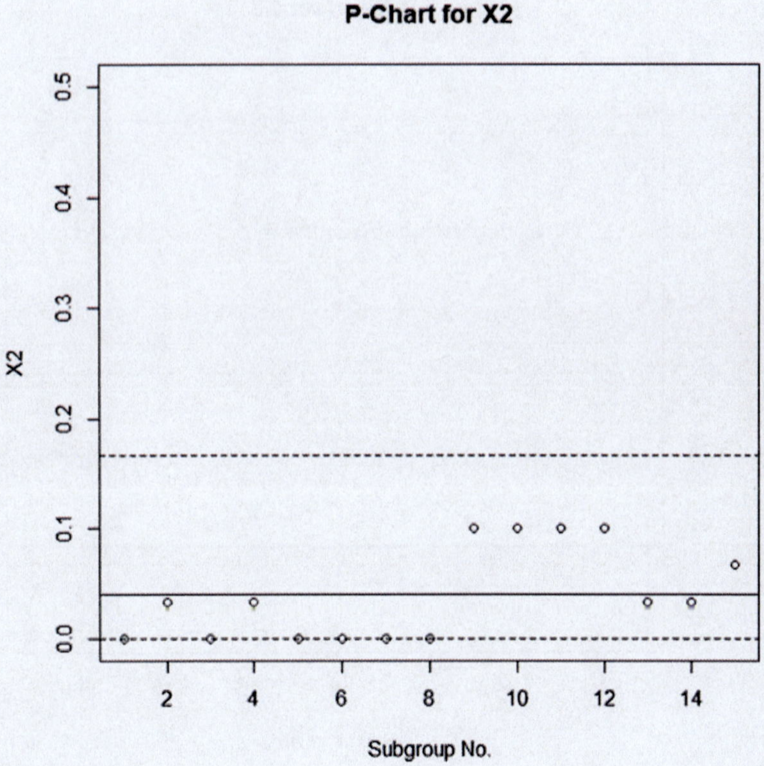

Fig. 7.14 P-chart for *X*2

The vector-values function $\boldsymbol{h}$ expresses any constraints on the input variables. So, for example, suppose a setting, such as a speed (call it $X1$), cannot be less than some minimum, $S_{\min}$, and cannot be greater than some maximum, $S_{\max}$. Then there could be two "component" functions of the vector $\boldsymbol{h}$:

$$h_{\min} : X1 \geq S_{\min} \Rightarrow X1 - S_{\min} = 0$$

$$h_{\max} : X1 \leq S_{\max} \Rightarrow S_{\max} - X1 = 0$$

It is advisable that in constructing *f*, only include in any component function those input variables that are statistically significant or in some way meaningful.

To learn more about methods of optimization, especially in the face of constraints, see Reklatis et al. (1983).

There are two basic types of models we will consider:

1. Static
2. Dynamic, $y_t = f(y_{t-\tau}, \varepsilon_{t,}, \varepsilon_{t-\tau}, \boldsymbol{C})$

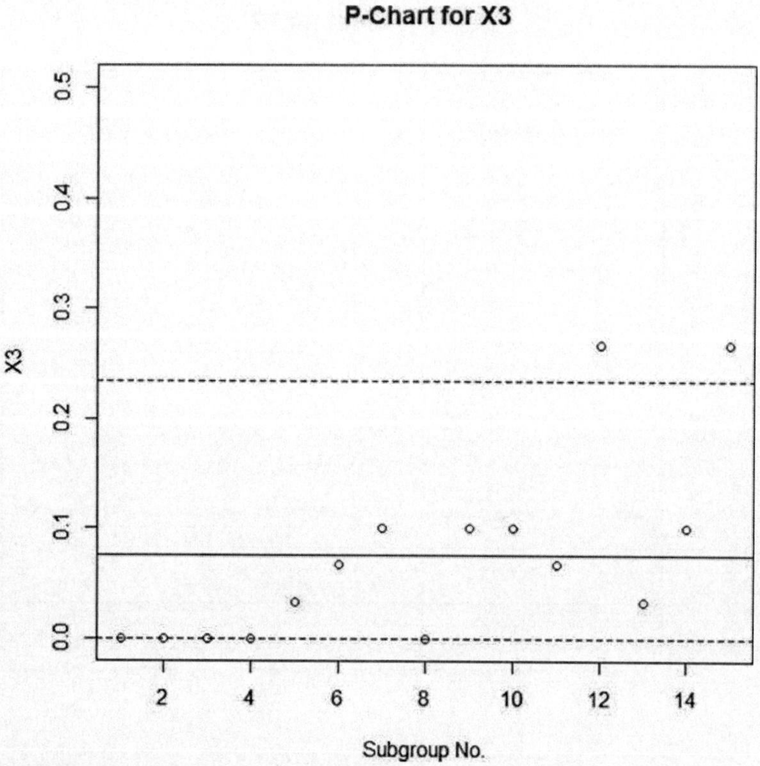

Fig. 7.15 P-chart for *X*3

The former type is often expressed as a polynomial in the control variables, e.g.:

$$y_t = \gamma^T \boldsymbol{C} + \varepsilon_t$$

The idea is that the response variable, which is the variable to be controlled, is a static function of some controlling variables, represented by a vector, $\boldsymbol{C}$. There is some stochastic, or random, effects, represented by ε.

In the dynamic models, the response at time t is in part a function of previous values of the response and previous values of the random effect, ε. It may also be a function of time itself, as well as a function of control variables, such as:

$$y_t = f(t) + \beta_1 y_{t-1} + \alpha_1 \varepsilon_{t-1} + \varepsilon_t + \gamma^T \boldsymbol{C}$$

The function $f(t)$ is referred to as a "secular trend," a function of time only. The model is referred to as an autoregressive, moving average (ARMA) model of order (1,1). That is to say, the model is in terms of the response at one "period" previous to

```
setwd("<your path here>")
df1 <- read.csv("20220502 Example 7.2 P-Chart.csv")
#
# Variables
#
# Subgroup
# X1
# X2
# X3
#
#
n.groups <- max(df1$Subgroup)
s.size <- nrow(df1) / n.groups
s.group <- seq(from=1,to=n.groups,by=1)
sub.size <- tapply(df1$X1,df1$Subgroup,FUN=length)
#
Center.P1 <- mean(df1$X1)
Center.P2 <- mean(df1$X2)
Center.P3 <- mean(df1$X3)
Subgroup.P1 <- tapply(df1$X1,df1$Subgroup,FUN=mean)
Subgroup.P2 <- tapply(df1$X2,df1$Subgroup,FUN=mean)
Subgroup.P3 <- tapply(df1$X3,df1$Subgroup,FUN=mean)
#
SE1 <- sqrt(Subgroup.P1*(1-Subgroup.P1)/s.size)
SE2 <- sqrt(Subgroup.P2*(1-Subgroup.P2)/s.size)
SE3 <- sqrt(Subgroup.P3*(1-Subgroup.P3)/s.size)
#
LL1 <- qbinom(p=0.005,size=s.size,prob=Center.P1)/s.size
UL1 <- qbinom(p=0.995,size=s.size,prob=Center.P1)/s.size
#
LL2 <- qbinom(p=0.005,size=s.size,prob=Center.P2)/s.size
UL2 <- qbinom(p=0.995,size=s.size,prob=Center.P2)/s.size
#
```

Fig. 7.16 R code for making P-charts

```
LL3 <- qbinom(p=0.005,size=s.size,prob=Center.P3)/s.size
UL3 <- qbinom(p=0.995,size=s.size,prob=Center.P3)/s.size
#
plot(x=s.group,y=Subgroup.P1,main="P-Chart for X1",ylim=c(0,0.5),xlab="Subgroup
No.",ylab="X1")
abline(h=Center.P1,lty=1)
abline(h=LL1,lty=2)
abline(h=UL1,lty=2)
dev.new()
plot(x=s.group,y=Subgroup.P2,main="P-Chart for X2",ylim=c(0,0.5),xlab="Subgroup
No.",ylab="X2")
abline(h=Center.P2,lty=1)
abline(h=LL2,lty=2)
abline(h=UL2,lty=2)
#
dev.new()
plot(x=s.group,y=Subgroup.P3,main="P-Chart for X3",ylim=c(0,0.5),xlab="Subgroup
No.",yla="X3")
abline(h=Center.P3,lty=1)
abline(h=LL3,lty=2)
abline(h=UL3,lty=2)
#####
```

Fig. 7.16 (continued)

the current period and also a function of whatever noise (model errors) existed in the previous time period.

Static models are simpler. The static models are usually treated as multiple regressions and are most frequently "linear" models (i.e., the unknown coefficients to be estimated are multipliers for each respective term in the model).

As an example, suppose that a response variable, Y, is a function of a control variable, $C1$. The control variable can be adjusted to change the response. An experiment can be performed to obtain a model of the form:

$$Y = \beta_0 + \beta_1 C1 + \varepsilon$$

Suppose it is desired to keep the response (at least on the average) constant, at a value of Y_0. Then on the average, we would have:

$$Y_0 = \beta_0 + \beta_1 C1$$

Thus, to make the response equal to the desired value, we would set the control variable, $C1$, as:

$$C1_0 = \frac{Y_0 - \beta_0}{\beta_1}$$

In an experiment, the value of $C1$ could be varied between two values, so that estimates of β_0 and β_1 could be obtained. Then the process would be run at $C1 = C1_0$, and the data would be plotted against the time.

Suppose that the desired value of the response is $Y_0 = 5.5$. To "simulate" the effects of changing $C1$, the data could be "adjusted" using the formula:

$$Y_{\text{adj}} = Y_{\text{actual}} - Y_{\text{predicted}} + Y_0$$

Figure 7.17 shows R code for fitting such a model.

The results of the model fit are shown in Fig. 7.18.

Residual standard error: 0.3065 on 78 degrees of freedom

Multiple R-squared: 0.9519, **Adjusted R-squared: 0.9513**

F-statistic: 1544 on 1 and 78 DF, p-value: < 2.2e-16

Figure 7.19 shows the data, together with the model's predicted values, from the experiment with $C1$ set to two different values (3 and 6). The response, Y, is plotted over time.

From the regression:

$$\widehat{C1}_0 = \frac{Y_0 - \widehat{\beta}_0}{\widehat{\beta}_1} \approx \frac{5.5 - 0.01050}{0.89750} \approx 6.11$$

Figure 7.20 shows a plot of the "adjusted" response values, together with 95% confidence bounds on the predicted value at $Y_0 = 5.5$.

The plot indicates that the response with $C1$ set at the level of 6.11 should yield values in the interval (6.11, 4.89) about 95% of the time (the value of $C1 = 6.11$ and the prediction upper bound of 6.11 are the same by coincidence).

As mentioned earlier, dynamic models can be much more complicated. For example, suppose the model for the response is of the form:

$$y_t = f(t) + \gamma^T \mathbf{C} + \varepsilon$$

That is to say, the dynamic, time-varying part of the response is a secular trend. If the situation is simplified so that there is only one control variable, C1, then the model becomes:

$$y_t = f(t) + \gamma C1 + \varepsilon$$

```
setwd("<your path here>")
df1 <- read.csv("20220427 Example 7.1 Control Variable.csv")
#
# Variables
#
# Y
# Time
# C1

#
#
#
attach(df1)
n.obs <- length(C1)
SS.C1 <- (n.obs-1)*var(C1)
C1.mean <- mean(C1)
Y.nom <- 5.5000
Y.mod <- lm(Y ~ C1) #this is a static model in C1; Y is not a function of Time.
#                   However, Y is observed over Time.
sd.residuals <- sd(Y.mod$residuals)
beta0 <- Y.mod$coefficients[1]
beta1 <- Y.mod$coefficients[2]
C1.nom <- (Y.nom - beta0) / beta1

se.reverse <- sd.residuals*sqrt(1 + 1/n.obs + ((C1.nom-C1.mean)**2)/SS.C1)
LPL <- Y.nom - qt(p=0.975,df=n.obs-2)*se.reverse
UPL <- Y.nom + qt(p=0.975,df=n.obs-2)*se.reverse
tol <- (Y.nom - LPL)/Y.nom
LTL <- (1-tol)*Y.nom
UTL <- (1+tol)*Y.nom
Y.nom.txt <- as.character(round(Y.nom,2))
```

Fig. 7.17 Code for controlling with static model

```
LPL.txt <- as.character(round(LPL,2))
UPL.txt <- as.character(round(UPL,2))
Y.predicted <- Y.mod$fitted.values
Adjustment <- Y.predicted - Y.nom
Adjusted <- Y - Adjustment
Y.mid <- mean(Adjusted)
Y.mid.txt <- as.character(round(Y.mid,2))
#
plot(x=Time,y=Y.mod$fitted.values,
   main="Predicted and Actual Values for Y",xlab="Time",ylab="Y",pch=1,type="p")
points(x=Time,y=Y,pch=2,type="p")
legend(x=10,y=4.25,legend=c("Predicted","Actual"),pch=c(1,2))
dev.new()
plot(x=Time,y=Adjusted,main="Adjusted Response Values using Model
Predictions",ylab="Adjusted = Y - Prediction + Nominal",ylim=c(4.5,6.5))
abline(h=LPL,lty=2)
abline(h=UPL,lty=2)
#abline(h=LTL,lty=3)
#abline(h=UTL,lty=3)
abline(h=Y.mid,lty=1)
text(x=3,y=Y.nom+0.05,labels=Y.nom.txt)
#text(x=3,y=Y.mid+0.05,labels=Y.mid.txt)
text(x=3,y=LPL+0.05,labels=LPL.txt)
text(x=3,y=UPL+0.05,labels=UPL.txt)
#
df.output <- c(Y,Time,C1,Y.predicted,Adjusted)
write.csv(df.output,"20220427 Example 12.1 Control Variable Output.csv")
#####
```

Fig. 7.17 (continued)

The function $f(t)$ may be a fixed function of time, with no estimation of parameters required, or a parametric form where parameters must be estimated by a regression technique. Then the predicted values of the response would be:

$$\widehat{y}_t = f(t) + \widehat{\gamma} C1$$

If it was desired for the response to be equal to some fixed value, y_0, then:

Fig. 7.18 Model fit statistics

```
Call:
lm(formula = Y ~ C1)

Residuals:
    Min      1Q  Median      3Q     Max
-0.5955 -0.2055  0.0095  0.2014  0.7070

Coefficients:
             Estimate  Std. Error  t value   Pr(>|t|)
(Intercept)   0.01050     0.10835    0.097    0.923
C1            0.89750     0.02284   39.291   <2e-16 ***
---
Signif. codes:  0 '***' 0.001 '**' 0.01 '*' 0.05 '.' 0.1 ' ' 1
```

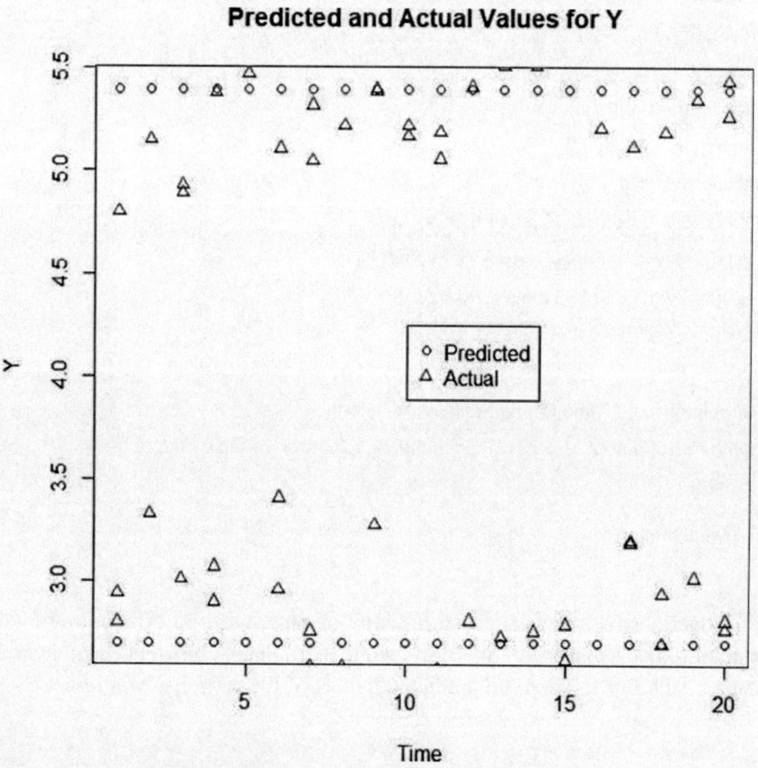

Fig. 7.19 Data and predicted values from $C1$ experiment

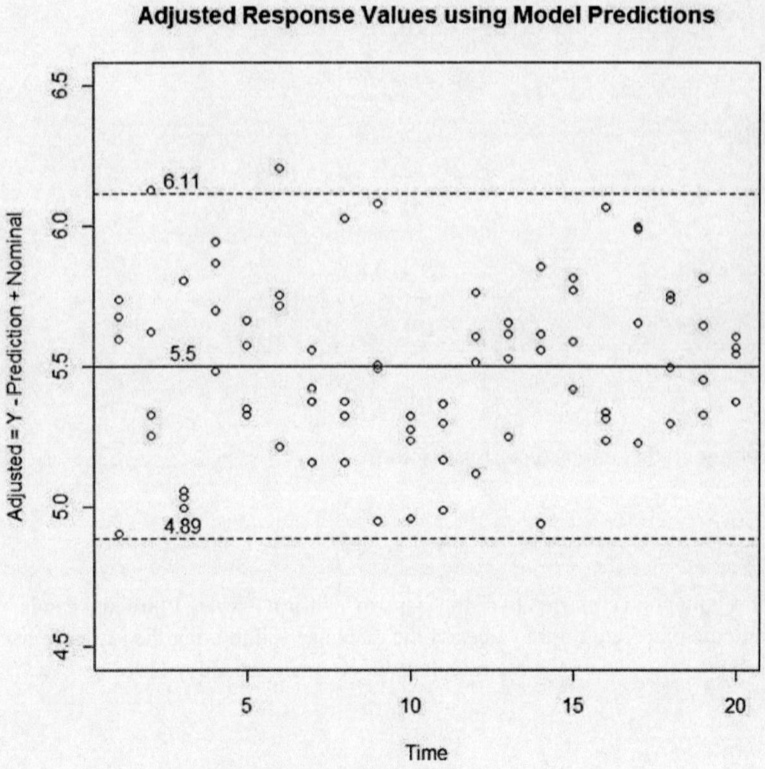

Fig. 7.20 Adjusted response values with prediction limits

$$\widehat{C1}_t = \frac{y_0 - f(t)}{\widehat{\gamma}}$$

This equation implies that the control variable values would change with time, based on the function *f*(*t*). That would mean the controller would need to be adjusted as time moved on.

Autoregressive responses may require some more involved methods. For example, a model of the form:

$$y_t = \beta_2 y_{t-2} + \beta_1 y_{t-1} + f(t) + \varepsilon$$

could be expressed as:

$$y_{t-2} - \alpha_1 y_{t-1} - \alpha_0 y_t = g(t, \varepsilon)$$

Then let:

$$x_{t-1}^{(2)} = y_{t-2} = \alpha_1 y_{t-1} + \alpha_0 y_t + g(t, \varepsilon)$$
$$x_{t-1}^{(1)} = x_t^{(2)} = y_{t-1}$$

Then:

$$x_t^{(1)} = y_t$$

So:

$$x_{t-1}^{(2)} = \alpha_1 x_t^{(2)} + \alpha_0 x_t^{(1)} + g(t, \varepsilon)$$
$$x_{t-1}^{(1)} = x_t^{(2)}$$

Putting all this into a vector/matrix form:

$$\begin{bmatrix} x_{t-1}^{(1)} \\ x_{t-1}^{(2)} \end{bmatrix} = \begin{bmatrix} 0 & 1 \\ \alpha_0 & \alpha_1 \end{bmatrix} \begin{bmatrix} x_t^{(1)} \\ x_t^{(2)} \end{bmatrix} + \begin{bmatrix} 0 \\ g(t, \varepsilon) \end{bmatrix}$$

This equation is referred to as the state-space form (Cryer, 1986) and can be used to compute predicted future values of the response and to bring the response back to its desired value, using the methodology of Kalman and Bucy (1961).

7.7 Summary

Statistical process control is about two things:

1. Monitoring a process to determine whether it is in a state of statistical control
2. Using data to returning a process to the state of statistical control

Generally, the process is monitored through estimating parameters, most commonly means, standard deviations, coefficients of variation, and proportions of nonconforming items, over time. At discrete time points, a sample of items is obtained, and a sample statistic, generally a parameter estimate, is computed. Each sample is referred to as a subgroup. The subgroup estimates are plotted against time. Statistical limits are computed, based on the aggregate of all the subgroups. The number of subgroups with estimates falling outside the limits is counted. If this number is greater than what would be expected, the process is said to be out of statistical control. Otherwise, the process is said to be in statistical control, meaning that departures from the aggregated parameter estimate are within what would be expected due to random fluctuations.

If a process is found to be out of control, then there are a number of possible ways to bring the process into a state of statistical control. The methods described in this chapter were based on doing experiments with the response and a set of control variables, i.e., variables whose values can be altered by process operators, and which will have an effect on the response. The experimental data are used to fit a predictive model for the response, and this model is in turn used to determine the values to which the control variables should be set.

References

Cryer, J. D. (1986). *Time series analysis*. Duxbury Press.

Grant, E. L., & Leavenworth, R. S. (1980). *Statistical quality control* (5th ed.). McGraw-Hill.

Kalman, R. E., & Bucy, R. S. (1961). New results in linear filtering and prediction theory. *Journal of Basic Engineering, 83*, 95–108.

Kang, C. W., Lee, M. S., Seong, Y. J., & Hawkins, D. M. (2007). A control chart for the coefficient of variation. *Journal of Quality Technology, 39*(2), 151–158.

Reklatis, G. V., Ravindran, A., & Ragsdell, K. M. (1983). *Engineering Optimization: Methods and Applications*. John Wiley and Sons, New York

Schilling, E. G. (1982). *Acceptance Sampling in Quality Control*. Marcel Dekker, Inc., ASQC Press

Shewhart, W. A. (1931). *The economic control of quality of manufactured product*. D. Van Nostrand, Inc.

Chapter 8
Inspection and Acceptance Sampling

Abstract Acceptance sampling is used as a method for making lot/product release decisions. The notion of the operating characteristic curve is introduced. Sampling by attributes and by variables are described.

8.1 Inspection and Acceptance Sampling

Despite the notion that you cannot sample quality into a product, the need for acceptance sampling persists. The core idea is that first there is some idea of how good a population must be in order to be acceptable. Then, a sampling plan is formulated, usually consisting of a sample size for a random sample, and some acceptance criterion for a sample statistic. After sampling and computing the statistic, if the statistic's value satisfies the criteria, and then the population is considered to be acceptable. If the statistic's value does not satisfy the criteria, then the population is "rejected," i.e., it is not considered acceptable without some kind of intervention.

Acceptance sampling can be classified into two different categories:

1. Attributes – generally the data are discrete binary variables, where the states are classified as either "good" or "bad." The test statistic is either the number of "good" results (or the number of "bad" results) out of a sample of n observations.
2. Variables – the statistic is a continuously valued function of (usually) a continuously valued variable. The statistic is often a function of two other statistics, e.g., sample mean and standard deviation.

Attribute sampling plans will be discussed first.

8.2 Attribute Sampling Plans

8.2.1 Risk

The simplest attribute plans use a simple random sample of n items. Before sampling, a rule must be established for deciding whether or not the population from

S. A. Pardo, *Statistical Methods and Analyses for Medical Devices*,
https://doi.org/10.1007/978-3-031-26139-8_8

which the n items were chosen is acceptable or not, with respect to some characteristic. Acceptability means that either the proportion of "good" items in the populations is sufficiently high or, conversely, the proportion of "bad" items is sufficiently low. High and low are relative concepts, so the definition of "sufficient" is context-sensitive. There is no universal definitions for sufficiently high or low. Sometimes they are defined by a regulatory authority, such as the Food and Drug Administration (FDA), as dictated by the Food, Drug, and Cosmetic Act (U.S. Food and Drug Administration, 2022). Often, however, the definition is either determined historically or in a very conservative fashion. That is to say, rather than deciding that X% is sufficiently high or low, the decision-makers fix the minimum number of "good" items that must be observed in a sample to decide that the population is acceptable to be 100% of the sample, i.e., n.

The question is, "What are the consequences of deciding the population is acceptable if we sample n items and we must have n "good" times in the sample?" The answer is dependent upon what proportion of the populations is actually "good."

You may have already concluded a few things. First, are there an infinite number of items in the population, and if so, what does it even mean that a proportion of them are "good"? Secondly, even if the population is infinite, is it possible to find the only n "good" times in the whole population, just by luck?

The answer to the first question is that we cannot truly expect a "proportion" to apply to an infinite population. Rather, we talk about the probability that a randomly selected item will be "good." Our intuition about probability may lead us to think of sampling a very large number of items and counting the number of good items. The problem is that if there are an infinite number of items in the population, then there are potentially an infinite number of "good" items. If there were a finite number of good items, then computing the ratio of good items found to total number of items counted would eventually, if we continued counting items, result in a number very close to zero.

OK, so let us presume that there are an infinite number of items in the population and an infinite number of good items in the population and that there is a number between 0 and 1 which indicates something about whether the next item we choose will be "good" or "bad."

There is a fairly simple formula for computing the probability of getting some fixed number of "good" items out of n items sampled randomly form an infinite population. If X is the number of good times in the sample, and you want that number to be at least x_c before concluding that the population is acceptable, and p represents the probability that a randomly selected item is "good," then the formula is:

$$\Pr\{X \geq x_c | n, p\} = \sum_{k=x_c}^{n} \binom{n}{k} p^k (1-p)^{n-k} = 1 - \sum_{k=0}^{x_c-1} \binom{n}{k} p^k (1-p)^{n-k}$$

This is a single equation in three unknowns: n, p, x_c. Usually, that value of n is chosen initially based on cost/availability. Often the value of x_c is chosen to be some proportion of n (often x_c is chosen to be n). The value of p is sometimes also chosen a

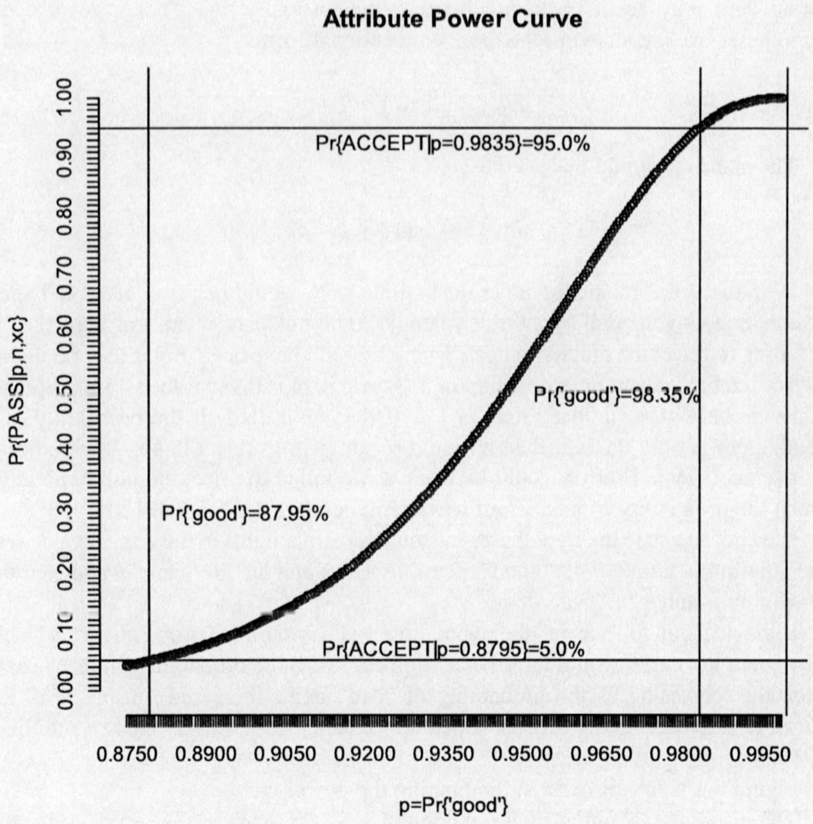

Fig. 8.1 Power curve for attribute sampling plan ($n = 50, x_c = 48$)

priori, often a number "close" to 1 (e.g., 0.90, 0.95, 0.99). This discussion may seem familiar; in fact, it is exactly the same situation as described in Chap. 4. Figure 8.1 shows the curve of the probability of observing at least $x_c = 48$ "good" items out of $n = 50$ items, sampled from an infinite population, as a function of p, the probability that a randomly selected item is "good."

This power curve relates to the hypotheses:

$$H_0 : \Pr\{'\text{good}'\} < p_0$$

Versus the alternative:

$$H_1 : \Pr\{'\text{good}'\} \geq p_0$$

If Pr{"good"} $= p_0 = 0.8795$, then there is a probability of making Type I error approximately equal to 5%. Similarly, there is Type II error probability of approximately 95% if Pr{"good"} $= 0.9835$. The statements about Type I and Type II errors here may seem backward because they actually are. That is to say, the hypotheses are not stated in the more conventional forms:

$$H_0 : \Pr\{'\text{bad}'\} = 1 - p_0$$

The alternate would be stated as:

$$H_1 : \Pr\{'\text{bad}'\} \leq 1 - p_0$$

With these formulations, the critical number, x_c, would be $50 - 48 = 2$. Type I error is rejecting the null hypothesis when it should not be rejected, and Type II error is failing to reject the null hypothesis when it should have been. So, if the population is "acceptable" when the probability of a "good" item is 0.9835, then it is acceptable if the probability of a "bad" item is $1 - 0.9835 = 0.0165$. If the probability of a "bad" item is only 0.0165, then it would be an error to reject this null hypothesis. Conversely, Type II error would be made if we failed to reject the null hypothesis when the probability of a bad item was as high as $1 - 0.8795 = 0.1205$.

We chose to state the hypotheses in a more positive light, so the critical value was the minimum number of "good" items to be found in the sample and not the maximum number of "bad" items.

If the decision to "accept" the population was based on the probability of a "bad" item, then the curve would have been "flipped," with the probability of an "accept" decision decreasing as the probability of "bad" items increased. In that form the curve is often referred to as an "operating characteristic" (OC) curve (Schilling, 1982).

Figure 8.2 shows R code for computing the power curve.

The discussion about attribute sampling has so far made the assumption that the population from which samples are taken is of infinite size. When the population is finite, a different distribution should be used to compute OC curves, the hypergeometric. The hypergeometric probability mass function:

$$\Pr\{X = k | n, p, N\} = \frac{\binom{pN}{k}\binom{N - pN}{n - k}}{\binom{N}{n}}$$

$N =$ population size
$n =$ sample size
$p =$ population proportion of items having the characteristic of interest

Keep in mind that each item is either "interesting" or "not interesting" and $X =$ the number of interesting items in the sample of size n.

The "population" is often a lot of product, or the total number of items in a production run. Even if the population size is relatively small (say 500 and greater), the binomial approximation is often quite close.

Let $k =$ the maximum number of nonconforming items allowed in the sample. Figure 8.3 has a script for computing and comparing OC curves with both

```
setwd("<your path here>")
#df1 <- read.csv("20151228 Binomial Inputs.csv")
#
# Inputs:
# n
# p
# Xc
#
Xc <- 48
n <- 50
#attach(df1)
alpha <- 0.05
pr <- seq(from=0.875,to=0.999,by=0.0005)
probpass <- 1-pbinom(Xc-1,size=n,prob=pr)
plot(pr,probpass,type="b",main="Attribute Power
Curve",xlab="P=Pr{'good'}",ylab="Pr{PASS|p,n,xc}",axes=FALSE,ylim=c(0.00,1.01))
axis(side=1,at=pr)
axis(side=2,at=seq(from=0.00,to=1.00,by=0.01))
text(x=0.8975,y=0.30,labels="Pr{'good'}=87.95%")
text(x=0.9679,y=0.50,labels="Pr{'good'}=98.35%")
text(x=0.9370,y=0.0700,labels="Pr{ACCEPT|p=0.8795}=5.0%")
text(x=0.9370,y=0.9250,labels="Pr{ACCEPT|p=0.9835}=95.0%")

abline(v=1.00)
abline(h=0.95)
abline(v=0.8795)
abline(v=0.98350)
abline(h=0.05)
q95 <- qbinom(p=0.95,size=n,prob=pr,lower.tail=TRUE)
q05 <- qbinom(p=0.05,size=n,prob=pr,lower.tail=TRUE)
dev.new()
plot(x=pr,y=q95,type="b",main="95th Percentiles for Binomial Distributions",ylab="95th
Percentile")

df2 <- cbind(n,pr,Xc,probpass,q95,q05)
write.csv(df2,"20210811 Attribute Power Curve No to Pass Outputs.csv")
#####
```

Fig. 8.2 Power curve attribute sampling plan

```
#setwd("<your path here>")
hcdfobs <- c()
hcdfless <- c()
bcdfobs <- c()
bcdfless <- c()
#
N <- 500 #population size
p.pop <- seq(from=0.01,to=0.99,by=0.01)
D <- p.pop*N

sampsize <- 30
observed <- 1

hcdfobs <- phyper(q=observed,m=D,n=N-D,k=sampsize,lower.tail=TRUE) # Pr{no.
interesting items in sample <= observed}
hcdfless <- phyper(q=observed-1,m=D,n=N-D,k=sampsize,lower.tail=TRUE)
hpdf <- hcdfobs - hcdfless # Pr{no. interesting items in sample also in the D items =
observed}
#
# Binomial approximations
#
bcdfobs <- pbinom(q=observed,size=sampsize,prob=p.pop,lower.tail=TRUE)
bcdfless <- pbinom(q=observed-1,size=sampsize,prob=p.pop,lower.tail=TRUE)
bpdf <- bcdfobs - bcdfless
#
plot(x=p.pop,y=100*hcdfobs,main="OC Curves: Hypergeometric and
Binomial",pch=1,xlab="Population Pr{non-conforming}",ylab="Pr{X<=k}(%)")
points(x=p.pop,y=100*bcdfobs,pch=2)
legend(x=0.5,y=60,legend=c("hypergeometric","binomial"),pch=c(1,2))
#####
```

Fig. 8.3 Computing OC curves with hypergeometric and binomial distributions

hypergeometric and binomial distributions. Figure 8.4 shows the OC curves with $N = 500$, $n = 30$, $k = 1$.

The closeness of the binomial approximation to the hypergeometric depends on N, n, p, and k. In general, the approximation will get worse as the sample size, n, gets closer to the population size, N.

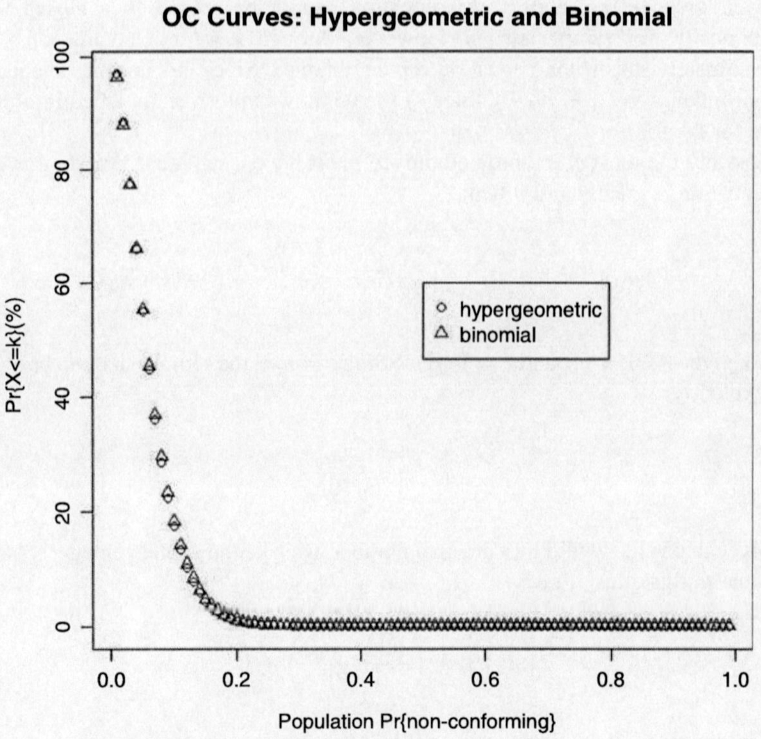

Fig. 8.4 OC curves with hypergeometric and binomial distributions

8.2.2 *Sample Size Considerations for Attribute Plans*

For an attribute sampling plan $(c, n) = (c =$ max. no. nonconforming in the sample, $n =$ sample size), the probability of *pass*ing given $p = \Pr\{$an individual item is in-spec$\}$ is given by the binomial formula. If $p_0 =$ the maximum allowable probability of finding a nonconforming item in the population (also called acceptable quality limit, or AQL), then:

$$\Pr\{X \leq k|n, p_0\} = \sum_{i=0}^{k} \binom{n}{i} p_0^i (1 - p_0)^{n-i}$$

Suppose further we wanted some specific level of assurance, call it P, that the test would result in a *pass* if in fact the probability of a nonconforming item in the population was p_0. Then we would have the equation:

$$\Pr\{X \leq c|n, p_0\} = \sum_{i=0}^{c} \binom{n}{i} p_0^i (1 - p_0)^{n-i} = P$$

Even given c, p_0, and P, this equation cannot be solved in a closed form, algebraically, for sample size n. However, through iteration, a value of n that approximately satisfies the equation can be obtained. Since this is a single equation in four unknowns (c, n, p_0, P), three of the unknowns must be specified in order to solve for the fourth.

Possibly the most commonly employed plans have c set equal to 0. In this case, the equation is greatly simplified:

$$\Pr\{X \leq c|n, p_0\} = \sum_{i=0}^{c} \binom{n}{i} p_0^i (1-p_0)^{n-i} = P = (1-p_0)^n$$

So, given AQL $= p_0$, and $P = \Pr\{\text{PASSing}|p = p_0\}$, the sample size can be solved for explicitly:

$$n = \frac{\ln(1-p_0)}{\ln(P)}$$

Since this value would be a decimal number, we take the greatest integer less than or equal to this, plus 1, i.e.:

$$n = \text{trunc}\left(\frac{\ln(1-p_0)}{\ln(P)}\right) + 1$$

Sample size calculations for variables sampling plans are more complicated. There are some approximations, but none are exact as in the case of the binomially based attribute sampling plans.

8.3 Variable Sampling Plans and *Cpk*

8.3.1 The Population Cpk

Suppose some measurement, dimension, or another continuously valued quantity, X, has acceptable tolerance limits (L, U). Furthermore, suppose that X is normally distributed with mean μ and standard deviation σ. Then we would be able to compute a new variable, Z, which would also have a normal distribution, with $\mu = 0$ and $\sigma = 1$:

$$Z = \frac{X - \mu}{\sigma}$$

This variable is referred to as a standard normal variable. The value of Z at a given value of $X = x$ is called the z-score for x. This standardized normal variable was very useful in the days before computers were so ubiquitous. Values of the cumulative probability distribution for the standard normal could be generated and tables made. Then, provided the values of μ and σ were known, values for the CDF of any normally distributed variable could be computed since, for example:

$$\Pr\{L \leq X \leq U\} = \Pr\{z_L \leq Z \leq z_U\}$$

$$z_L = \frac{L - \mu}{\sigma}$$

$$z_U = \frac{U - \mu}{\sigma}$$

It turns out that for any normally distributed variable:

$$\Pr\{\mu - 3\sigma \leq X \leq \mu + 3\sigma\} \approx 0.9973$$

That means:

$$\Pr\{-3 \leq Z \leq +3\} \approx 0.9973$$

So, if:

$$L \leq \mu - 3\sigma$$

and:

$$U \geq \mu + 3\sigma$$

Then there is at least a 99.73% chance that a randomly selected item would have a value of *X* between the limits *L* and *U*. In other words:

$$z_L = \frac{L - \mu}{\sigma} \leq -3$$

and:

$$z_U = \frac{U - \mu}{\sigma} \geq +3$$

For the moment, suppose that having a 99.73% chance that a randomly selected item would have an *X* value in the range (*L*, *U*) would be acceptable. Dividing the two quantities by 3 would yield:

$$\frac{z_L}{3} = \frac{L - \mu}{3\sigma} = -1$$

and:

$$\frac{z_U}{3} = \frac{U - \mu}{3\sigma} = +1$$

If the sign on z_L were reversed, then the two quantities would become:

$$C_{pL} = \frac{-z_L}{3} = \frac{\mu - L}{3\sigma} = +1$$

and:

$$C_{pU} = \frac{z_U}{3} = \frac{U - \mu}{3\sigma} = +1$$

For some reason, the value +1 seemed more appealing than the range (−3,+3), or simply +3. Nevertheless, presuming that the tolerance limits may not be symmetrically placed around the mean, μ, we might say that the population is acceptable if:

$$C_{pk} = \text{Min}\left(C_{pL}, C_{pU}\right) \geq +1$$

So, if we were to sample n items and measure the continuous variable X on each of them, we might want to use those data to test the hypothesis:

$$H_0 : C_{pk} < +1$$

against the alternative:

$$H_1 : C_{pk} \geq +1$$

8.3.2 Sample Cpk *and Hypothesis Tests*

Sample C_{pk}, or $\widehat{C}_{pk}$, is the lesser of two quantities:

$$\widehat{C}_{pL} = \frac{\overline{X} - L}{3S}$$

and:

$$\widehat{C}_{pU} = \frac{U - \overline{X}}{3S}$$

where:

$$\overline{X} = \frac{1}{n}\sum_{i=1}^{n} x_i$$

$$S = \sqrt{\frac{\sum_{i=1}^{n}\left(x_i - \overline{X}\right)^2}{n-1}}$$

That is:

$$\widehat{C}_{pk} = \min\left\{\widehat{C}_{pL}, \widehat{C}_{pU}\right\}$$

Of course, $x_i =$ the result obtained with item i, $i = 1, n$.

The population passes the test if the sample statistics, $\widehat{C}_{pL}$ and $\widehat{C}_{pU}$, both equal or exceed some (minimum) critical value for $\widehat{C}_{pk}$.

It can been shown (Kushler & Hurley, 1992; Pardo, 2014) that the sampling distribution of either $\widehat{C}_{pL}$ or $\widehat{C}_{pU}$ is that of a noncentral T distribution, divided by $3\sqrt{n}$, with noncentrality parameter $\delta = 3\sqrt{n}K_0$, where K_0 is $\frac{1}{3}$ the desired absolute value of the z-score for the limits L or U (in this case, the null distribution is symmetric about the target, T, so that $|z_L| = |z_U|$). A hypothesis test would have the following null and alternative hypotheses:

$$H_0 : C_{pk} < K_0$$

versus the alternative:

$$H_1 : C_{pk} \geq K_0$$

So, in the previously mentioned hypotheses, $K_0 = +1$. The first problem is to find a value, call it c, such that, for a given sample size, n:

$$Pr\left\{\widehat{C}_{pk} \geq c | K_0 = +1, n\right\} \approx 1 - \alpha$$

That is to say, if it is really true that $K_0 = +1$, we want a fairly high chance that the sample *Cpk* will be high enough to reject H_0 in favor of H_1. In other words, if the sample *Cpk* is greater than or equal to c, then we will reject H_0. More often than not, the value of α is chosen to be 0.05.

The critical value for $\widehat{C}_{pL}$ or $\widehat{C}_{pU}$ should be the 100(0.05)th percentile of a noncentral T with $n - 1$ of freedom and non-centrality parameter $\delta = 3\sqrt{n}K_0$.

The first problem is that we cannot solve a single equation with two unknowns, namely, c and n. First presume that n is chosen to satisfy some external requirement, such as regulatory guidance documents, or perhaps based on test fixture capacities. Then we must vary c to find the value that gives us the desired probability:

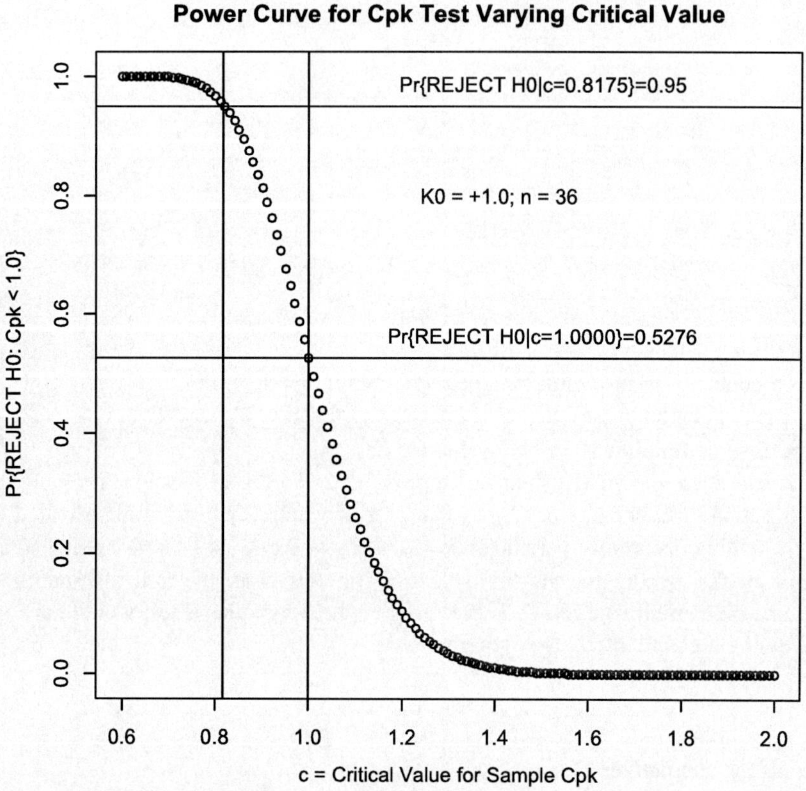

Fig. 8.5 Power curve for Cpk test, varying critical values

$$Pr\left\{\widehat{C}_{pk} \geq c | K_0 = +1, n\right\} \approx 1 - 0.05 = 0.95$$

For the sake of illustration, suppose $n = 36$. Figure 8.5 shows the power curve for different critical values, c, for sample *Cpk*.

If the population's *Cpk* is 1.0, then setting the critical value to 1.0 would yield a relatively low chance of rejecting H_0, i.e., passing the test. In particular, if the critical value, c, is equal to 1.0, and the population *Cpk* is also 1.0, then with $n = 36$, there is approximately a 52.67% chance that the sample *Cpk* will exceed 1.0. That is to say, if we set the critical value of sample *Cpk* to be 1.0, then we have an approximately 52.76% chance of passing the test when the population is actually acceptable.

Conversely, if the critical value for sample *Cpk* is 0.8175, then if the population *Cpk* is 1.0, there is an approximately 95% chance of passing the test.

Since we do not actually know what the population's *Cpk* is, we can only determine the likelihood of passing, i.e., getting a sample *Cpk* greater than or equal to the critical value, under different possible values of the population *Cpk*.

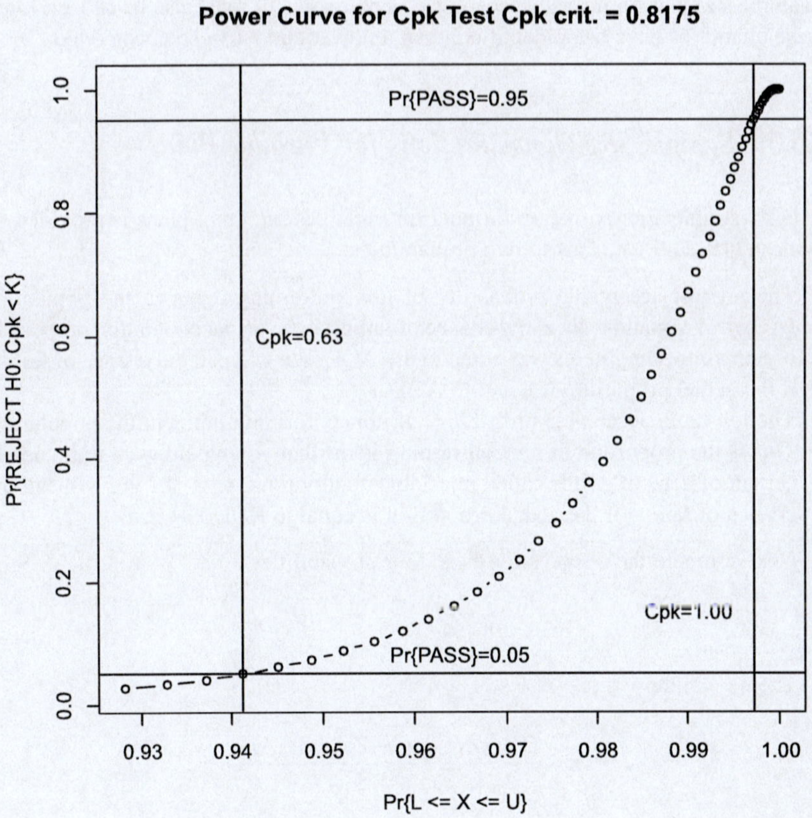

Fig. 8.6 Power curve with critical value at sample $Cpk = 0.8175$

So, with $n = 36$, and the critical value fixed at 0.8175, we can compute the probability of passing (sample *Cpk* greater than or equal to 0.8175).

Figure 8.6 shows the power as a function of the probability that the variable X is within specification limits (L, U). Of course, $\Pr\{L \le X \le U\}$ has a one-to-one, onto relationship to the population *Cpk*. For example, a *Cpk* of 0.63 corresponds to $\Pr\{L \le X \le U\} \approx 0.9412$ and a *Cpk* of 1.00 corresponds to $\Pr\{L \le X \le U\} \approx 0.9973$.

With $n = 36$, and a critical value of $c = 0.8175$, there is about a 5% chance that the sample *Cpk* $(\widehat{Cpk})$ would be at least 0.8175 if the population $Cpk = 0.63$. Similarly, there is about a 95% chance that the sample *Cpk* would be at least 0.8175 if the population $Cpk = 1.00$.

In summary, a "variables" acceptance sampling plan involves a continuously valued measurement, X, with some specification limits on its value. The parameter *Cpk* is used to indicate the probability that a randomly selected item would have an X value within the specification limits. A critical value is chosen for the sample *Cpk*. If the sample *Cpk* meets or exceeds the critical value, then the population is considered acceptable and passes the test. In addition to choosing a critical value, a sample size and a minimally acceptable population *Cpk* must also be chosen. Once these quantities have been identified, then a power curve can be constructed.

8.3.3 Sample Size Considerations for Variable Plans

A fairly simple approximation formula for variables sampling plans is provided by Duncan (1965). First, specify two probabilities:

1. The highest acceptable probability of nonconforming items in the population (AQL). We would want a low chance of failing if the actual population proportion of nonconforming items was equal to the AQL. We will call the chance of failing if the actual proportion is equal to AQL α.
2. The lowest unacceptable probability of nonconforming items in the population. This is the proportion of nonconforming items that you would want to be nearly certain of detecting. It is called rejectable quality limit, or RQL. We will call the chance of failing if the actual proportion is equal to RQL $1-\beta$.

Next compute the z-scores for these four probabilities:

$$Z_{1-\text{AQL}}, Z_{1-\text{RQL}}, Z_{1-\alpha}, Z_{1-\beta}$$

Then compute:

$$\kappa = \frac{Z_{1-\alpha}Z_{1-\text{RQL}} + Z_{1-\beta}Z_{1-\text{AQL}}}{Z_{1-\alpha} + Z_{1-\beta}}$$

Finally, compute:

$$n = \text{trunc}\left[\left(1+\frac{\kappa^2}{2}\right)\left(\frac{Z_{1-\alpha}+Z_{1-\beta}}{Z_{1-\text{AQL}}-Z_{1-\text{RQL}}}\right)^2\right]+1$$

Keep in mind that this is an approximation. Your results may vary.

8.4 Confidence Limits for *Cpk*

There are several formulas for approximate confidence intervals for Cpk. Here is the one used in this book (Kushler & Hurley, 1992):

$$\widehat{C}_{pk}(1-\alpha) = \widehat{C}_{pk} - z_{1-\alpha}\sqrt{\frac{1}{9n}+\frac{\widehat{C}_{pk}^2}{2(n-1)}}$$

The quantity $z_{1-\alpha}$ is the $100(1-\alpha)$th percentile of a standard normal distribution.

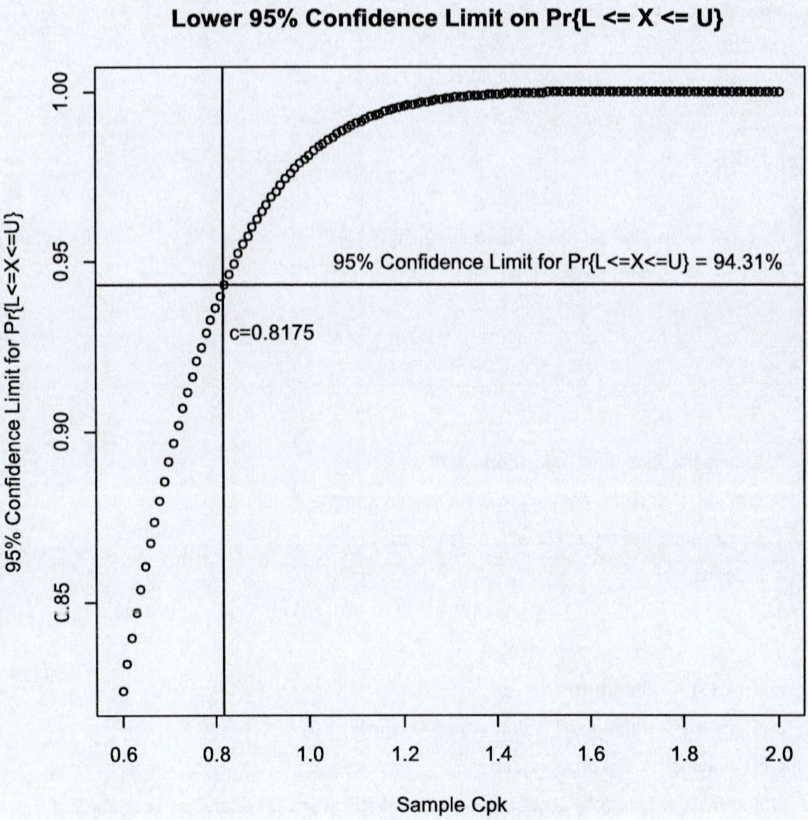

Fig. 8.7 Lower 95% confidence limits for $\Pr\{L \leq X \leq U\}$ based on confidence limits for *Cpk*

Inasmuch as the larger *Cpk* is, the better, only a lower confidence limit will be considered.

The lower confidence limit on *Cpk* can then be used to compute a lower confidence limit on the probability $\Pr\{L \leq X \leq U\}$:

$$P_{1-\alpha} = 1 - 2\Phi\left(-3\widehat{C}_{pk}(1-\alpha)\right)$$

and $\Phi(z)$ is the cumulative distribution function for a standard normal variable.

Figure 8.7 shows the lower 95% ($\alpha = 0.05$) limits on $\Pr\{L \leq X \leq U\}$ as a function of potential values for sample *Cpk*, with $n = 36$. Note that if the sample *Cpk* is equal to 0.8175 (the critical value chosen earlier), then the lower 95% confidence limit on $\Pr\{L \leq X \leq U\}$ would be approximately 94.31%.

Figure 8.8 shows R code for computing the curves shown in Figs. 8.2 and 8.3.

```
setwd("<your path here>")
n <- 36
n.tests <- 1
alpha_prime <- 0.05/n.tests
K0 <- 1.00
#
Cpk_crit <- seq(from=0.60,to=2.00,by=0.01)
#df_Cpk <- cbind(n,prob_inspec,Z0,K_list,Cpk_crit)

#
#
#
# This is the main part of the code.
# crit_val_Cpk is the critical value (minimum sample
# Cpk required to reject H0 with 95% probability
# if Cpk = K0)
#
prob_null <- 1 - 2*pnorm(-3*K0)
noncenT.1 <- (3*sqrt(n))*Cpk_crit
prob_reject <- pt(noncenT.1,df=n-1,ncp=3*sqrt(n)*K0)
prob_accept.1 <- 1 - prob_reject
plot(x=Cpk_crit,y=prob_accept,type="b",main="Power Curve for Cpk Test Varying
Critical Value",xlab="c = Critical Value for Sample Cpk",ylab="Pr{REJECT H0: Cpk <
1.0}")
abline(v=K0)
abline(v=0.8175)
abline(h=0.52756)
abline(h=0.95)
text(x=1.4,y=0.80,labels="K0 = +1.0; n = 36")

text(x=1.5,y=0.9675,labels="Pr{REJECT H0|c=0.8175}=0.95")
text(x=1.5,y=0.5450,labels="Pr{REJECT H0|c=1.0000}=0.5276")
#
#
#
dev.new()
```

Fig. 8.8 R code for variable (*Cpk*) sampling plan power curves

```
c.fixed <- 0.8175
noncenT.2 <- (3*sqrt(n))*c.fixed
prob_reject.2 <- pt(noncenT.2,df=n-1,ncp=3*sqrt(n)*Cpk_crit)
prob_accept.2 <- 1 - prob_reject.2

prob_within <- 1 - 2*pnorm(-3*Cpk_crit)
prob_good <- 1 - 2*pnorm(-3*1.0)

plot(x=prob_within,y=prob_accept.2,type="b",main="Power Curve for Cpk Test Cpk crit.
= 0.8175",xlab="Pr{L <= X <= U}",ylab="Pr{REJECT H0: Cpk < K}")
abline(h=0.05)
abline(h=0.95)
abline(v=prob_good)
abline(v=0.9412)
text(x=0.9475,y=0.60,labels="Cpk=0.63")
text(x=0.9900,y=0.15,labels="Cpk=1.00")
text(x=0.9650,y=0.9700,labels="Pr{PASS}=0.95")
text(x=0.9650,y=0.0700,labels="Pr{PASS}=0.05")
#compute approximate lower confidence limit for Cpk given Cpk_crit
CpkLim <- Cpk_crit - qnorm(.95)*sqrt((1/(9*n)) + (Cpk_crit**2)/(2*(n-1)))
prob_Lim <- 1 - 2*pnorm(-3*CpkLim)

#
# prob_accept = probability of PASSing for Cpk plan

#
dev.new()
plot(x=Cpk_crit,y=prob_Lim,main="Lower 95% Confidence Limit on Pr{L <= X <=
U}",xlab="Sample Cpk",ylab="95% Confidence Limit for Pr{L<=X<=U}")
abline(v=0.8175)
abline(h=0.9431)
text(x=0.9200,y=0.9300,labels="c=0.8175")
text(x=1.52,y=0.9500,labels="95% Confidence Limit for Pr{L<=X<=U} = 94.31%")
#
df3 <- cbind(n,K0,Cpk_crit,prob_within,prob_accept.1,prob_accept.2,CpkLim,prob_Lim)

#write.csv(df3,"20210818 Cpk and Power.csv")
#####
```

Fig. 8.8 (continued)

```
#
# prob_accept = probability of PASSing for Cpk plan

#
dev.new()
plot(x=Cpk_crit,y=prob_Lim,main="Lower 95% Confidence Limit on Pr{L <= X <=
U}",xlab="Sample Cpk",ylab="95% Confidence Limit for Pr{L<=X<=U}")
abline(v=0.8175)
abline(h=0.9431)
text(x=0.9200,y=0.9300,labels="c=0.8175")
text(x=1.52,y=0.9500,labels="95% Confidence Limit for Pr{L<=X<=U} = 94.31%")
#
df3 <- cbind(n,K0,Cpk_crit,prob_within,prob_accept.1,prob_accept.2,CpkLim,prob_Lim)

#write.csv(df3,"20210818 Cpk and Power.csv")
#####
```

Fig. 8.8 (continued)

There are other variable acceptance sampling plans. The grandfather of this *Cpk*-based plan is very similar. Instead of *Cpk*, the parameters are:

$$z_L = \frac{\mu - L}{\sigma}$$

and:

$$z_U = \frac{U - \mu}{\sigma}$$

Notice that the quantity z_L is just the z-score for the lower specification limit, L, multiplied by -1. So, the description and explanation of the *Cpk* plan applies to the original plan, without the factor of 1/3. The plan was codified in a government document and called MIL-STD-414. Later, the plan was adopted by the American National Standards Institute (ANSI) and the American Society for Quality (ASQ), and renamed ANSI/ASQ Z1.9. The standard was adopted by the International Standards Organization (ISO) and called ISO 3951. The principals are largely the same. The big difference is that the acceptance criteria (critical values) are based directly on the probability that the variable falls outside the specification limits (L, U). The *Cpk* plan has critical values that are scaled z-scores, i.e., z-scores divided by 3.

All these variable plans may also be used if there is only a one-sided specification, i.e., either *L* or *U*. In the case of the *Cpk* plans, *Cpk* will be either *CpL* or *CpU*, depending on which limit is present.

8.5 Double Sampling Plans: Attributes

Sometimes it seems that a second chance ought to be given before a failing grade is given. It is possible to have a sampling plan that allows for a second sample to be drawn. Such a plan is called a double sampling plan. It works like this:

1. Sample *n1* items. There are three possible outcomes:

   (a) PASS
   (b) FAIL
   (c) RESAMPLE

2. If the resample decision was reached, sample other *n2* items. Based on the combined samples (*n1* + *n2*), there are now only two possibilities:

   (a) PASS
   (b) FAIL

The PASS, RESAMPLE, or FAIL decisions are based on some variable, *X*. Here we will deal with a discrete *X*, namely, the number of conforming (or nonconforming) items.

Consider this example:

1. Sample $n1 = 36$ items. Let $X1 =$ the number of conforming items in this sample. The rules are:

   (a) If $X1 \geq 34$, then PASS.
   (b) If $X1 \leq 31$, then FAIL.
   (c) If $32 \leq X1 \leq 33$, then RESAMPLE.

2. If the RESAMPLE decision was reached, then sample other $n2 = 36$ items. If the total number of conforming items (call it $X1 + X2$) is at least 68, then PASS. Otherwise, FAIL.

If in the first sample 32 out of $n1 = 36$ were conforming, then $X2 = 36$ out of the *n2* items must be conforming to PASS. If, however, 33 items out of $n1 = 36$ were conforming, then at least $X2 = 35$ out of $n2 = 36$ must also be conforming.

The primary question to ask is, what are the risk characteristics of this plan? How easy/difficult is it to PASS on the first sample? How likely is it to RESAMPLE? To PASS on the second sample?

The graph in Fig. 8.9 is the operating characteristic/power curve for the first sample of $n1 = 36$ items.

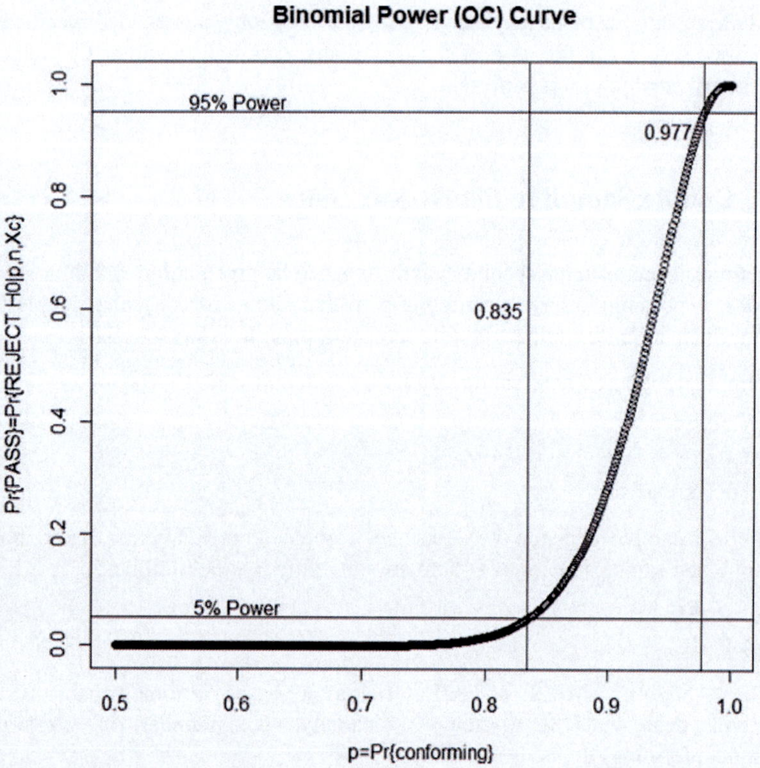

Fig. 8.9 Power (operating characteristic) curve: $n1 = 36$, $X1 = 34$

For the first sample, there is approximately a 95% chance of *pass*ing if the probability of a conforming item is about 97.7%. Conversely, there is only about a 5% chance of *pass*ing if the probability of a conforming item is only 83.5%.

This is only for the first sample. The probability of a FAIL on the first sample is shown in Fig. 8.10.

Let $c1 = 34$, $c3 = 68$, $c2 = c3 - n2 = 68 - 36 = 32$

The probability of drawing a second sample is given in Fig. 8.11.

For this sampling plan, there is at most a 23.39% chance that a RESAMPLE decision would be reached after the first sample, if Pr{conforming} = 91.7%. Suppose that $X1 = 32$ or $X1 = 33$. Then what are the chances of *pass*ing with the second sample added to the first?

The probability to PASS after the second sample would be:

$$\Pr\{\text{PASS}|n1 + n2 = 72\} = \Pr\{\text{PASS}|n1\} + \Pr\{\text{RESAMPLE}|n1\} \\ * \Pr\{X2 \geq 68 - X1|X1, n1, n2\}$$

Then the formula becomes:

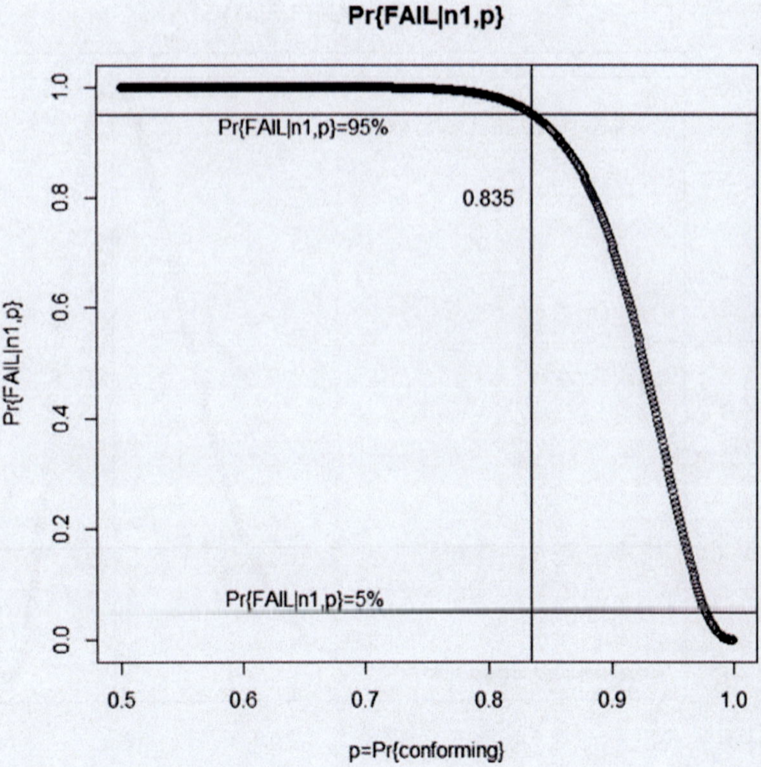

Fig. 8.10 Pr{FAIL|$n1 = 36, p$}

$$\Pr\{\text{PASS}|n1,n2,p\} = \Pr\{\text{PASS}|n1,p\} + \sum_{i=1}^{c1-c2} \times \left(\Pr\{X1 = c2 + i - 1|n1,p\} * \sum_{j=1}^{i} \Pr\{X2 = c3 - c2 - j + 1|n2,p\} \right)$$

and:

$$\Pr\{X = x|n,p\} = \binom{n}{x} p^x (1-p)^{n-x}$$

Figure 8.12 shows the power/OC curve for the overall sampling plan.

A brief note should be made concerning double variable sampling plans. While such plans do exist (Schilling, 1982), computing the OC curve is much more complicated.

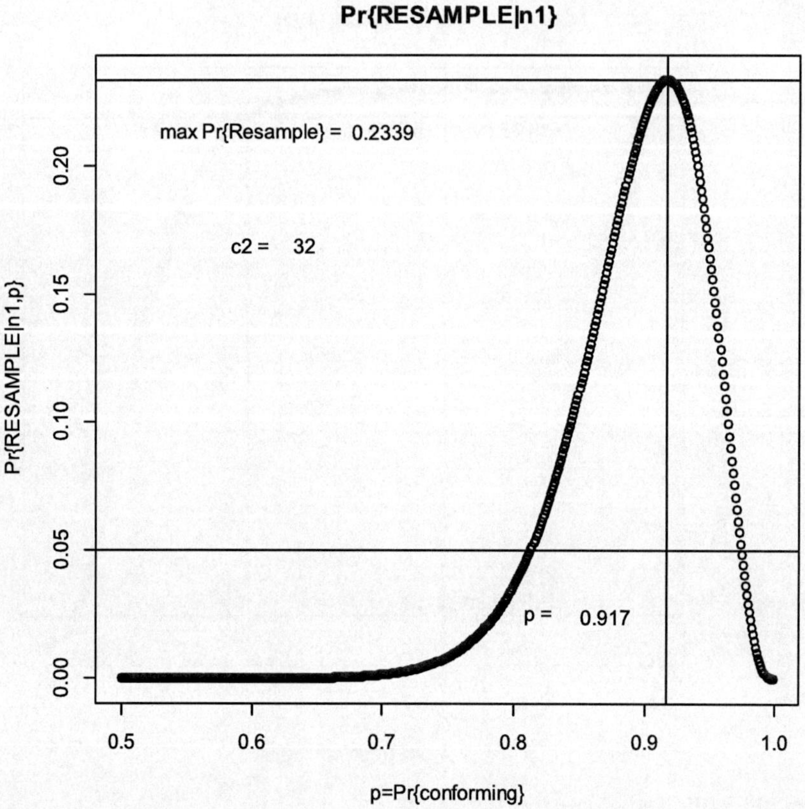

Fig. 8.11 Resampling probabilities, $32 \leq X1 \leq 33$

Furthermore, they are not frequently employed. As such, we will not address them in this work.

8.6 Sampling Plans for Precision Parameters

8.6.1 Introduction

Sometimes a process or lot of products must meet some requirement or specification for precision. That is to say, it is desired for the variability in some critical quality variable to be kept relatively low. There are only two parameters that will be addressed here:

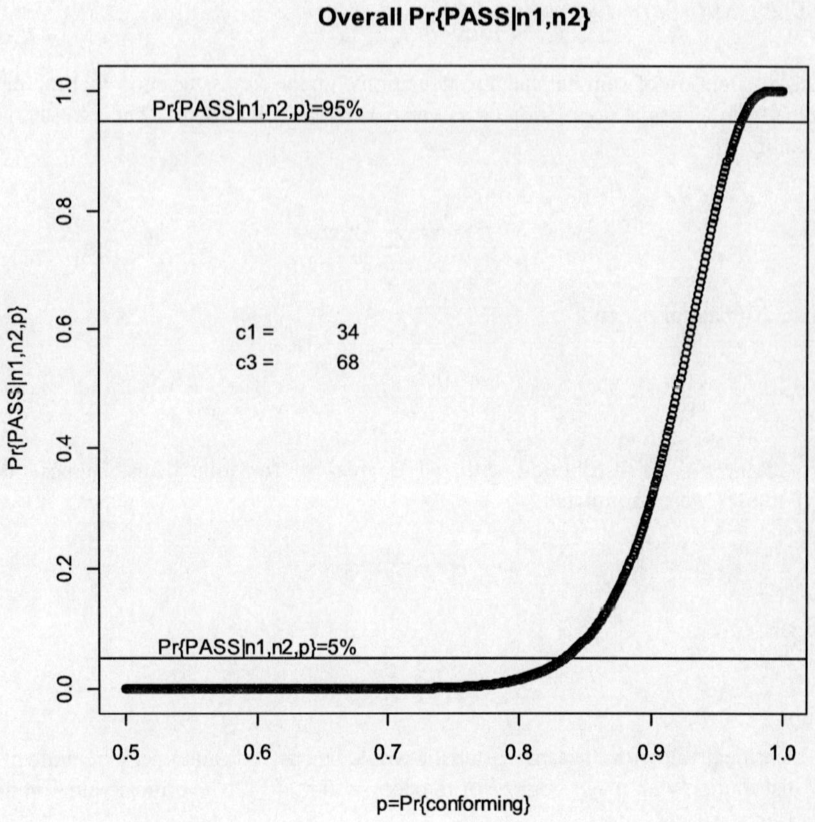

Fig. 8.12 Power curve: overall

1. Standard deviation (SD)
2. Coefficient of variation (CV)

The SD of a population is frequently symbolized with the Greek letter σ. For CV, we will use the letter c to represent the population value.

As described earlier, the coefficient of variation is a unitless quantity and is often expressed as a percentage:

$$c = \frac{\sigma}{\mu} * 100\%$$

8.6.2 Standard Deviation

The distribution of sample standard deviations, under the assumption that the data come from a normal population or universe, is more well known. That is to say, the quantity:

$$s^2 = \frac{1}{n-1}\sum_{i=1}^{n}(y_i - \bar{y})^2$$

is an estimate of σ^2, and:

$$\frac{(n-1)s^2}{\sigma^2}$$

would have a χ^2 distribution with n-1 degrees of freedom. Thus, suppose the hypotheses were formulated:

$$H_0 : \sigma^2 > \sigma_0^2$$

versus:

$$H : \sigma^2 \leq \sigma_0^2$$

A critical value for the sample statistic would be based on an upper percentile of a χ^2 distribution with $n - 1$ degrees of freedom; call it $\chi^2_{1-\alpha}$. The critical value for the statistic s^2 would be:

$$s^2_{\text{crit}} = \frac{\sigma_0^2}{n-1}\chi^2_{1-\alpha}$$

To compute an OC or power curve, all we need to do is compute:

$$\Pr\{s > s|n\} = \Pr\left\{\chi^2 > \frac{(n-1)s^2_{\text{crit}}}{\sigma_a^2}|n\right\}$$

For various values of $\sigma_a^2 \geq \sigma_0^2$.

Figure 8.13 shows an R script for computing the power (OC) curve for a sampling plan of SD where $n = 36$. Example hypotheses for the curve are:

$$H_0 : \sigma^2 > 1.59^2$$

versus:

```
setwd("<your path here>")
#
SDcrit <- 1.9
alpha<- 0.05
Calt <- c()
ncpalt <- c()
power <- c()
n <- 36
SDcrit.text <- as.character(round(SDcrit,digits=2))
#
sigalt <- seq(from=sig0,to=sig0+4,by=0.01)

power = pchisq(q=((n-1)*(SDcrit**2))/(sigalt**2),df=n-1)
sigalpha <- sigalt[which(power>0.946 & power<0.952)]
sigalpha.txt <- as.character(round(sigalpha,3))
sigbeta <- sigalt[which(power>0.049 & power < 0.051)]
sigbeta.txt <- as.character(round(sigbeta,3))
df2 <- cbind(sigalt,power,n,sig0,SDsamp)
plot(sigalt,power,axes=F,xaxt="n",yaxt="n",main="Operating Characteristic Curve for
SD",xlab="Population Sigma",ylab="Pr{PASS|Pop.Sigma}",type="b")
axis(side=1,at=seq(from=sig0,to=sig0+4,by=0.01))
axis(side=2,at=seq(from=0.0,to=1.03,by=0.01))
abline(h=0.95)
abline(h=0.05)
abline(v=sigalpha)
abline(v=sigbeta)
text(x=1.2,y=0.932,labels="95%")
text(x=1.2,y=0.075,labels="5%")
text(x=2.75,y=0.20,labels="Sigma= ")
text(x=3.20,y=0.20,labels=sigbeta.txt)
text(x=1.00,y=0.70,labels="Sigma= ")
text(x=1.40,y=0.70,labels=sigalpha.txt)
write.csv(df2,"20220202 Power for SD test.csv")
#
# s.crit should be very close to SDcrit
#
vcrit <- qchisq(0.95,df=n-1)*(sigalpha**2)/(n-1)
s.crit <- sqrt(vcrit)
s.crit
#####
```

Fig. 8.13 R script: computing an OC curve for standard deviation

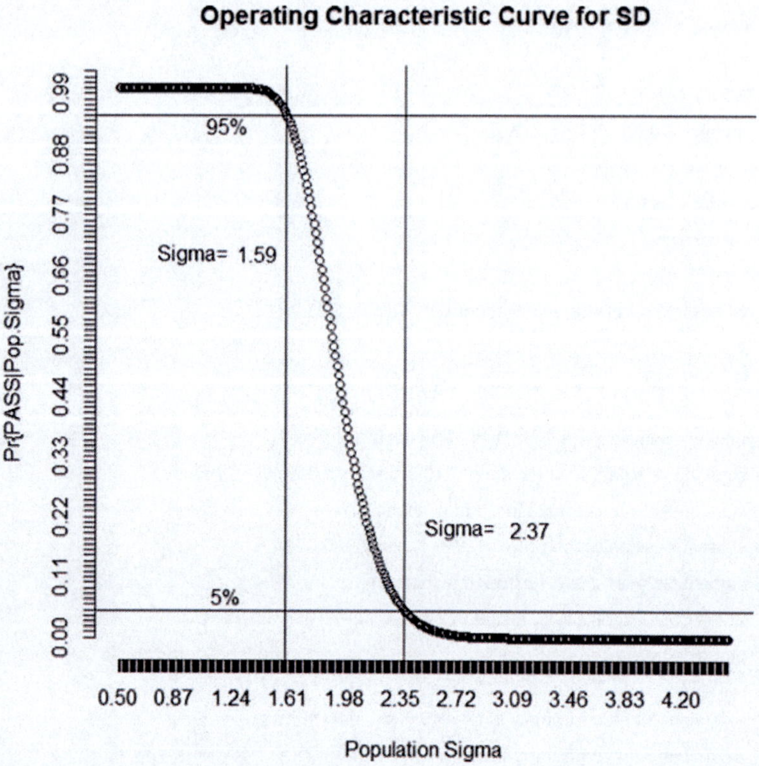

Fig. 8.14 Power (OC) curve for standard deviation plan

$$H : \sigma^2 \leq 1.59^2$$

The critical value for sample SD is approximately 1.9, which corresponds to an approximately 95% chance that a sample SD would be at or below if the population standard deviation is equal to $\sigma_0 = 1.59$.

Figure 8.14 shows the example curve with $n = 36$ and $\sigma_0 = 0.50$.

In this example, the population standard deviation, σ, must be no greater than 1.59 in order to have at least a 95% chance of a PASS, i.e., getting a sample SD less than or equal to the critical value, 1.9.

As in the case of any acceptance sampling plan, there is a rule for the number of items to sample, a statistic used to judge the sample, and a critical value for the statistic based on a hypothesis. A power or OC curve can be constructed to show the chances of *pass*ing or *fail*ing under various alternative hypothetical values of a parameter or even a set of parameters. One of the most difficult parts of creating a sampling plan is making the computations for constructing the power curve. In the cases of Cpk and SD, the determination of the distribution of the sample statistics was not difficult. In the case of CV, however, that distribution must be approximated.

8.6.3 Coefficient of Variation

Kang et al. (2007) approximated the distribution of the sample CV inversely proportional to a non-central t distribution:

$$\mathrm{cv} = \frac{\sqrt{n}}{T'}$$

T' has a non-central t distribution with degrees of freedom (df) = $n - 1$ and non-centrality parameter:

$$\mathrm{nct} = \frac{\sqrt{n}}{c_0}$$

The value of c_O is the "null" or target value of the population coefficient of variation. That is to say, c_O is the hypothetical value of CV:

$$H_0 : CV > c_0$$

This hypothesis is tested against the alternative:

$$H_1 : CV \leq c_0$$

Rejecting the null hypothesis, H_O, means a PASS.

The critical value for sample CV is computed to be that value for which there is a $1 - \alpha$ chance that the sample CV would be at or below if the population CV is equal exactly to c_O.

Figure 8.15 shows an R script for computing the OC curve for CV. The value of n is set to 36, and c_O = 1.75%. The critical value at α = 0.05 (5%) is approximately 2.08%. That is to say, if a sample CV is less than or equal to 2.08, the result is PASS. There is approximately a 95% chance of *pass*ing if the population CV actually equals 1.75%. Conversely, there is only about a 5% chance of *pass*ing (i.e., observing a sample CV less than or equal to the critical value of 2.08%) if the population CV is as high at 2.62%. Figure 8.16 shows the OC curve for this plan.

```
setwd("<your path here>")
#df1 <- read.csv("20181109 Two Sample Mean Equivalence.csv")
#
#
# Kang,C.W.,  Lee, M.S., Seong, Y.J., Hawkins, D.M. (2007)
# "A Control Chart for the  Coefficient of Variation",
# JQT, April 2007, Vol. 39, No. 2
#
# Mark Vangel (1996), "Confidence Intervals for a Normal Coefficient of Variation",
# American Statistician, Vol. 15, No. 1, pp. 21-26.
#
#
#conf <- 0.65
#conf <- 0.999975
conf <- 0.95

alpha <- 1 - conf #this has no effect in this version.  Critical Value
# is chosen external to any risk calculation
k <- 1
alpha.prime <- 1-conf**(1/k)

#alpha.prime <- alpha / k
#
# Note that in this version, alpha, alpha.prime, and conf are not used in
# Power calculations.
# C.crit is set without consideration for Type I error risk
#
n <- 36
C0 <- 1.75 #CV in %
C <- C0 / 100
C0.text <- as.character(round(C0,4))
NCT <-sqrt(n) / C #This is the noncentrality parameter
#
C.crit <- 100*sqrt(n) / qt(alpha.prime,df=n-1,ncp=NCT)
#C.crit <- 2.6 #This is an arbitrary Critical Value Specification
T.quant <- sqrt(n) / (C.crit/100)
```

Fig. 8.15 R script: computing OC curve for CV sampling plan

```
C.crit.text <- as.character(round(C.crit,2))
C.alt <- seq(from=0.01,to=0.0575,by=0.0001)
C.alt.pct <- C.alt*100
NCT.alt <- sqrt(n) / C.alt
Power <- pt(T.quant,df=n-1,ncp=NCT.alt)
PrPass <- 1 - Power
plot(x=C.alt.pct,y=Power,main="Power Curve for Testing CV Hypothesis",xlab="Pop.
CV (%)",ylab="Power=Pr{Fail}")
#
C.alt.95 <- C.alt.pct[which(Power>.949 & Power<0.952)] #This is the pop. CV that yields
95% power (Pr{Fail})
C.alt.95.txt <- as.character(round(C.alt.95,4))
text(x=4.0,y=0.50,labels="Crit. CV(%) =")
text(x=4.85,y=0.50,labels=C.crit.text)
text(x=4.0,y=0.40,labels="Target Pop. CV(%) =")
text(x=4.85,y=0.40,labels=C0.text)
abline(v=C.alt.95)
#abline(v=1.7)
abline(h=0.95)
abline(h=alpha)
text(x=2.0,y=0.975,labels="Pr{FAIL} = 95%")
text(x=4.0,y=0.03,labels="Pr{FAIL} = 5%")
text(x=3.7,y=0.875,labels="CV%(95) =  ")
text(x=4.3,y=0.875,labels=C.alt.95.txt)
#abline(h=0.05)
dev.new()
plot(x=C.alt.pct,y=PrPass,main="OC Curve (Pr{PASS}) for CV Sampling
Plan",xlab="Pop. CV (%)",ylab="Pr{PASS}")
abline(h=conf)
abline(h=alpha.prime)
abline(v=C0)
abline(v=C.alt.95)
text(x=2.85,y=0.25,labels=C.alt.95.txt)
text(x=4.0,y=0.975,labels="Pr{PASS} = 95%")
text(x=4.0,y=0.07,labels="Pr{PASS} = 5%")
text(x=4.0,y=0.50,labels="Crit. CV(%) =")
text(x=4.85,y=0.50,labels=C.crit.text)
```

Fig. 8.15 (continued)

```
text(x=4.0,y=0.40,labels="Target Pop. CV(%) =")
text(x=4.85,y=0.40,labels=C0.text)

CV.UCL <- 100*sqrt(n) / T.quant
CV.UCL
C.alt.95
#
# Vangel Modified McKay Limit (for lower limit, use qchisq(1-alpha.prime,df=n-1)
#
CV.UCL.V <- 100*C / sqrt(((qchisq(alpha.prime,df=n-1)/n) - 1)*C**2 +
qchisq(alpha.prime,df=n-1)/(n-1))
CV.UCL.V
#
# This computes an upper confidence limit for standard deviation:
#
S <- 1.3
Sig.UCL <- S*sqrt((n-1)/qchisq(p=alpha.prime,df=n-1))
Sig.UCL
df2 <- cbind(NCT.alt,C.alt,Power)
#write.csv(df2,"20211025 Power for CV test.csv")
#####
```

Fig. 8.15 (continued)

8.7 Summary: A Final Word

Sampling plans, or acceptance sampling plans, are used in a wide variety of contexts. They are often used to assess the percent of conforming items in a lot or batch, or to assess the probability that a process would produce a conforming item. More specifically, sampling plans can be used to assess whether to believe that the probability of finding a conforming item is sufficiently high. The sufficiency is context-sensitive; there are no global values or rules about how high the probability of finding a conforming item needs to be. There is likewise no global standard for the degree of variability allowed among items, or whatever the observational unit happens to be.

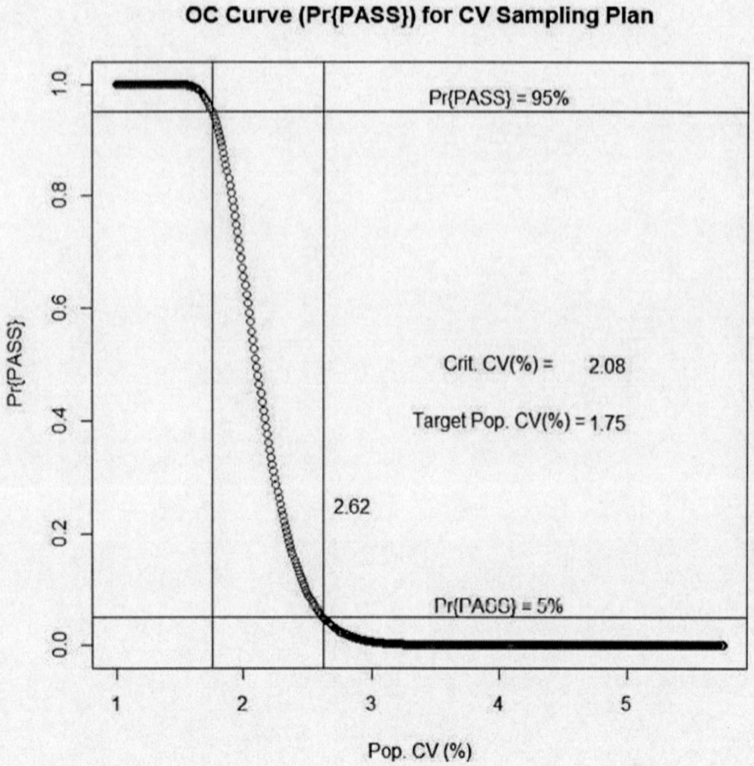

Fig. 8.16 Operating characteristic curve for a CV sampling plan

References

Duncan, A. J. (1965). *Quality control and industrial statistics* (3rd ed.). Richard D. Irwin, Inc.

Kang, C. W., Lee, M. S., Seong, Y. J., & Hawkins, D. M. (2007). A control chart for the coefficient of variation. *Journal of Quality Technology, 39*(2), 151–158.

Kushler, R. H., & Hurley, P. (1992). Confidence bounds for capability indices. *Journal of Quality Technology, 24*(4), 188–195.

Pardo, S. A. (2014). *Equivalence and noninferiority tests for quality, manufacturing, and test engineers*. Chapman & Hall/CRC.

Schilling, E. G. (1982). *Acceptance sampling in quality control*. Marcel Dekker, Inc., ASQC Press.

U.S. Food and Drug Administration. (2022). *Food, drug, and cosmetic act, chapter V, title 21, U.S. code.*

Chapter 9
Reliability, Life Testing, and Shelf Life

Abstract Shelf life, stability, and reliability are described as applications of life testing methods. An experimental approach to fitting models to non-exponential time-to-event data is presented.

9.1 The Reliability and Related Functions

Everything will fail, eventually. Reliability is a probability that a system (or a component) will fail no sooner than t time units from the time it begins operating. Reliability has many manifestations, or one might consider reliability as a special case of probabilities that some specific type of event will occur no sooner than t units from some initial reference time. Other manifestations include survival of patients having a particular disease, or the shelf life of drugs. For simplicity, we will refer to reliability in terms of time-to-failure, with the understanding that this time could be the time from a reference point to the occurrence of some specific type of event (such as death, progression of disease, a drug concentration in a human body drops below some threshold, or potency/reactivity loss). The point is to derive a model by which the probability of interest can be predicted and to incorporate design parameters into this model. In this way, we hope to help design a system so that it will have a desired reliability for a particular time-to-event.

To begin, we borrow from elementary chemical kinetics (Whitten et al., 2004) and consider a first-order system. Suppose we can measure a response variable that indicates the degree to which a system is operating properly (we realize this could be challenging in many cases, but it is our point of departure). Let $Y(t)$ be the response variable observed at time t. A first-order differential equation model that describes the change in this response over time can be expressed as:

$$\frac{dY(t)}{dt} = -\lambda Y(t)$$

With initial condition $Y(0) = y_0$, the solution to the equation is:

S. A. Pardo, *Statistical Methods and Analyses for Medical Devices*,
https://doi.org/10.1007/978-3-031-26139-8_9

$$Y(t) = y_0 e^{-\lambda t}$$

Some may recognize this as the equation governing first-order chemical reaction kinetics (Whitten et al., 2004), or the equation for radioactive decay (Rutherford, 1900). The parameter λ is called the rate parameter (as in the rate of reaction). If we presume that the response variable *Y(t)* is a decreasing function of time, as in the case of analyte concentrations in chemical reactions, then it is a maximum at $t = 0$, and its maximum value is y_0. Thus, we can express the response as a proportion of its initial value, namely:

$$\frac{Y(t)}{y_0} = e^{-\lambda t}$$

The proportion can be thought of at a probability, namely, the probability that the value of the response is not zero after *t* time units. So, if we replace *Y*(*t*) with a new variable, namely, $T =$ the time at which *Y*(*t*) is zero, then we can express the value of *Y*(*t*) as a proportion of its initial value with $\Pr\{T \geq t\}$, or:

$$\Pr\{T \geq t\} = e^{-\lambda t}$$

It turns out that this is the complement of the cumulative distribution function of an exponential random variable. That is:

$$\Pr\{T \leq t\} = 1 - \Pr\{T \geq t\} = 1 - e^{-\lambda t}$$

In the language of reliability, the parameter λ is called the failure rate. The exponential time-to-failure variable is characterized by a constant failure rate. That is to say, potentially the rate of failure could change with time, *t*. If *h*(*t*) represents the failure rate, then for the exponential time-to-failure variable:

$$h(t) = \lambda \forall t$$

The exponential time-to-failure variable is related to a discrete random variable with a Poisson distribution. The variable, *X*, is the number of failures (or events) within a fixed length of time. It has a Poisson distribution if:

$$\Pr\{X = x | \lambda, t\} = \frac{(\lambda t)^x e^{-\lambda t}}{x!}$$

The value of *t* is a fixed length of time, and λ is the (constant) failure rate.

Suppose that the failure rate actually changes with time. For example, the rate of failure could be fairly high initially, then drop after a "burn-in" period to a constant, and then climb back up after the product reaches a "wear-out" time. Such a failure rate function is sometimes called a "bath-tub" curve, as illustrated in Fig. 9.1.

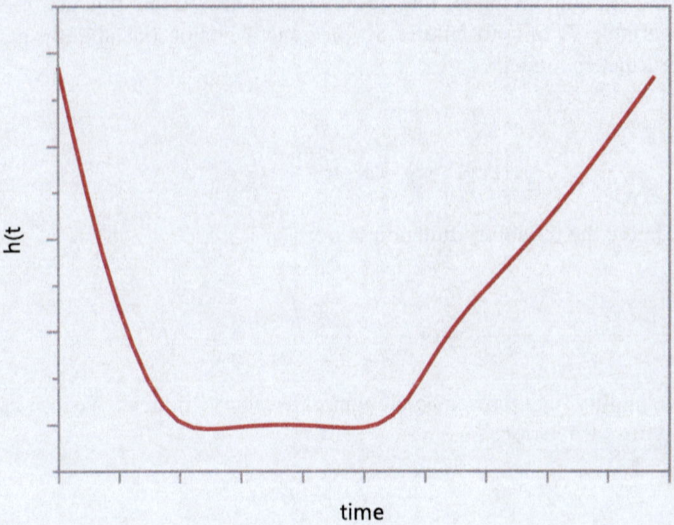

Fig. 9.1 "Bath-tub" failure rate curve

The failure rate function is also called the hazard rate function.

To generalize our initial first-order differential equation model for the response variable *Y*(*t*), we could replace the constant λ with *h*(*t*):

$$\frac{dY(t)}{dt} = -h(t)Y(t)$$

Again using the initial condition $Y(0) = y_0$, the solution is:

$$Y(t) = y_0 e^{-\int_0^t h(\tau)d\tau}$$

Of course, if $h(t) = \lambda$ (a constant), then:

$$\int_0^t h(\tau)d\tau = \lambda t$$

The function:

$$H(t) = \int_0^t h(\tau)d\tau$$

is called the cumulative failure rate or cumulative hazard rate function for our new random variable *T*, time-to-failure. So, in general, using our slightly generalized first-order kinetics model:

$$\Pr\{T \geq t\} = e^{-\int_0^t h(\tau)d\tau} = e^{-H(t)}$$

If we define the reliability function to be:

$$R(t) = \Pr\{T \geq t\} = e^{-\int_0^t h(\tau)d\tau} = e^{-H(t)}$$

then the reliability function can be thought of as a curve in time. We also have some potentially useful relationships:

$$-\ln R(t) = H(t)$$

$$h(t) = \frac{dH(t)}{dt}$$

Another related pair of functions is:

$$F(t) = 1 - R(t) = 1 - e^{-H(t)}$$

$$G(t) = -\ln F(t)$$

F(*t*) is the cumulative distribution function for time-to-failure.

Thus, we have the relationship:

$$R(t) = 1 - e^{-G(t)}$$

The design problem is to state either the reliability function or cumulative failure rate function in terms of design parameters and then choose values of those parameters that yield the desired reliability curve.

9.2 Obtaining an Empirical Reliability Model

Suppose the designer does not know the specific form of *R*(*t*), *H*(*t*), or *h*(*t*). She or he could perform an experiment to obtain a polynomial approximation. We will call the experiment a life test. In this experiment, *n* items (devices, systems, components) will be "started" and allowed to run until failure. The elapsed time from start to failure will be recorded. Suppose the *n* times to failure are ordered form shortest to longest. Call these times t_1, t_2, ..., t_k, ..., t_n. These times are referred to as "order statistics" (Conover, 1999). Compute the empirical reliability function:

$$\widehat{R}(t_k) = 1 - \frac{k}{n} = \frac{n-k}{n} \quad k = 1, n$$

or the empirical cumulative distribution function:

$$\widehat{F}(t_k) = 1 - \widehat{R}(t_k) = \frac{k}{n} \qquad k = 1, n$$

For a lack of a better term, we will call this formula, or estimator, the empirical maximum likelihood (EML) estimator. Now suppose that the cumulative hazard rate function, $H(t)$, can be approximated by a low-order polynomial in t, for example:

$$H(t) = \beta_0 + \beta_1 t + \beta_2 t^2$$

Given the relation:

$$-\ln R(t) = H(t)$$

it may be advantageous to approximate $G(t)$ as a second-order polynomial, i.e.:

$$G(t) = \beta_0 + \beta_1 t + \beta_2 t^2$$

The designer can now obtain via least squares an estimate of the parameters, β_k, and thus an approximation formula for $R(t)$. That is to say, the estimated approximation formula would be:

$$\tilde{R}(t) = 1 - \exp(-G(t)) = 1 - \exp\left(-\widehat{\beta}_0 - \widehat{\beta}_1 t - \widehat{\beta}_2 t^2\right)$$

The $\widehat{\beta}_i$ are the least squares estimates of the $H(t)$ approximation formula parameters. The approximation formula could be used to interpolate values of $R(t)$, but interpolation is not its most important use. More importantly, it can be used as a design tool.

9.3 Relating the β_i to Design Parameters

Consider a system with design factors or parameters, x_1, x_2, ...,x_j, ..., x_m. The system designer could perform a designed experiment in these factors, where the response is T = time-to-failure. At each run in the experiment, the reliability function approximation can be fit to the data. In this way, the reliability parameters can be

Table 9.1 2^3 Factorial experiment in hip replacement prototypes

Prototype	Metal	PE Pin	Lube
1	Steel	Cone	Albumin
2	Steel	Cone	GG
3	Steel	Cylinder	Albumin
4	Steel	Cylinder	GG
5	Aluminum	Cone	Albumin
6	Aluminum	Cone	GG
7	Aluminum	Cylinder	Albumin
8	Aluminum	Cylinder	GG

Table 9.2 Time-to-failure (years) data from simulator experiments

Rhat	*P*1	*P*2	*P*3	*P*4	*P*5	*P*6	*P*7	*P*8
0.9	16.23	18.37	16.36	14.30	21.48	24.05	19.71	22.28
0.8	18.25	18.56	18.51	16.84	22.01	24.63	20.93	22.57
0.7	18.95	18.72	18.72	17.76	22.90	26.86	21.43	22.95
0.6	19.27	19.21	18.95	19.86	23.20	27.35	21.67	24.17
0.5	19.41	20.94	18.97	20.38	23.86	29.09	23.53	25.29
0.4	20.05	21.83	19.59	20.63	24.08	29.51	24.07	25.66
0.3	20.42	22.01	19.79	20.77	24.71	30.10	24.57	25.66
0.2	20.91	22.62	20.37	21.99	25.57	30.50	24.93	26.12
0.1	21.19	22.80	20.71	23.57	25.59	31.00	25.40	26.53
0	21.95	24.05	21.21	24.10	26.64	31.51	25.56	27.63

related to the design factors. That is to say, we want to obtain an approximation formula that allows us to predict each of the parameters, β_0, β_1, and β_2, as polynomial functions of the design factors $x_1, x_2, \ldots, x_j, \ldots, x_m$. These approximating polynomials are obtained once again via least squares. Pardo (2009) describes how this methodology can be used for designing solid dosage forms for pharmaceuticals.

An example will help. Suppose the objective is to design an artificial hip joint replacement. The design will be a metal-on-polyethylene system. The questions are which metal material (steel or aluminum), the shape of the polyethylene (PE) pin (cylinder or truncated cone), and the type of lubricant (albumin or gamma globulin) to use. Normally, hip joint replacements are intended to last 15–20 years with a high degree of probability. That is to say, the expectation is that the reliability at 15 years should be at least 90%. The engineering team decides to perform a 2^3 factorial experiment in the three factors, metal, PE, and lube. Table 9.1 shows the eight different prototypes to be constructed.

A simulator device was constructed, so that years of wear could be achieved without actually implanting the devices in people and waiting for them to fail. A total of $n = 10$ units of each prototype were constructed and tested on the simulator until they failed.

Table 9.2 shows the time-to-failure data from the simulator. For each prototype, the times have been sorted from shortest to longest, and the empirical reliability has been calculated.

Table 9.3 Parameter estimates for second-order polynomial fits to *H*(t)

	$G(t) = B0 + B1 * t + B2 * t^2$						
Prototype	Metal	PE Pin	Lube	*B*0	*B*1	*B*2	*R*(20)
P1	Steel	Cone	Albumin	15.11	−1.04	0.02	0.459
P2	Steel	Cone	GG	28.92	−2.38	0.05	0.610
P3	Steel	Cylinder	Albumin	16.06	−1.08	0.02	0.351
P4	Steel	Cylinder	GG	9.65	−0.67	0.01	0.506
P5	Aluminum	Cone	Albumin	58.80	−4.43	0.08	0.973
P6	Aluminum	Cone	GG	16.97	−0.88	0.01	0.975
P7	Aluminum	Cylinder	Albumin	29.81	−2.21	0.04	0.869
P8	Aluminum	Cylinder	GG	41.50	−2.92	0.05	0.974

For each of the prototypes, a second order polynomial approximation was fit to:

$$\widehat{G}(t) = -\ln \widehat{F}(t) = \beta_0 + \beta_1 t + \beta_2 t^2 + \epsilon$$

Table 9.3 shows the resulting estimates of the parameters. In addition, the predicted reliability at $t = 20$ years is computed for each prototype.

If the design objective is to achieve a 95% reliability at $t = 20$ years, it is clear that steel is not adequate. What is not clear is whether there is an interaction between The PE pin and the lube. It appears that the cone shape may be superior to cylinder, but it may be that cylinder with gamma globulin (GG) is at least as good as cone with albumin.

An ANOVA could be performed to determine which factors have an effect on time-to-failure, at least on the average. Figure 9.2 shows an R script using the function ANOVA in package *car*.

The output is shown in Fig. 9.3. The interaction plot for the metal x lube term is shown in Fig. 9.4.

The only statistically significant (i.e., probably repeatable) interaction effect was between metal and lube.

Table 9.4 shows the mean and standard deviation (SD) for time-to-failure and the mean predicted reliability at 20 years for each level of the three factors. The maxima for both mean failure time and reliability are shown in bold font. Note that the conditions that yield maximal values are aluminum, cone, and GG (P6). The average time-to-failure for this condition is 28.46 years, and the predicted reliability at 20 years is 0.975. A close second place is aluminum, cylinder, and GG (P8) with a predicted reliability of 0.974. A very close third place is aluminum, cone, and albumin (P5) with predicted $R(20) = 0.973$. The lower SD for P5 indicates that it may yield more consistent wear and hence more predictable time-to-failure. The lower limit of the 95% confidence interval for predicted reliability at 20 years for P5 is 0.9585, which actually exceeds our design criterion (0.950).

The ANOVA has indicated several things we might not have known without this experiment:

```
setwd("<your path here>")
df1 <- read.csv("20220130 SMAMD Example 14.1 Hip Replacement Survival.csv")
#
# Variables
# Prototype
# Metal
# PE.Pin
# Lube
# Time
#
attach(df1)
library(car)
library(lsmeans)
Time.mod <- lm(Time~Metal+PE.Pin+Lube+Metal:PE.Pin+Metal:Lube+PE.Pin:Lube)
Anova(Time.mod,type="III")
lsmeans(Time.mod,"Metal",by="Lube")
interaction.plot(x.factor=Metal,trace.factor=Lube,response=Time,fun=mean,type="b")
#####
```

Fig. 9.2 ANOVA for hip replacement time-to-failure

1. Metal, pin, and lube types all significantly affect the time-to-failure.
2. Although related, the probability that time-to-failure exceeds 20 years and the mean time-to-failure are not surrogates for each other.
3. The effect of lube may in fact depend on the type of pin; the interaction effect may be more apparent in terms of consistency of wear.

Had we not constructed the low-order polynomial approximations for $G(t)$, we would not have been able to generate predictions for reliability at 20 years. Had we not used a designed experiment, we might not have realized that all three factors actually do affect time-to-failure. Furthermore, without both the designed experiment and the models for $G(t)$, we might not have discovered that the pin and lube types interact with each other in terms of reliability, even though they do not appear to interact in terms of mean time-to-failure.

Fig. 9.3 ANOVA output for hip replacement time-to-failure data

```
Anova Table (Type III tests)

Response: Time
               Sum Sq  Df   F value      Pr(>F)
(Intercept)    6737.1   1  1523.8627   < 2.2e-16 ***
Metal           159.5   1    36.0874   6.768e-08 ***
PE.Pin           12.6   1     2.8509     0.09559 .
Lube            101.7   1    22.9946   8.348e-06 ***
Metal:PE.Pin     12.5   1     2.8269     0.09697 .
Metal:Lube       22.2   1     5.0160     0.02816 *
PE.Pin:Lube      13.6   1     3.0716     0.08387 .
Residuals       322.7 73

---
Signif. codes:  0 '***' 0.001 '**' 0.01 '*' 0.05 '.' 0.1 ' ' 1
> lsmeans(Time.mod,"Metal",by="Lube")
Lube = Albumin:
 Metal       lsmean    SE  df  lower.CL upper.CL
 Aluminum      23.6  0.47  73      22.7     24.5
 Steel         19.5  0.47  73      18.6     20.4
Lube = GG:
 Metal       lsmean    SE  df  lower.CL upper.CL
 Aluminum      26.7  0.47  73      25.7     27.6
 Steel         20.5  0.47  73      19.5     21.4

Results are averaged over the levels of: PE.Pin
Confidence level used: 0.95
```

9.4 Censored Time-to-Failure

Sometimes limits are placed on the length of time over which experimental units will be observed or on the number of units (out of a sample of size n) that will be allowed to fail before the test is stopped. Such restrictions are called censoring. Stopping the test at a fixed time is referred to as type I censoring (Mann et al., 1974), and stopping the test after a fixed number of units fail is called type II censoring (Mann et al., 1974) The question is how to estimate reliability in the face of such censoring. We will address type I censoring first.

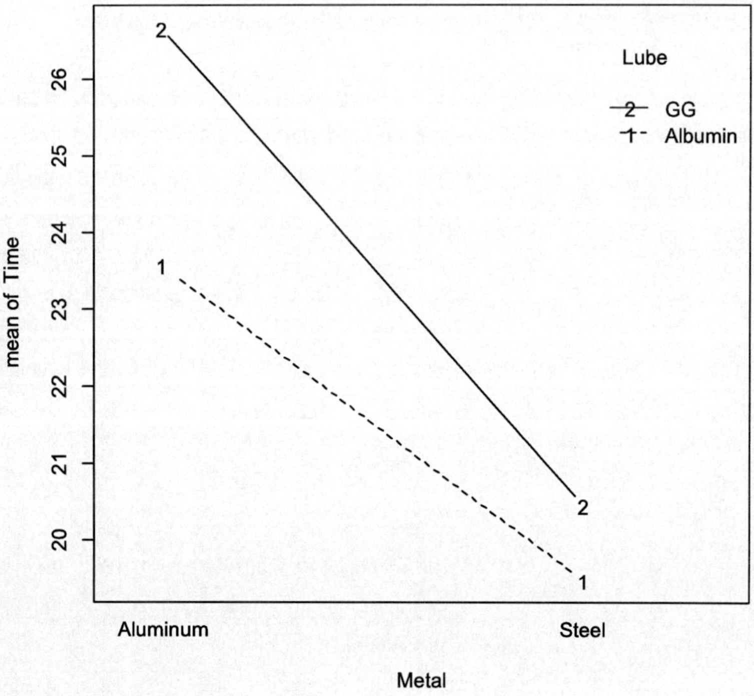

Fig. 9.4 Interaction between metal and lube

Table 9.4 Mean time-to-failure and predicted R(20)

Time-to-failure									
Prototype	Metal	PE pin	Lube	*N*	Mean	SD	Pred. R(20)	LCL-R(20)	UCL-R(20)
P5	Aluminum	Cone	Albumin	10	24.00	1.66	0.973	0.9585	0.9826
P6	Aluminum	Cone	GG	10	28.46	2.62	0.975	0.9117	0.9930
P7	Aluminum	Cylinder	Albumin	10	23.18	2.08	0.869	0.8290	0.9001
P8	Aluminum	Cylinder	GG	10	24.89	1.81	0.974	0.9039	0.9930
P1	Steel	Cone	Albumin	10	19.66	1.64	0.459	0.3844	0.5239
P2	Steel	Cone	GG	10	20.91	2.06	0.610	0.4510	0.7229
P3	Steel	Cylinder	Albumin	10	19.32	1.37	0.351	0.2158	0.4627
P4	Steel	Cylinder	GG	10	20.02	3.02	0.506	0.4434	0.5610

Suppose the team did a life test with a sample of n experimental units and stopped the test after T_{max} time units. Out of the n units on test, only $r < n$ actually failed. The remaining $n - r$ units were still functioning at time T_{max}. Suppose further that the r failure times are sequenced from shortest to longest and that $t_1, t_2, t_3, \ldots, t_r$ represent those order statistics. So the empirical reliability function could be represented as it was when there was no censoring:

$$\widehat{R}(t_k) = 1 - \frac{k}{n} = \frac{n-k}{n} \quad k = 1, r$$

This is the EML with right-censoring (EMLC). This computed empirical reliability function can be treated in the exact same way it was when there was no censoring. There is another method for computing an empirical reliability function at times other than those which were explicitly observed failure times. The method is by Kaplan and Meier (1958) and is described by Lee (1992). The formula for the Kaplan-Meier (K-M) estimator is:

$$\widehat{R}(t) = \prod_{t_k \le t} \frac{n-k}{n-k+1}$$

where t_k are the failure times for uncensored observations. At $t = t_k$, this formula can be written as:

$$\widehat{R}(t_k) = \widehat{R}(t_{k-1}) \frac{n-k}{n-k+1}$$

The K-M estimator is particularly useful if censoring can occur even though the time has not exceeded $T_{\max}$. For example, a unit may fail during the test, but due to some cause other than the particular failure mode of interest.

The variance, and thus standard error, of the K-M reliability estimate can be approximated (Lee, 1992) by:

$$V\left(\widehat{R}(t_k)\right) \approx \widehat{R}^2(t_k) \sum_{i=1}^{k} \frac{1}{(n-i)(n-i+1)}$$

The approximate standard error is the square root of this variance approximation.

Using either the EML, EMLC, or K-M estimators, the highest reliability estimate is at failure time t_1, and it is $(n - 1)/n$. The lowest reliability value would be at tr. If $r = n$, then EML = EMLC, and at t_r, $R(t_r) = 0$. For the K-M estimator, this is not the case if there are any intermediate censored failures (i.e., if a unit fails before $T_{\max}$ in some mode other than the one(s) of interest).

As an example, consider testing a circuit board. The engineer is concerned about a particular component failing. She puts $n = 15$ boards on test and sets $T_{\max} = 1000$ hours. Some of the boards fail before $T_{\max}$, and some do not. Of the boards that fail, some of them had a failure in a component other than the one of interest. Those boards that failed due to other components are censored. Table 9.5 shows the failure times, whether the failures were censored or not, the intermediate calculation $(n - k)/(n - k + 1)$, the K-M estimator for $R(t_k)$, and an approximate 95% confidence interval (LCL, UCL) for the reliability at each uncensored failure time. The column labeled "factor" is the summation term:

Table 9.5 Censored failure times with Kaplan-Meier estimates

k	Time-to-failure	Censored?	$(n - k)/(n - k + 1)$	$R(t_k)$ $K - M$	Factor	SE	LCL	UCL
1	870	Uncensored	0.9333	0.9333	0.00476	0.06441	0.8071	1.0000
2	872	Uncensored	0.9286	0.8667	0.01026	0.08777	0.6946	1.0000
3	884	Uncensored	0.9231	0.8000	0.01667	0.10328	0.5976	1.0000
4	889	Uncensored	0.9167	0.7333	0.02424	0.11418	0.5095	0.9571
5	909	Uncensored	0.9091	0.6667	0.03333	0.12172	0.4281	0.9052
6	915	Uncensored	0.9000	0.6000	0.04444	0.12649	0.3521	0.8479
7	916	Uncensored	0.8889	0.5333	0.05833	0.12881	0.2809	0.7858
8	932	Uncensored	0.8750	0.4667	0.07619	0.12881	0.2142	0.7191
9	937	Censored						
10	939	Uncensored	0.8333	0.3889	0.10952	0.12870	0.1366	0.6411
11	951	Uncensored	0.8000	0.3111	0.15952	0.12426	0.0676	0.5547
12	962	Uncensored	0.7500	0.2333	0.24286	0.11499	0.0080	0.4587
13	979	Uncensored	0.6667	0.1556	0.40952	0.09955	0.0000	0.3507
14	1000	Censored						
15	1000	Censored						

$$\sum_{i=1}^{k} \frac{1}{(n-i)(n-i+1)}$$

Note that for censored observations, no computation is made. The ninth failure time was censored, as this unit failed in a mode that was not of interest to the engineer.

Regardless of how the reliability estimates were obtained (EML, EMLC, K-M), the computations for $\widehat{F}(t_k)$, the polynomial approximation of $G(t)$, and the computation of predicted values for $R(t)$ follow the same procedures.

9.5 Accelerated Life Tests

In the case of the hip replacement problem, the design team had a simulator that could simulate years of life in a fairly short time. Sometimes, the simulation of life is performed by subjecting experimental units to some condition which is presumed to accelerate the failure process in such a way as to allow the designers to predict the increase in the failure rate. Commonly, temperature is used as the accelerating condition. Presuming that the increase in failure rate is proportional to an increase in the rate of a chemical reaction, a model called the Arrhenius reaction rate law (Mann, et al., 1974) is employed to relate the failure rate to temperature. The Arrhenius model is:

$$\lambda_P = A \exp\left(\frac{-E/K}{P}\right)$$

The constant A is specific to the particular materials and reactions that underlie the failure mode. The constant E is called the energy of activation and is also specific to the materials and chemical reactions involved in failure. The letter K stands for Boltzmann's constant, and P is the temperature in degrees Kelvin (we use the letter P, for "parameter", so as to not confuse it with T for time-to-failure). So λ_P is the average failure rate at temperature P. The usual presumption is that the time-to-failure follows an exponential distribution, so at temperature P, the reliability function is given by:

$$R(t|P) = e^{-\lambda_P t}$$

The question is how to determine the degree of acceleration achieved by exposing experimental units to a particular temperature, say P_A. The first problem is to estimate the parameters A and $B = -E/K$ (note that B is just a normalized energy of activation). The answer depends on the nature of the data. Life tests may be performed by placing n units into a temperature chamber, at a fixed temperature, P_A, for a given time, T. At time T, the units are taken out and inspected or tested, and the number of units that "survive," $S = s$, is counted. Assuming that the time-to-failure is exponentially distributed, and an estimate of the reliability at time T is $p_A = \frac{s}{n}$, then the reliability is given by:

$$\widehat{R}(T|P_A) = e^{-\lambda_A T} = p_A = \frac{s}{n}$$

Solving for λ_A gives an estimate for the failure rate:

$$\widehat{\lambda}_A = \frac{-\ln(p_A)}{T} = \frac{-\ln\left(\frac{s}{n}\right)}{T}$$

Suppose an experiment was performed where n_1 units were put on test for T time units at temperature P_1 and another n_2 units put on test for T time units at temperature P_2. Then we would have two equations in the two unknowns A and B:

$$-\ln\left(\frac{s_1}{n_1}\right) = TA \exp\left(\frac{-B}{P_1}\right)$$

$$-\ln\left(\frac{s_2}{n_2}\right) = TA \exp\left(\frac{-B}{P_2}\right)$$

Taking logs on both sides yields:

$$\ln\left(-\ln\left(\frac{s_1}{n_1}\right)\right) = \ln T + \ln A - \frac{B}{P_1}$$

$$\ln\left(-\ln\left(\frac{s_2}{n_2}\right)\right) = \ln T + \ln A - \frac{B}{P_2}$$

These in turn yield the solutions for the estimates:

$$\widehat{B} = P_1\left[\ln\widehat{A} - \ln\left(-\ln\frac{s_1}{n_1}\right) + \ln T\right]$$

$$\ln\widehat{A} = \frac{P_2}{P_2 - P_1}\left[\ln\left(-\ln\frac{s_2}{n_2}\right) + \frac{P_1}{P_2}\ln\left(-\ln\frac{s_1}{n_1}\right) + \frac{P_1}{P_2}\ln T\right]$$

Thus, an expression for the estimate of parameter B in only known quantities is:

$$\widehat{B} = \begin{array}{l} P_1\left[\frac{P_2}{P_2 - P_1}\left[\ln\left(-\ln\frac{s_2}{n_2}\right) + \frac{P_1}{P_2}\ln\left(-\ln\frac{s_1}{n_1}\right) + \frac{P_1}{P_2}\ln T\right]\right. \\ \left. -\ln\left(-\ln\frac{s_1}{n_1}\right) + \ln T\right] \end{array}$$

These estimates will allow the designers to determine how much acceleration was achieved at any given temperature, P_2, compared to a lower temperature, P_1. The estimates of B and A may be useful for future experiments or tests, provided that the materials involved in such tests are at least similar if not identical to those used to obtain the estimates.

Suppose that the team has at least a hypothetical failure rate desired at say 25 ° C = 25 + 273 = 298°K = P_0. Such a failure rate might be determined by having a specification or requirement that the reliability at time T_0 must be at least r_0. Assuming the exponential time-to-failure, the failure rate is given by:

$$\lambda_0 = \frac{-\ln(r_0)}{T_0}$$

Now suppose that the designers want to accelerate the failure process k times, so that the actual test time would need to be $T_a = \frac{T_0}{k}$ (“a” stands for “accelerated”). This would also mean that the failure rate under the accelerated conditions would need to be:

$$\lambda_a = k\lambda_0$$

The team must choose a temperature, $P_a > P_0$, to achieve the desired acceleration.

Using the Arrhenius equation:

$$k = \frac{\exp(-B/P_a)}{\exp(-B/P_0)} = \exp\left(B\left(\frac{1}{P_0} - \frac{1}{P_a}\right)\right)$$

Since k is actually given (i.e., the desired acceleration to allow the test to occur in a short enough time), and P_0 is known, the equation can be solved for P_a:

$$P_a = \frac{BP_0}{B - P_0 \ln k}$$

The only thing required is a value for B, the normalized energy of activation. A simple experiment as described earlier can be used to obtain an estimate of B.

Recall that test temperatures should be expressed in degrees Kelvin (°K) when using these equations.

There are two potential drawbacks to the procedures described for determining parameters of an accelerated life test. First, we have assumed that the time-to-failure is affected by temperature in the way described by the Arrhenius equation. Second, we have assumed that time-to-failure has an exponential distribution. Both of these assumptions stem from a more fundamental assumption that the failure process is related to a first-order chemical reaction (Chow, 2007). While these assumptions may not be completely valid, they may provide at least a practical approach to determining the amount of acceleration achieved by putting units on test at temperature P_a for time T_0.

As an example, consider a life test with $P_0 = 25$ °C = (25 + 273 = 298)°K and P_a is 40 °C = (40 + 273 = 313)°K. At P_0, a test with $n_0 = 30$ units is performed for $T_0 = 720$ hours. At P_a, the test is also performed for $T_0 = 720$ hours with $n_2 = 30$ units. At P_0, the number of "surviving" units was $s_0 = 29$. At P_a, the number of operating units after 720 hours was $s_a = 20$. Figure 9.5 shows some R code, together with session window output, for computing the estimates of A, B, and k.

Note that due to the vectorized nature of the computations in R, and the fact that the data were entered with separate rows for the 25 °C and 40 °C inputs, R computes the parameter estimates as vectors of length 2.

In this example, increasing temperature from 25 to 40 °C results in an acceleration of about 18.8 times. Thus, the 720 hours at 40 °C is equivalent to 18.8 * 720 = 13,536 hours at 25 °C, or about 1.54 years. If the designers wished to simulate 2 years = 17,532 hours with a test of 720 hours, then they would need a test temperature that would yield an acceleration of approximately 17,532/ 720 = k = 24.35 times. Using the equation for Pa, given B and k, yields:

$$P_a = \frac{BP_0}{B - P_0 \ln k} = \frac{18231.39 * 298}{18231.39 - 298 * \ln(24.35)} \approx 314.1\,^\circ K = 41.1\,^\circ \mathrm{C}$$

```
setwd("<your path here>")
df1 <- read.csv("20141110 accelerated life test.csv")
attach(df1)
# inputs:
# Test = index
# TempC = test temp in degrees C
# Time = test time in hours
# n = number of units on test
# s = number surviving units after test
#
# the R function "log" is the natural logarithm
#
TK1 <- TempC[1] + 273
TK2 <- TempC[2] + 273
n1 <- n[1]
n2 <- n[2]
s1 <- s[1]
s2 <- s[2]
lnA <- (TK2)/(TK2 - TK1)*(log(-log(s2/n2)) + TK1*log(-log(s1/n1))/TK2 + log(Time))
B <- TK1*(lnA - log(-log(s1/n1)) + log(Time))
K <- exp(-B/TK2) / exp(-B/TK1)
df2 <- cbind(lnA,B,K)
write.csv(df2,file="20141110.acceleration.parameters.csv")
```

Fig. 9.5 R code for accelerated life test parameter estimation

So, it turns out that a relatively small change (1.1 °C) in the test temperature would give the desired acceleration.

Finally, the team may want to have an estimate of the failure rate at the new test temperature of 41.1 °C. Using the Arrhenius equation:

$\lambda_P = A\exp\left(\frac{-B}{P}\right) \approx 0.0012$, or approximately 0.0012 failures per hour at 41.1 °C. The estimated failure rate at 25 °C would then be:

$$\lambda_0 = \frac{\lambda_a}{k} \approx \frac{0.0012}{24.35} \approx 0.000049$$

The estimated probability that the device would last 2 years is given by the exponential reliability function:

$$R(2\text{ years}|25\,^\circ\text{C}) = e^{-0.000049*17532} \approx 0.4236,$$

or about a 42.36% chance that the device will not fail before 2 years. Recall that at 25 °C, there were 29 out of 30 parts that survived after only 720 hours. It does not seem unreasonable to expect such a low reliability at 2 years. The question of why it required a designed experiment in the features the design team determines may have a critical impact on reliability. Armed with the Arrhenius parameter estimates, they may perform the experiment at 41.1 °C for 720 hours per each experimental set of prototypes. If the only information per unit tested is whether at 720 hours it was or was not operational, logistic regression methods may well be appropriate to determine which device features are most critical and what particular choices for each critical feature may be optimal.

Logistic regression (Hosmer & Lemeshow, 1989) is a form of generalized linear modeling where the response is most commonly binary, and there are several factors or regressors that may affect the probability of obtaining one or the other response values. That is to say, the response is one of two possible states, call them + and −. Say the probability of state "+" is the thing of interest (it could either be failure or non-failure). The notion is that first, the response is coded as:

$$Y = \begin{cases} 1 & \text{if state is} + \\ 0 & \text{if state is} - \end{cases}$$

Then suppose there are a set of p regressors that can potentially affect the probability that $Y = 1$. The model is:

$$\Pr\{Y=1|x_1, x_2, \ldots, x_p\} = h(x_1, x_2, \ldots, x_p) = \frac{\exp\left(\beta_0 + \sum_{i=1}^{p} \beta_i x_i\right)}{1 + \exp\left(\beta_0 + \sum_{i=1}^{p} \beta_i x_i\right)}$$

Note that:

$$1 - h(x, x, \ldots, x) = \Pr\{Y=0|x_1, x_2, \ldots, x_p\} = \frac{1}{1 + \exp\left(\beta_0 + \sum_{i=1}^{p} \beta_i x_i\right)}$$

So that:

$$r(x, x, \ldots x) = \ln\left(\frac{h(x_1, x_2, \ldots, x_p)}{1 - h(x_1, x_2, \ldots, x_p)}\right) = \ln\left(\exp\left(\beta_0 + \sum_{i=1}^{p} \beta_i x_i\right)\right) = \beta_0 + \sum_{i=1}^{p} \beta_i x_i$$

The function $r(x_1, x_2, \ldots, x_p)$ is called the log odds (Fleiss et al., 2003). Using log odds, the model looks like a multiple linear regression. Inasmuch as the log odds cannot be directly computed with data, parameters are usually estimated via a maximum likelihood approach. For more details, see Hosmer and Lemeshow (1989).

9.6 Stability and Shelf Life

Shelf life is the amount of time an item may remain unused and still perform adequately. Chemical products, such as pharmaceuticals, paint, solvents, and even food items, are subject to the problem of estimating, controlling, and elongating shelf life. Shelf life is closely related to reliability, although the two are not identical concepts. We will treat the case where there is a continuously valued quality response variable. Furthermore, we will assume that this variable is monotonic with respect to time spent "on the shelf" and that it is continuously degrading. As in the cases of Chaps. 5, 6, 7, and 8, the shelf-life variable can be optimized over the list of critical features, components, or factors. We will concentrate on how to estimate shelf life, following methods described by Chow (2007).

Assume that the quality variable, Y, is linearly related to shelf time, S, i.e.:

$$Y = \beta_0 + \beta_1 S + \varepsilon$$

There are two sorts of shelf-life problems: (1) find the time, S_e, such that there is a $100p\%$ chance that the value of Y will still be acceptable, and (2) determine the probability, p, that Y is acceptable at a pre-determined "warranty" time, S_w. For both of these problems, a prediction interval approach will be employed.

Recall that earlier, the standard error of a predicted value from a polynomial regression model was presented. As a special case of a linear model in a single regressor, namely, time, the formula for a future predicted value at time $= s_k$ is:

$$\mathrm{SE}\left(\widehat{Y}|s_k\right) = \sigma\sqrt{1 + \frac{1}{n} + \frac{(s_k - \overline{s})^2}{\sum_{i=1}^{n}(s_i - \overline{s})^2}}$$

The standard deviation of the noise variable, ε, namely, σ, is estimated by the root mean square error of the regression fit, $\widehat{\sigma}$. The lower confidence limit for a predicted value of Y at time $= s_k$ is then:

$$\widehat{Y}_L = \widehat{Y}_{s_k} - t_{1-\alpha}(n-2)\widehat{\sigma}\sqrt{1 + \frac{1}{n} + \frac{(s_k - \overline{s})^2}{\sum_{i=1}^{n}(s_i - \overline{s})^2}}$$

To solve problem 1, we will employ the inverse regression approach (Draper & Smith, 1981). To find the time, s_k, at which it is expected the response variable, Y, would be no lower than the lower specification limit, L, with probability $1 - \alpha$, solve the equation for s_k:

$$L = \widehat{Y}_{s_k} - t_{1-\alpha}(n-2)\widehat{\sigma}\sqrt{1 + \frac{1}{n} + \frac{(s_k - \overline{s})^2}{\sum_{i=1}^{n}(s_i - \overline{s})^2}}$$

Solving yields:

$$s_k = \overline{s} + \sqrt{\left[\frac{\left(L - \widehat{Y}_{s_k}\right)^2}{t^2\widehat{\sigma}^2} - 1 - \frac{1}{n}\right]\sum_{i=1}^{n}(s_i - \overline{s})^2}$$

where t is the $100(1 - \alpha)$ percentile of a t distribution with $n - 2$ degrees of freedom.

Problem 2 is simpler. Presume that there is a desired warranty time, call it s_w, and we want to know how likely it is that the product will have an adequate value of Y at that time. If the predicted value of Y at s_w is $\widehat{Y}_{s_w}$, then the lower confidence limit for the predicted value at time s_w is:

$$\widehat{Y}_L = \widehat{Y}_{s_w} - t^*\widehat{\sigma}\sqrt{1 + \frac{1}{n} + \frac{(s_w - \overline{s})^2}{\sum_{i=1}^{n}(s_i - \overline{s})^2}}$$

Setting $\widehat{Y}_L = L$ and solving for t^* gives:

$$t^* = \frac{\widehat{Y}_{s_w} - \widehat{Y}_L}{\widehat{\sigma}\sqrt{1 + \frac{1}{n} + \frac{(s_w - \overline{s})^2}{\sum_{i=1}^{s}(s_i - \overline{s})^2}}}$$

The probability that Y will not fall below Y_L before time s_w, $\Pr\{Y \geq Y_L | s_w\}$, is $\Pr\{T \leq t^* | n - 2\}$, the probability that T, a Student's t random variable with $n - 2$ degrees of freedom, will be less than t^*.

Generally, design problems cannot afford to wait for product to be put on test until the expiration or warranty time has elapsed. Thus, some form of accelerated test is generally required. Presume that the accelerating parameter is temperature. The question is how to determine the temperature and the degree to which acceleration is achieved at that temperature. The methods for determining the Arrhenius parameters described earlier could be employed. However, since the response variable, Y, is continuously valued, and is assumed here to be monotonic and linear with respect to time, there is another possible approach. Perform the test at a fixed set of times, say $s_1, s_2, \ldots s_n$, at two temperatures, P_1 and P_2, both of which are greater than the "nominal" temperature, P_0 (typically $P_0 = 25$ °C), and $P_2 > P_1$. Obtain slope estimates for each set of data. Call the estimates b_1 (at temperature P_1) and b_2

(at temperature P_2). An estimate of the acceleration rate for a temperature difference of $\Delta_P = P_2 - P_1$ is $k_{\Delta P} = b_2/b_1$. If the designers desire to simulate T_0 time units with $T_a < T_0$ time units, they would require $k_a \approx T_0/T_a$. Let $\Delta_a = P_a - P_0$ be the temperature difference from nominal required to achieve acceleration k_a. Then:

$$\frac{\Delta_a}{\Delta_P} = \frac{k_a}{k_{\Delta P}} \Rightarrow \Delta_a = \frac{k_a}{k_{\Delta P}} \Delta_P$$

So, to simulate T_0 time units at nominal temperature P_0, test over T_a time units at temperature $P_a = P_0 + \Delta_a$. The number of time points at which to measure Y should be at least 3, 2 more than the highest order term in the model, but 4 or 5 would be better. For the purposes of assessing the adequacy of the model (which we have assumed was first order), replication at each time point in the form of multiple units measured is highly recommended.

The values of s_k or t^* depend on the product, which is in turn dependent upon the chemical or other components. If prototype products are formulated in a designed experimental fashion, then either s_k or t^* could act as response variables. In that way, a desired prototype formulation could be found.

In the shelf-life discussion so far, we have only dealt with a case where the quality response variable is monotonically decreasing and that there was a lower limit of acceptability. The case where Y is monotonically increasing, with an upper limit of acceptability, is analogous.

9.7 Summary

The purpose of this section was to provide some ideas concerning shelf-life testing and to encourage the use of designed experiments in order to achieve a desired shelf-life with some stated level of probability. Shelf life is a fairly broad topic, and those who are interested are encouraged to read the book by Professor Chow (2007).

References

Conover, W. J., (1999). *Practical Nonparametric Statistics*, 3rd Ed. John Wiley and Sons, New York.

Chow, S.-C. (2007). *Statistical design and analysis of stability studies*. Chapman and Hall/CRC.

Drape, N. R., & Smith, H. (1981). *Applied regression analysis* (2nd ed.). Wiley, New York.

Fleiss, J. L., & Levin, B., Paik, M. C. (2003). *Statistical Methods for Rates and Proportions*, 3rd Ed., John Wiley and Sons, New York.

Hosmer, D. W., & Lemeshow, S. (1989). *Applied logistic regression*. John Wiley and Sons, Inc..

Kaplan, E. L., & Meier, P. (1958). Nonparametric estimation from incomplete observations. *Journal of the American Statistical Association, 53*, 457–481.

Lee, E. T. (1992). *Statistical methods for survival data analysis*. John Wiley and Sons.

Mann, N. R., Schafer, R. E., & Singpurwalla, N. D. (1974). *Methods for statistical analysis of reliability and life data*. John Wiley and Sons.

Pardo, S. (2009). A differential equation model for dissolution profiles, and its uses in designing dosage forms. *Statistics in Biopharmaceutical Research, 1*(2), 194–200.

Rutherford, E. (1900). A radioactive substance emitted from thorium compounds. *Philosophical Magazine Series 5, 49*, 1–14.

Whitten, K. W., Davis, R. E., Peck, M. L., & Stanley, G. G. (2004). *General chemistry* (7th ed.). Brooks/Cole.

Chapter 10
Diagnostics: Sensitivity and Specificity

Abstract Sensitivity and specificity are two fundamental issues in diagnostic tests. Experimental methods for generating the receiver operating curve (ROC) are described, and the ROC is used to provide estimates of sensitivity and specificity.

10.1 Receiver Operating Characteristic (ROC) Curves

A topic that is somewhat related to reliability/survival curves is the receiver operating characteristic (ROC) curve (Pepe, 2003). Many analytic tests yield a binary response, i.e., + or −. The thing is, the chemical or other reactions used by the test actually is a value of a continuous variable. The resulting + or − is based on whether an individual test yields a value for the continuous response that is either above or below some threshold value. So, the question is, "How do we set the threshold?"

Usually, an experiment is conducted where the value of the continuous variable, call it Y, is measured on a sample of n experimental units who are known to be positive for some characteristic (usually either a disease state or a condition such as pregnancy). Then, for a set of possible threshold values for Y, call them T_k, the proportion of units that have a Y value at or above T_k is computed. Ostensibly, as the value of T_k is increased, the proportion of Y values at or above T_k decreases. In addition to known positives, a similar curve can be constructed for known negatives. Usually the question is the complement to that for known positives, namely, the probability that a true negative would yield a result below the threshold.

As an example, suppose the data were gathered (really simulated) on a continuous response for some analytic test. Call it Y.sim, with $n = 200$ experimental units. The R code is given in Fig. 10.1.

Figure 10.2 shows the histogram of Y.sim for the known positives, and Fig. 10.3 shows the ROC + curve (Fig. 10.3).

Figure 10.4 shows the histogram of simulated response values for the "true negatives." Figure 10.5 shows the ROC − curve.

A balance between a low chance of false positives and false negatives would potentially be achieved if the ROC + and ROC − curves intersected. One way to discover this threshold is to find the point at which the difference in the two curves is

S. A. Pardo, *Statistical Methods and Analyses for Medical Devices*,
https://doi.org/10.1007/978-3-031-26139-8_10

```
setwd("<your path here>")
#
prop.pos <- c()
prop.neg <- c()
set.seed(26)
Y.sim.pos <- rgamma(n=200,shape=5,scale=10)+20
Y.sim.neg <- rgamma(n=200,shape=5,scale=2.5)
hist(Y.sim.pos)
dev.new()
hist(Y.sim.neg)
Low.lim.pos <- min(Y.sim.pos)
High.lim.pos <- max(Y.sim.pos)
Tk <- seq(from=Low.lim.pos-14,to=High.lim.pos+14,by=2)
n.length <- length(Tk)
for (i in 1:n.length) {
 Y.sim.pos.sub <- Y.sim.pos[Y.sim.pos >= Tk[i]]
 num.pos <- length(Y.sim.pos.sub)
 prop.pos[i] <- num.pos / 200
 Y.sim.neg.sub <- Y.sim.neg[Y.sim.neg <= Tk[i]]
 num.neg <- length(Y.sim.neg.sub)
 prop.neg[i] <- num.neg / 200
 }
#
df2 <- cbind(prop.pos,prop.neg,Tk)
dev.new()
plot(Tk,prop.pos,type="b",main="Receiver Operating Characteristic (ROC) Curve
+",xlab="Threshold, Tk",ylab="Pr{Y.sim >= Tk}")
abline(h=0.95)
#
threshold.p <- Tk[which(prop.pos>0.949 & prop.pos<0.951)]
thresh.p.txt <- as.character(round(threshold.p,2))
abline(v=threshold.p)
text(x=60,y=0.970,labels="Pr{Y.sim>=Tk} = 0.95}")
text(x=47,y=0.20,labels=thresh.p.txt)
#
```

Fig. 10.1 ROC computations

```
dev.new()
plot(Tk,prop.neg,type="b",main="Receiver Operating Characteristic (ROC) Curve -
",xlab="Threshold, Tk",ylab="Pr{Y.sim >= Tk}")
abline(h=0.95)
threshold.n <- Tk[which(prop.neg>0.949 & prop.neg<0.951)]
thresh.n.txt <- as.character(round(threshold.n,2))
#
abline(v=threshold.n)
text(x=60,y=0.960,labels="Pr{Y.sim<=Tk} = 0.95}")
text(x=36,y=0.85,labels=thresh.n.txt)
#
write.csv(df2,"20220109 ROC Curve Results.csv")
#####
```

Figure 10.2 shows the histogram of Y.sim for the known positives, and Figure 10.3 shows the ROC + curve.

Fig. 10.1 (continued)

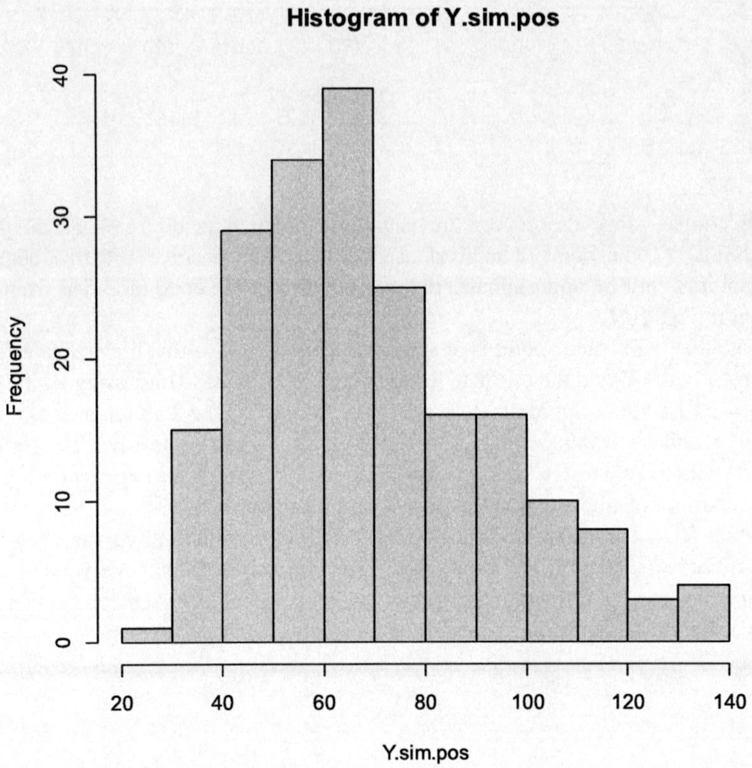

Fig. 10.2 Histogram for Y.sim +

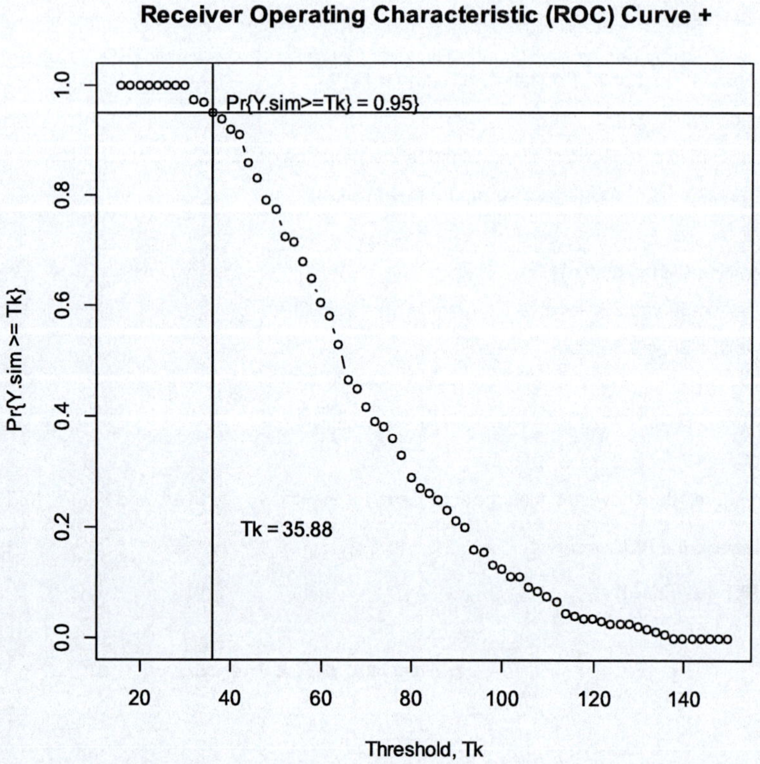

Fig. 10.3 ROC curve +

0. Of course, since the curves are actually computed at discrete values of the thresholds, T_k, the point of equivalence between true positive and true negative probabilities will be approximate. Figure 10.6 shows the code modified from that shown in Fig. 10.1.

Note that the balance point is at approximately 31.88, so that the estimate for Pr{True +} = 0.975 and the estimate for Pr{True −} = 0.985. Thus, using a threshold of $T_k = 31.88$ yields approximately a $1 - 0.975 = 0.025$, or 2.5% chance of a false negative, and a $1 - 0.985 = 0.015$, or 1.5% chance of a false positive. The script in Fig. 10.6 includes a test to see if the threshold identifies the "true negatives" (at least, the proportion of true negatives expected in the sample).

While ROC curves are not the same as reliability or survival curves, they have certain characteristics in common. Both are monotonic. Both have *y*-axis values varying between 0 and 1. Both are used to estimate probabilities. Of course, there are many differences (censoring perhaps the most significant).

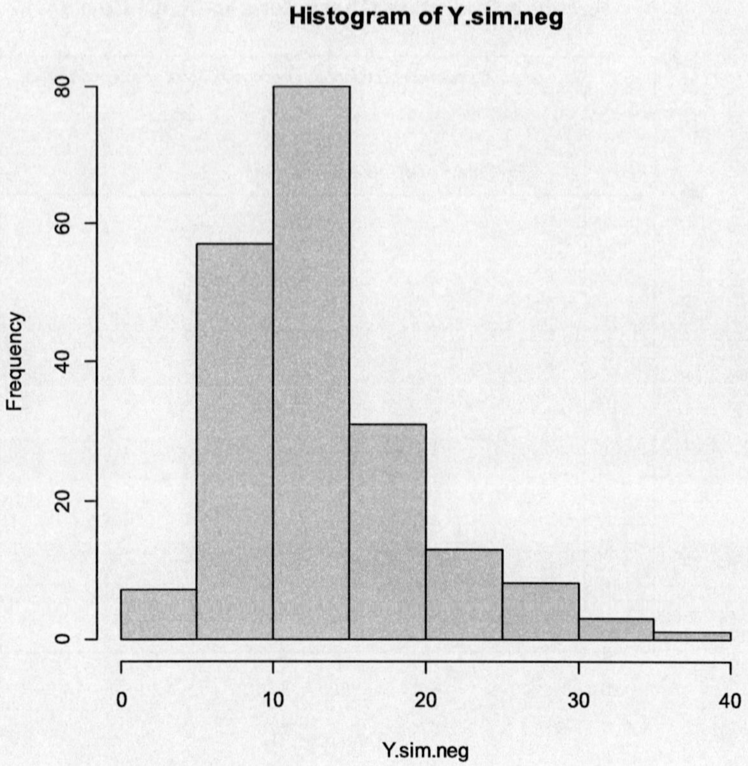

Fig. 10.4 Histogram for Y.sim –

10.2 Sensitivity and Specificity

Sensitivity is the probability of obtaining a "+" test result given that the sample is in fact +. Specificity is the probability of obtaining a "−" test result when the sample is truly −. The ROC curves were used to determine the values of responses to define a + or − test.

Once a diagnostic test is developed, so that the threshold level of response for a + test is determined, it is generally required to assess the sensitivity and specificity.

In the example used to illustrate ROC curves, there were $n = 200$ true + and $n = 200$ true −. With the threshold value of 31.88, there were 197 true + that tested +, or 98.5%. Similarly, there were 195 true – that tested −, or 97.5%. Thus, the estimate of sensitivity is:

$$\widehat{\Pr}\{+|+\} = \frac{197}{200} = 0.985$$

The estimate of specificity is:

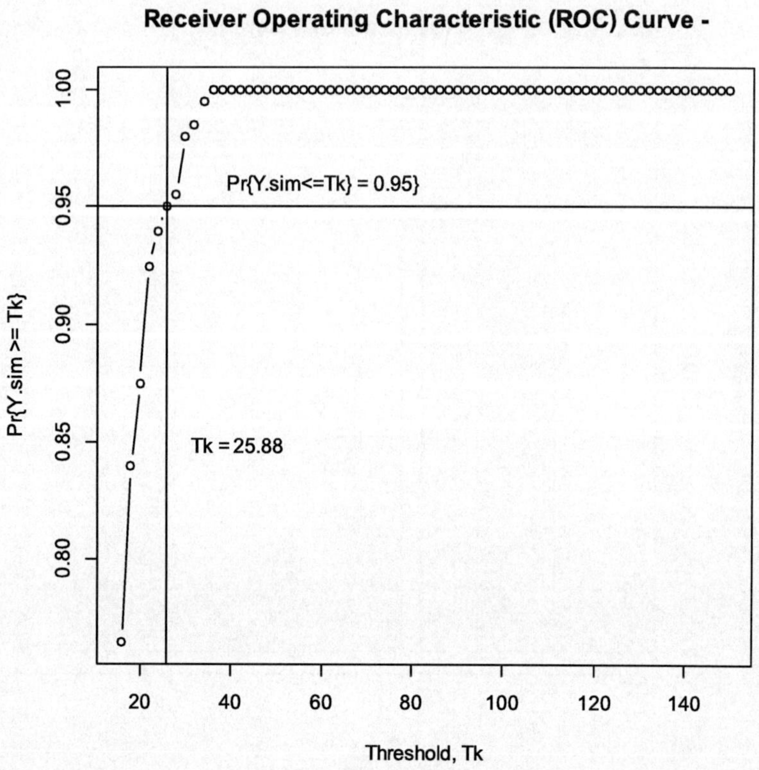

Fig. 10.5 ROC curve –

$$\widehat{\Pr}\{-\mid-\}=\frac{195}{200}=0.975$$

Confidence intervals can be computed for both sensitivity and specificity. For sensitivity, the Clopper-Pearson 95% interval is 0.9568–0.9969. For specificity, the interval is 0.9426–0.9918.

10.3 Summary

Diagnostic devices often rely on a chemical reaction to determine the presence or absence of some condition in human tissue. Often the products that are detectable exist in individuals without the condition of interest. As a result, the level that the diagnostic device uses to indicate the presence of the condition must be determined experimentally. The receiver operating characteristic (ROC) curve is a tool for finding such thresholds. The degree to which the diagnostic is helpful is quantified

```
setwd("<your path here>")
#
sample.size <- 200
prop.pos <- c()
prop.neg <- c()
set.seed(26)
Y.sim.pos <- rgamma(n=sample.size,shape=5,scale=10)+20 #these are the true
positives
Y.sim.neg <- rgamma(n=sample.size,shape=5,scale=2.5) #these are the true negatives
hist(Y.sim.pos)
abline(v=31.88)
dev.new()
hist(Y.sim.neg)
abline(v=31.88)
Low.lim.pos <- min(Y.sim.pos)
High.lim.pos <- max(Y.sim.pos)
Tk <- seq(from=Low.lim.pos-14,to=High.lim.pos+14,by=2) #these are potential threshold
values
n.length <- length(Tk)
for (i in 1:n.length) {
 Y.sim.pos.sub <- Y.sim.pos[Y.sim.pos >= Tk[i]]
 num.pos <- length(Y.sim.pos.sub)
 prop.pos[i] <- num.pos / sample.size
 Y.sim.neg.sub <- Y.sim.neg[Y.sim.neg <= Tk[i]]
 num.neg <- length(Y.sim.neg.sub)
 prop.neg[i] <- num.neg / sample.size
 }
#
pos.vec <- prop.pos
neg.vec <- prop.neg
diff <- pos.vec - neg.vec
level <- Tk[diff>-0.0105 & diff<0.0105]
dev.new()
plot(Tk,prop.pos,type="b",main="Receiver Operating Characteristic (ROC) Curve
+",xlab="Threshold, Tk",ylab="Pr{Y.sim >= Tk}")
```

Fig. 10.6 ROC curve code with balance point calculations

```
abline(h=0.95)
thresh.all <- level[1]
threshold.p <- Tk[which(prop.pos>0.949 & prop.pos<0.951)]
thresh.p.txt <- as.character(round(threshold.p,2))
abline(v=threshold.p)
#abline(v=thresh.all)
pos.best.pos <- prop.pos[which(Tk==thresh.all)]
text(x=60,y=0.970,labels="Pr{Y.sim>=Tk} = 0.95}")
text(x=47,y=0.20,labels="Tk = ")
text(x=57,y=0.20,labels=thresh.p.txt)
#
level
#
pos.best.pos
dev.new()
plot(Tk,prop.neg,type="b",main="Receiver Operating Characteristic (ROC) Curve -
",xlab="Threshold, Tk",ylab="Pr{Y.sim >= Tk}")
abline(h=0.95)
threshold.n <- Tk[which(prop.neg>0.949 & prop.neg<0.951)]
thresh.n.txt <- as.character(round(threshold.n,2))

abline(v=threshold.n)
#abline(v=thresh.all)
pos.best.neg <- prop.neg[which(Tk==thresh.all)]
pos.best.neg
text(x=60,y=0.960,labels="Pr{Y.sim<=Tk} = 0.95}")
text(x=36,y=0.85,labels="Tk = ")
text(x=46,y=0.85,labels=thresh.n.txt)
#
dev.new()
all.cases <- c(round(Test.sim.pos,0),round(Test.sim.neg,0))
F.all <- ecdf(all.cases)
prob.detect <- F.all(level[1])
level.txt <- as.character(round(level[1],2))
```

Fig. 10.6 (continued)

```
prob.detect.txt <- as.character(round(100*prob.detect,2))
LL <- min(all.cases)
UL <- max(all.cases)
hist.all <- hist(all.cases, plot=FALSE)
hist.all$density <- 100*hist.all$counts/sum(hist.all$counts)
plot(hist.all,freq=FALSE,col="light grey",xaxt="n",main="Histogram of Response for All
Subjects",xlab="Response",ylab="Frequency")
axis(side=1,at=seq(from=LL,to=UL,by=5))
text(x=level[1]-8,y=25,labels=level.txt)
#text(x=110,y=100*prob.detect+1.3,labels=prob.detect.txt)
#text(x=115,y=100*prob.detect+1.3,labels="%")
#abline(h=100*prob.detect)

abline(v=level[1])
#
df2 <- cbind(prop.pos,prop.neg,diff,Tk)#diff is the difference between prop.pos and
prop.neg at each value of Tk
#
write.csv(df2,"20220130 ROC Curve Results.csv")
#
# Here is a test of the overall threshold.
# In this simulated sample, 20 out of 200 are simulated
# from the "negative" population, and 180 from
# the "positive" population.
# The two samples are combined, and then the proportion
# of the combined sample that is negative is computed
#
#
Test.sim.pos <- rgamma(n=180,shape=5,scale=10)+20
Test.sim.neg <- rgamma(n=20,shape=5,scale=2.5)
#
Test.All <- c(Test.sim.pos,Test.sim.neg)
F.check <- ecdf(Test.All)
#####
```

Fig. 10.6 (continued)

as sensitivity and specificity. Sensitivity is the probability of the diagnostic detecting the condition when it is in fact existent, i.e., a "true" positive. Specificity is the probability that the diagnostic indicates the condition does not exist when in fact it does not, i.e., a true negative. The ROCV curves are tools for setting thresholds for controlling sensitivity and specificity.

Reference

Pepe, M. S. (2003). *The statistical evaluation of medical tests for classification and prediction*. Oxford University Press.

Chapter 11
Equivalence and Noninferiority

Abstract Equivalence and noninferiority testing and hypothesis formulation are explained. In particular, product design or performance specifications are used in the context of hypothesis tests to demonstrate equivalence in the context of validation.

11.1 Introduction

There is much literature on equivalence tests in the clinical bioavailability world. Unfortunately, this paradigm and its associated statistical methods presume that the "limits of equivalence" are actually undesirable values for the variables or parameters under consideration. Thus, for example, if a government regulation states that on the average, a generic version of a drug must have a time to maximum blood concentration that is within ±20% of that for the brand name drug, the regulation actually implies that the difference in those averages must be much less than 20%. Thus, statistical equivalence tests for such bioavailability parameters are constructed to *fail* with near certainty (95% probability) if the actual difference is 20%. An engineer is not likely to specify tolerances or limits on a parameter that are actually unacceptable. The engineer would more likely specify the tolerance around a parameter to be such that fit, form, and function of the product or system or process are not adversely affected as long as the parameter remains within the tolerance limits. So, the methods for "bioequivalence," as they are described by bioavailability writers and researchers, would be useless to the engineer. In this work, the engineer will find methods for demonstrating that a product or process meets the design specifications required to ensure quality.

The need for equivalence and noninferiority tests in engineering and quality control often arises in situations known as validation, process validation, or design verification. In Chap. 6, the emphasis is on finding an "effect," that is, deciding whether a response differs in a repeatable way when a "factor" is set to two (or more) different levels. In the cases of validation or verification, the goal is to demonstrate no effect or, at least, no meaningful effect.

In the more conventional hypothesis testing paradigm, the null hypothesis is that there is no effect. Rejecting the null means we will now believe that there is a

S. A. Pardo, *Statistical Methods and Analyses for Medical Devices*,
https://doi.org/10.1007/978-3-031-26139-8_11

repeatable effect; failure to reject the null means we do not have sufficient evidence to believe that a repeatable effect exists. This is similar to an American court of law. A suspect is considered not guilty *unless* the prosecution has sufficient evidence to reject the null hypothesis of not guilty. If the prosecution fails to provide enough evidence to reject the hypothesis of not guilty, the suspect is released and found *not guilty*, but that that does not mean the suspect is *innocent*. Rather, it means there was not sufficient evidence to reject the null. So, failure to find an effect is not the same as concluding no effect exists.

In equivalence testing, the null hypothesis is that there is an effect that is minimally big enough to matter. Rejecting this null means we have sufficient reason to believe that any repeatable difference that exists will be small enough to not matter. The minimal size of an effect that matters is context sensitive; it cannot be determined by setting a threshold for a p-value, or setting a confidence level. Instead, that difference depends on the requirements for the device, process, or procedure, to either work sufficiently well or produce sufficiently acceptable results.

Finally, a word about the difference between "equivalence" and "noninferiority." "Equivalence" will be used when describing two-sided differences and "noninferiority" used for one-sided differences. For example, two methods of manufacture might be considered equivalent if the difference in the average response is within $\pm\delta$ units. That is to say, it does not matter which method produces a larger or smaller response, on the average, as long at the difference is within the range of $\pm\delta$ units. A new process might be considered noninferior to the current process if the new process produces no more than $100(p + \rho)\%$ nonconforming product, where p is the proportion of nonconforming produced by the current process.

Equivalence and noninferiority can be defined in terms of difference from a standard, or desired, value. For example, a device might be considered "equivalent" to an "ideal" if the average response value for the device is $\mu \pm \delta$, where μ is the desired nominal average value. Similarly, a process might be considered noninferior to an ideal if it produces no more than $100(p + \rho)\%$ nonconforming product, where p is the ideal proportion of non-defectives.

In this work, the two one-sided test (TOST) methodology of Schuirmann (1987) will be adopted for determining critical values for equivalence tests.

11.2 A Single Proportion

Suppose that it is desired for the proportion of conforming items produced by a process to be at least 0.99. That is to say, the process operators/designers wish to have no less than a 99% probability of producing "good" items. A conventional hypothesis test would more than likely be formulated for the proportion of nonconforming, or "bad" items, namely, $q = 1 - p = 1 - 0.99 = 0.01$, or 1%. The hypotheses (null and alternative) might be stated as:

$$H_0 : q = 0.01$$
$$H_1 : q > 0.01$$

So this calls for a one-sided hypothesis test. The sample statistic is:

X = no. "bad" items in a sample of size n

or alternatively:

$$\hat{q} = \frac{X}{n}$$

The critical values are either X_c, where:

$$\Pr\{X \geq X_c | q = 0.01, n\} = \sum_{k=X_c}^{n} \binom{n}{k} q^k (1-q)^{n-k} \approx 0.05$$

or alternatively:

$$q_c = \frac{X_c}{n}$$

In other words, if $X \geq X_c$, or $q \geq q_c$, then reject H_0.

If $X < X_c$, we fail to reject H_0. All we can say is that we are unable to say that the process probability of producing a nonconforming item is greater than 0.01, or 1%.

This could be formulated as a noninferiority test.

The hypotheses would be:

$$H_0 : q > 0.01$$
$$H_1 : q \leq 0.01$$

The critical value would be X_c:

$$\Pr\{X \leq Xc | q = 0.01, n\} = \sum_{k=0}^{X_c} \binom{n}{k} q^k (1-q)^{n-k} \approx 0.95$$

So for this formulation, if X is low enough, then H_0 is rejected, which is in fact the desirable result. Thus, in place of having a favorable result yield a conclusion that you cannot say the process in inadequate, the noninferiority formulation allows you to more definitively state that the probability of generating nonconforming items is at or below 1%.

With $n = 80$, the critical value for the conventional hypothesis test is $X_c = 3$:

$$\Pr\{X \geq 3 | q = 0.01, n = 80\} = \sum_{k=3}^{n} \binom{n}{k} q^k (1-q)^{n-k} \approx 0.04655$$

For the noninferiority test, the critical value is $X_c = 2$:

$$\Pr\{X \leq 2 | q = 0.01, n = 80\} = \sum_{k=0}^{2} \binom{n}{k} q^k (1-q)^{n-k} \approx 0.95345$$

It may be preferable to state the hypotheses in terms of the proportion of “good” items, p:

$$H_0 : p < 0.99$$
$$H_1 : p \geq 0.99.$$

The critical value, X_c, would be 78:

$$\Pr\{X \geq X_c | p = 0.99, n = 80\} = \sum_{k=X_c}^{n} \binom{n}{k} p^k (1-p)^{n-k} \approx 0.95345$$

11.3 Comparing Two Proportions

If the proportions of conforming items are to be compared, there are two ways in which to compare them:

1. A difference, $p_1 - p_2$
2. A ratio of some sort

For ratios, it is more common to compare the ratio of odds (Fleiss et al., 2003). If p represents the proportion of conforming items, or instances of the event of interest, then odds are:

$$O = \frac{p}{1-p}$$

An odds of 1 means that $p = 0.50$, that is, it is equally likely for the event of interest to occur or not occur. If $p = 1$, then the odds are infinite.

Odds are interpreted as the number of times it is as likely for the event to occur compared to it not occurring. So, if the odds of making a conforming item are 2.0, for example, then it is 2.0 times more likely for the process to make a conforming item compared to making a nonconforming item.

Odds ratios are a means of comparing odds. The odds ratio for two proportions is:

$$\mathrm{OR} = \frac{O_1}{O_2}$$

Another way to interpret odds is as a percent greater likelihood of the event of interest occurring. The formula is:

$$P = 100\left(1 - \frac{1}{O}\right)\%$$

So, if the odds of a conforming item are 2, then it is $P = 100\left(1 - \frac{1}{2}\right)\% = 50\%$ more likely that the process will produce a conforming item as not. The same transformation can be made for odds ratios, namely:

$$P = 100\left(1 - \frac{1}{\mathrm{OR}}\right)\%$$

In the case of odds ratios, it means that the odds of event occurrence is *P*% greater for group/condition 1 as it is for group/condition 2.

To define limits of equivalence or noninferiority, for differences, the two proportions would be considered equivalent if the difference in population proportions are within $\pm\Delta$ of each other. For OR, it might be that the population OR must be no greater than some threshold, say ρ. The hypotheses might be stated as:

$$H_0 : p_1 - p_2 < -\Delta \text{or } p_1 - p_2 > +\Delta$$

versus the alternative:

$$H_1 : |p_1 - p_2| \leq +\Delta$$

The limits of equivalence need not be symmetric about 0, that is, the low limit might be $-\Delta_1$ and the high limit $+\Delta_2$.

For OR, generally the limits are surrounding the value 1.0. That is to say, if OR = 1, then the two odds are identical. If the OR is less than 1, then the "numerator" group is less likely to yield the event of interest compared to the "denominator" group. Conversely, if OR > 1, then the "numerator" group is more likely to yield the vent compared to the "denominator" group. So, the limits of equivalence might be something like:

$$1 \pm \Delta$$

A test of equivalence could be performed using confidence intervals.

Unfortunately, unlike many tests of hypotheses, there is no simple formula for the sampling distributions of either $\widehat{p}_1 - \widehat{p}_2$ or $\widehat{\mathrm{OR}}$. The easiest method for testing hypotheses about these two parameters is via a resampling technique. We will discuss the use of the bootstrap (Efron, 1982).

Bootstrap sampling is a means of computing confidence intervals when no closed-form expression exists or is very difficult to compute. The procedure is fairly simple:

1. Compute the sample statistic, e.g., $\widehat{p}_1 - \widehat{p}_2$ or $\widehat{\text{OR}}$.
2. For N repetitions:

   2a. Randomly sample the data, separately from each group/condition and with replacement, and compute the sample statistics as done for the original sample.

   2b. Store these "resampled" statistics.

3. Find the 100α percentile and $100(1 - \alpha)$ percentile of the distribution of the resampled parameters, $p_1 - p_2$ or OR.

There are two ways to determine, based on the confidence limits, whether to reject H_0. The first is to determine if the confidence limits fall within the equivalence limits. This approach should be used if it is actually unacceptable for the parameters to equal the equivalence limits. This is the approach taken for bioequivalence studies.

If, however, it is perfectly acceptable for the parameters to be exactly at the equivalence limits, then if the lower confidence limit is less than or equal to the upper equivalence limit, and the upper confidence limit is greater than or equal to the lower equivalence limit, H_0 is rejected (Pardo, 2014).

For equivalence analyses, Schuirmann (1987) performed two separate comparisons, one for the lower limit and one for the upper limit. For each, the critical value was chosen with $100\alpha\%$ error and not $100(\alpha/2)\%$. Such a choice is referred to as two one-sided test (TOST).

As an example, consider a comparison of two alternative product designs. A total of $n = 80$ units of each type was sampled, and each unit was tested. The response was binary, i.e., either *pass* or *fail*. Table 11.1 shows the numbers of *pass* results for each product.

Table 11.2 shows the equivalence limits for both $p_1 - p_2$ and OR.

Figure 11.1 shows R code for computing bootstrap confidence intervals for both differences in the two proportions and the odds ratio. Figure 11.2 shows the

Table 11.1 Comparison of two proportions

Product	n	$X =$ *No. PASSed*
1	80	77
2	80	76

Table 11.2 Equivalence limits

Parameter	Low lim.	High lim.
$p_1 - p2$	−0.04	0.04
OR	0.90	5.21

```
options(na.action=na.omit)
#
# setwd tells R which folder to use for this script
#

setwd("<your path here>")

df1 <- read.csv("20220509 Ch 11 Two Proportions Summary Data.csv")
#
# Variables:
#  Group ( 1 and 2)
#  X
#  N
#
attach(df1)
OR <- c()
ODDS1 <- c()
ODDS2 <- c()
pDiff <- c()
samp1 <- c()
samp2 <- c()
P1 <- 0.99
P2 <- 0.95
Del <- abs(P1 - P2)
maxOR <- (P1/(1-P1))/(P2/(1-P2))
minOR <- 0.90
#
# Now expand the summary data into vectors of 0's and 1's
#
one.1 <- rep(x=1,times=X[1])
zero.1 <- rep(x=0,times=N[1] - X[1])
one.zero.1 <- c(one.1,zero.1)
#
# This shuffles the original (expanded) data
#
```

Fig. 11.1 R Code for bootstrap TOST confidence intervals: difference of proportions and odds ratio

```
rand.1 <- runif(length(one.zero.1),0,1)
rand.one.zero.1 <- one.zero.1[order(rand.1)]
#
one.2 <- rep(x=1,times=X[2])
zero.2 <- rep(x=0,times=N[2] - X[2])
one.zero.2 <- c(one.2,zero.2)
#
# This shuffles the original (expanded) data
#
rand.2 <- runif(length(one.zero.2),0,1)
rand.one.zero.2 <- one.zero.2[order(rand.2)]

#
bootn <- 2000

alpha <- 0.05
set.seed(26)
#

#
sampP1 <- X[1] / N[1]
sampP2 <- X[2] / N[2]
sampODDS1 <- sampP1 / ( 1 - sampP1)
sampODDS2 <- sampP2 / (1 - sampP2)
sampOR <- sampODDS1 / sampODDS2
sampDiff <- sampP1 - sampP2
#

for(i in 1:bootn) {
 samp1 <- sample(x=rand.one.zero.1,size=N[1],replace=TRUE)
 samp2 <- sample(x=rand.one.zero.2,size=N[2],replace=TRUE)

 permute1 <- mean(samp1)
 permute2 <- mean(samp2)
 if (permute1==1){
  ODDS1[i] <- 0
  }
 else {
  ODDS1[i] <- permute1 / (1 - permute1)
```

Fig. 11.1 (continued)

```
    }
   if (permute2==1) {
    ODDS2[i] <- 0
    }
   else {
    ODDS2[i] <- permute2 / (1 - permute2)
    }
   if (ODDS1[i]==0 | ODDS2[i]==0) {
    OR[i] <- NA
    }
   else {
    OR[i] <- ODDS1[i] / ODDS2[i]
    }
   pDiff[i] <- permute1 - permute2
   }
#
# this is a percentile-method bootstrap confidence interval
#
CL.pDiff <- quantile(pDiff,probs=c(0.05,0.95),type=8,na.rm=TRUE)
CL.OR <- quantile(OR,probs=c(0.05,0.95),type=8,na.rm=TRUE)
LCL.pDiff <- CL.pDiff[1]
UCL.pDiff <- CL.pDiff[2]
LCL.OR <- CL.OR[1]
UCL.OR <- CL.OR[2]

#
hist.OR <- hist(OR,plot=FALSE)

hist.diff <- hist(pDiff,plot=FALSE)
#
```

Fig. 11.1 (continued)

```
hist.OR$density <- 100*hist.OR$counts / sum(hist.OR$counts)
hist.diff$density <- 100*hist.diff$counts / sum(hist.diff$counts)

plot(hist.OR,freq=FALSE,col="light grey",main="Histogram of
OR",xaxt="n",yaxt="n",xlim=c(0,15))
axis(side=1,at=seq(from=0,to=13,by=1)) #controls x-axis
axis(side=2,at=seq(from=0,to=40,by=1)) #controls y-axis
abline(v=sampOR,lty=1)
abline(v=LCL.OR,lty=2)
abline(v=UCL.OR,lty=2)
abline(v=minOR,lty=3)
abline(v=maxOR,lty=3)
legend(x=6,y=27,legend=c("Sample OR","Conf.Limit","Equiv. Limit"),lty=c(1,2,3))
dev.new()
plot(hist.diff,freq=FALSE,col="light grey",main="Histogram of
Diff",xaxt="n",yaxt="n",xlim=c(-0.145,0.145))
axis(side=1,at=seq(from=-0.15,to=0.15,by=0.01)) #controls x-axis
axis(side=2,at=seq(from=0,to=40,by=1)) #controls y-axis
abline(v=sampDiff,lty=1)
abline(v=LCL.pDiff,lty=2)
abline(v=UCL.pDiff,lty=2)
abline(v=-Del,lty=3)
abline(v=Del,lty=3)
legend(x=0.062,y=27,legend=c("Sample Diff","Conf.Limit","Equiv. Limit"),lty=c(1,2,3))

#
df2 <- cbind(OR,pDiff) # these are bootstrap samples

write.csv(df2,"20220510 Bootstrap Samples OR and pDiff.csv")
#
LCL.pDiff
UCL.pDiff
LCL.OR
UCL.OR
Del
minOR
maxOR
#####
```

Fig. 11.1 (continued)

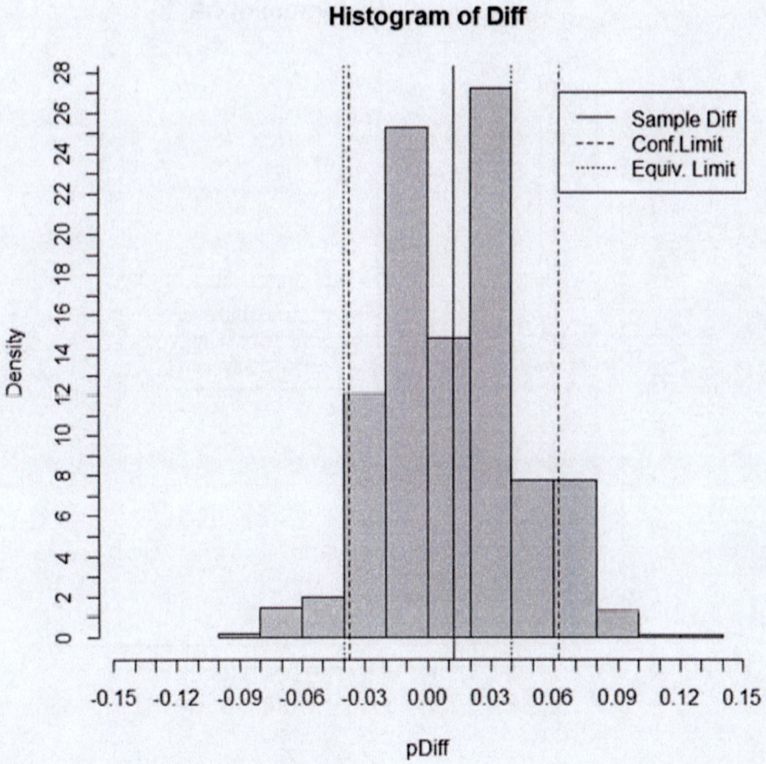

Fig. 11.2 Histogram of bootstrap samples, $\widehat{p}_1 - \widehat{p}_2$

bootstrap distribution for the differences in the proportions, together with the two one-sided confidence limits and the equivalence limits.

Figure 11.3 shows the histogram of the bootstrap samples of $\widehat{\text{OR}}$.

In both Figs. 11.2 and 11.3, the lower confidence limit is less than the high equivalence limit and the upper confidence limit is greater than the low equivalence limit.

A concept related to bootstrapping is called permutation tests (Good, 1994). First, the test statistic, either $\widehat{p}_1 - \widehat{p}_2$ or $\widehat{\text{OR}}$, is computed. Then a "null" distribution is formed by randomly sampling. With replacement, response values from each group and randomly assigning the values to one of the two groups. That means response values from one group originally may be assigned to the other. After this "permutation" of the response values is done, the test statistic(s) is computed and stored. The random sampling/permuting process is repeated N times, and the distribution of the test statistic(s) is formed. This is the null distribution, i.e., the distribution of the test statistic under the assumption that the grouping has no effect. The original test statistic is compared to the null distribution. If its value falls below or above some threshold tail (i.e., below the $100\alpha\%$ tail or above the $100(1-\alpha)\%$ tail), then the null

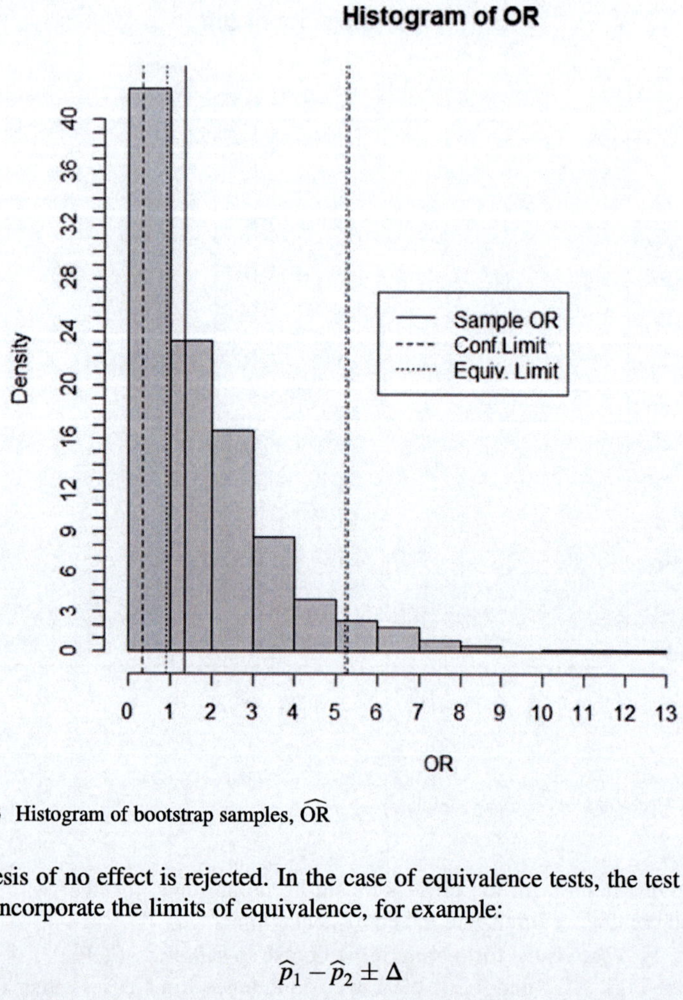

Fig. 11.3 Histogram of bootstrap samples, $\widehat{\text{OR}}$

hypothesis of no effect is rejected. In the case of equivalence tests, the test statistic would incorporate the limits of equivalence, for example:

$$\widehat{p}_1 - \widehat{p}_2 \pm \Delta$$

Figure 11.4 shows R code for doing a permutation equivalence test for both the statistics $\widehat{p}_1 - \widehat{p}_2$ or $\widehat{\text{OR}}$. Note that the p-values depend on whether the upper or lower limit applies. That is to say, if the sample OR is less than 1.0, we might be concerned that the population OR is lower than the low limit of equivalence. Conversely, if the sample OR is greater than 1.0, we are concerned that the population OR might be greater than the high equivalence limit.

Figure 11.5 shows the permutation distribution for $\widehat{p}_1 - \widehat{p}_2$ and Fig. 11.6 shows the permutation distribution for $\widehat{\text{OR}}$.

```
options(na.action=na.omit)
#
# setwd tells R which folder to use for this script
#

setwd("<your path here>")
rm(OR54,OR70,OR180,OR250)
df1 <- read.csv("20220509 Ch 11 Two Proportions Summary Data.csv")
#
# Variables:
#  Group ( 1 and 2)
#  X
#  N
#
#attach(df1)
OR <- c()
pDiff <- c()
maxOR <- 1.35
minOR <- 0.65
perm.reps <- 5000

alpha <- 0.05
set.seed(26)
#

#
```

Fig. 11.4 R code for permutation tests: $p_1 - p_2$ and OR

11.4 Comparing Two Means

In order for two means to be considered equivalent, the difference between them must be no greater, in absolute value, of some prespecified quantity, say Δ. The equivalence hypotheses could be stated as:

$$H_0 : \mu_1 - \mu_2 < -\Delta_0 \text{ or } \mu_1 - \mu_2 > +\Delta_0$$

versus the alternative:

```
sampP1 <- df1$X[1] / df1$N[1]
sampP2 <- df1$X[2] / df1$N[2]
sampODDS1 <- sampP1 / ( 1 - sampP1)
sampODDS2 <- sampP2 / (1 - sampP2)
sampOR <- sampODDS1 / sampODDS2
sampDiff <- sampP1 - sampP2
if (sampDiff < 0) {
 eq.diff.lim <- -0.04
 } else {
 eq.diff.lim <- 0.04
 }
#
if (sampOR >= 1.0){
 eq.OR.lim <- maxOR
 } else {
 eq.OR.lim <- -minOR
 }
#
# Now expand the summary data into vectors of 0's and 1's
#
one.1 <- rep(x=1,times=df1$X[1])
zero.1 <- rep(x=0,times=df1$N[1] - df1$X[1])
one.zero.1 <- c(one.1,zero.1)
#
# This shuffles the original (expanded) data
#
rand.1 <- runif(length(one.zero.1),0,1)
rand.one.zero.1 <- one.zero.1[order(rand.1)]
#
one.2 <- rep(x=1,times=df1$X[2])
```

Fig. 11.4 (continued)

```
zero.2 <- rep(x=0,times=df1$N[2] - df1$X[2])
one.zero.2 <- c(one.2,zero.2)
#
# This shuffles the original (expanded) data
#
rand.2 <- runif(length(one.zero.2),0,1)
rand.one.zero.2 <- one.zero.2[order(rand.2)]
#
# Now create a concatenated expanded dataset for doing the permutation sampling
#
big.expanded <- c(rand.one.zero.1,rand.one.zero.2)
for(i in 1:perm.reps) {
  samp1 <- sample(x=big.expanded,size=N[1],replace=TRUE)
  samp2 <- sample(x=big.expanded,size=N[2],replace=TRUE)
  permute1 <- mean(samp1)
  permute2 <- mean(samp2)
  ODDS1 <- permute1 / (1 - permute1)
  ODDS2 <- permute2 / (1 - permute2)
  OR[i] <- (ODDS1 / ODDS2) + eq.OR.lim
  pDiff[i] <- permute1 - permute2 - eq.diff.lim
  }
#
#
F.diff <- ecdf(pDiff)
F.OR <- ecdf(OR)
pval.Low.OR <- F.OR(sampOR)
pval.High.OR <- 1 - F.OR(sampOR)
pval.Low.diff <- F.diff(sampDiff)
pval.High.diff <- 1-F.diff(sampDiff)
sampOR
```

Fig. 11.4 (continued)

```
pval.Low.OR
pval.High.OR
sampDiff
pval.Low.diff
pval.High.diff
#
hist.OR <- hist(OR,plot=FALSE)

hist.diff <- hist(pDiff,plot=FALSE)
#
hist.OR$density <- 100*hist.OR$counts / sum(hist.OR$counts)
hist.diff$density <- 100*hist.diff$counts / sum(hist.diff$counts)


plot(hist.OR,freq=FALSE,col="light grey",main="Histogram of OR",xaxt="n",yaxt="n")
axis(side=1,at=seq(from=-13,to=13,by=1)) #controls x-axis
axis(side=2,at=seq(from=0,to=70,by=1)) #controls y-axis
abline(v=sampOR,lty=1)
dev.new()
plot(hist.diff,freq=FALSE,col="light grey",main="Histogram of Diff",xaxt="n",yaxt="n")
axis(side=1,at=seq(from=-0.30,to=0.30,by=0.01)) #controls x-axis
axis(side=2,at=seq(from=0,to=45,by=1)) #controls y-axis
abline(v=sampDiff,lty=1)
#####
```

Fig. 11.4 (continued)

$$H_1 : -\Delta_0 \leq \mu_1 - \mu_2 \leq +\Delta_0$$

Although we have stated the limits of equivalence in a symmetric fashion ($\pm\Delta$), it not need be the case.

The null hypothesis is rejected if:

$$\overline{X}_1 - \overline{X}_2 + t_{1-\alpha}\text{SE} \geq -\Delta_0$$

and

$$\overline{X}_1 - \overline{X}_2 - t_{1-\alpha}\text{SE} \leq +\Delta_0$$

$\overline{X}$ is the sample arithmetic mean and SE is the standard error of the differences:

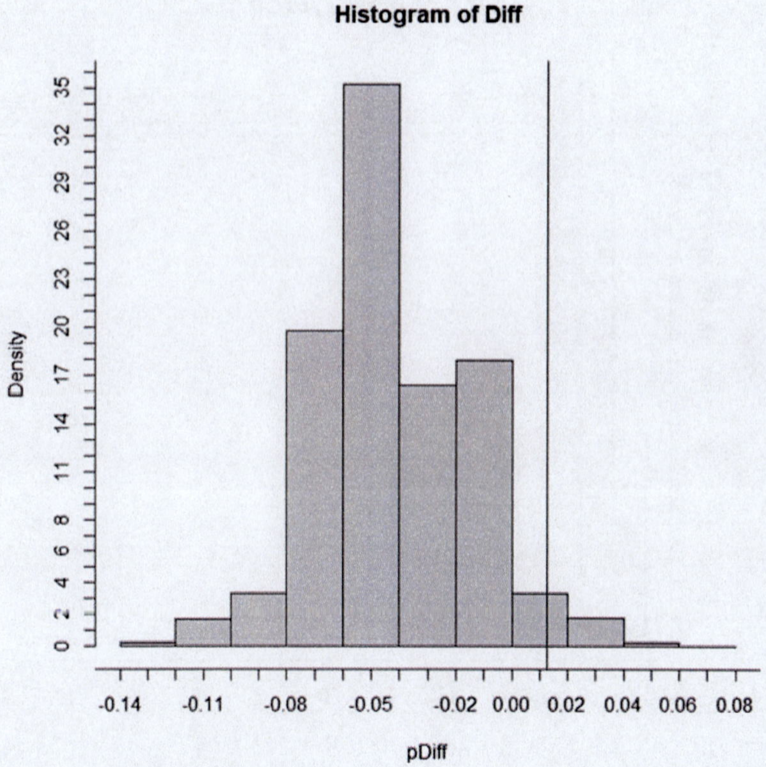

Fig. 11.5 Permutation distribution: $\widehat{p}_1 - \widehat{p}_2$

$$\mathrm{SE} = \sqrt{\frac{S_1^2}{n_1} + \frac{S_2^2}{n_2}}$$

$t_{1-\alpha}$ is the 100(1 − α) percentile of a Student's t distribution with df degrees of freedom. The value of df can be computed in two ways. First, assuming that the variance of the response is the same for the two groupings:

$$\mathrm{df} = n_1 + n_2 - 2$$

Welch (1947) devised a test that was similar to that of Student's t, without the assumption of equal variances. The degrees of freedom are more complicated, but not too difficult to compute. The formulas are:

$$W_1 = \left(\frac{S_1^2/n_1}{S_1^2/n_1 + S_2^2/n_2}\right)^2 / (n_1 - 1)$$

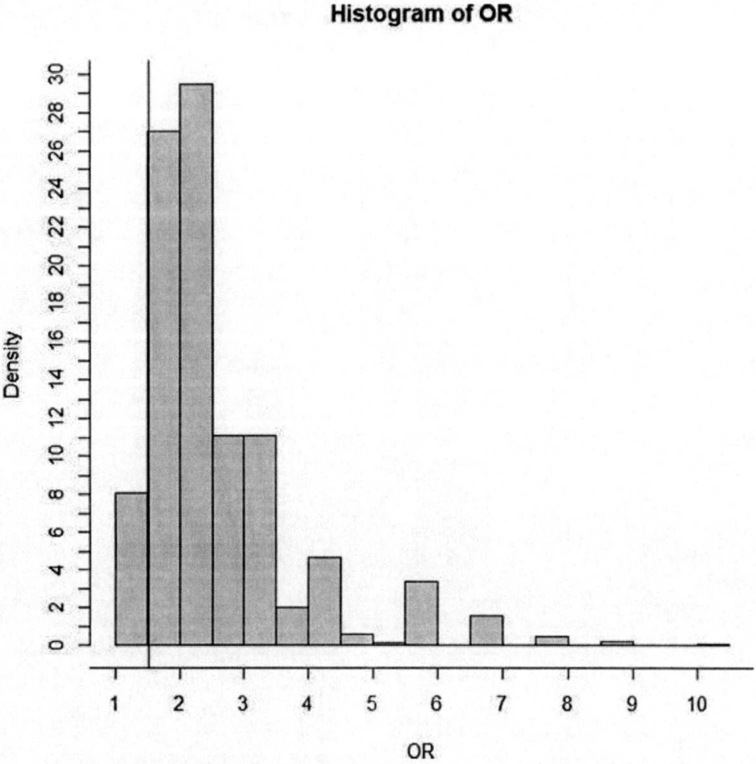

Fig. 11.6 Permutation distribution: $\widehat{\text{OR}}$

$$W_2 = \left(\frac{S_2^2/n_2}{S_1^2/n_1 + S_2^2/n_2} \right)^2 / (n_2 - 1)$$

$$\text{df} = \frac{1}{W_1 + W_2}$$

Hypothesis tests generally beg the question, "How likely will the null hypothesis be rejected?" The probability of rejecting the null hypothesis is called power and depends mostly on how much the truth departs from the null hypothesis and the sample size. What we would like to know is:

$$\Pr\left\{\overline{X}_1 - \overline{X}_2 + t_{1-\alpha}\text{SE} \geq -\Delta_0 \text{ and } \overline{X}_1 - \overline{X}_2 - t_{1-\alpha}\text{SE} \leq +\Delta_0 \middle| |\mu_1 - \mu_2| = \Delta_a \neq \Delta_0, n_1, n_2\right\}$$

The quantity:

$$\frac{\overline{X}_1 - \overline{X}_2 - (\mu_1 - \mu_2)}{\text{SE}}$$

has a Student's t distribution with df degrees of freedom. If the difference $\mu_1 - \mu_2$ were equal to Δ_0, then:

$$\frac{\overline{X}_1 - \overline{X}_2 - \Delta_0}{\text{SE}}$$

would have that Student's t distribution and:

$$\Pr\left\{\frac{\overline{X}_1 - \overline{X}_2 - \Delta_0}{\text{SE}} \leq t_{1-\alpha}\right\} = 1 - \alpha$$

Suppose that $\mu_1 - \mu_2 = \Delta_a > \Delta_0$. In that case the quantity:

$$\frac{\overline{X}_1 - \overline{X}_2 - \Delta_0}{\text{SE}}$$

has a non-central t distribution, with non-centrality parameter:

$$\delta = \frac{\Delta_a - \Delta_0}{\sqrt{\sigma_1^2/n_1 + \sigma_2^2/n_2}}$$

For the purposes of making probability calculations, we will make the simplification:

$$\sigma_1 = \sigma_2 = \sigma$$

and:

$$n_1 = n_2 = n$$

The non-centrality parameter simplifies to:

$$\frac{\Delta_a - \Delta_0}{\sqrt{2\frac{\sigma^2}{n}}} = \frac{\sqrt{n}(\Delta_a - \Delta_0)}{\sqrt{2}\,\sigma} = \sqrt{\frac{n}{2}}\frac{(\Delta_a - \Delta_0)}{\sigma}$$

So the probability of rejecting the null is:

```
setwd("<your path here>")

n <- 40 #assume same sample size per group
alpha<- 0.05
#sigma <- 2.694

#
#
# delta is the true parameter value:
#delta <- c(-1.0*sigma, -0.85*sigma, -0.75*sigma, -0.65*sigma, -0.5*sigma, -0.45*sigma,
-0.33*sigma, -0.25*sigma, -0.14*sigma, -0.10*sigma, -0.025*sigma, -0.005*sigma, 0.0,
0.005*sigma, 0.025*sigma, 0.1*sigma, 0.14*sigma, 0.15*sigma, 0.25*sigma,
0.33*sigma, 0.45*sigma, 0.5*sigma, 0.65*sigma, 0.75*sigma, 0.85*sigma, 1.0*sigma)
delsig <- seq(from=-3.0,to=3.0,by=0.01) #this is the difference between the two means
#                          per standard deviation (sigma)
#                          So, if delsig = 1, then the difference is 1 standard deviation.
#                          The null value is delsig = 0, i.e., no difference in the means of
the two groups.
# note that in general delta could be negative (mu1 < mu2), 0 (mu1 = mu2), or positive
(mu1 > mu2)
dfe <- 2*n - 2
# Tcrit is the two-sided (+/-) critical value for the t-test statistic:
Tcrit <- qt(p=1-alpha/2,df=dfe,ncp=0.0) #this is the 100(1-alpha/2) percentile of the t
distribution with 2n-2 d.f.
#
# ncp is called the non-centrality parameter; when the null hypothesis of zero difference
is true, then ncp = 0.0
# as the truth departs from the null (i.e., as delta differs from 0), the ncp changes
#
```

Fig. 11.7 R code for power of equivalence test for two means

$$\Pr\left\{\frac{\overline{X}_1 - \overline{X}_2 - \Delta_0}{\mathrm{SE}} \leq t_{1-\alpha} \middle| \, n, \frac{(\Delta_a - \Delta_0)}{\sigma}\right\}$$

Figure 11.7 shows R code for computing power in terms of $\frac{(\Delta_a - \Delta_0)}{\sigma}$. Figure 11.8 shows the power curve with $n_1 = n_2 = n = 40$.

```
# Note that in this formulation, delsig is the amount of difference beyond that which is
allowed.
# if delsig = 0, then it means the difference in means is exactly at the limit of
equivalence.
#
ncp.alt <- delsig*sqrt(n)/sqrt(2) #assume that sigma is the same in each group
# power is the probability of rejecting the null hypothesis: H0: mu1 - mu2 + e = 0, given
# that the true difference is delta.  The value of e is the allowable difference.
# Note that for equivalence test, if delta = 0, the power is equal to 1-alpha
power <- 1-(pt(q=Tcrit,df=dfe,ncp=ncp.alt,lower.tail=FALSE)+pt(q=-
Tcrit,df=dfe,ncp=ncp.alt,lower.tail=TRUE))

delsig90 <- delsig[which(power>=0.049 & power<=0.053)]
low.95 <- as.character(round(delsig90[1],2))
high.95 <- as.character(round(delsig90[2],2))
plot(delsig,power,ylim=c(0,1),main="Power Curve Two-Sample T-Test",xlab="(Delta.a-
Delta.0)/sigma",ylab="Power=Pr{Reject Ho | Ha}")
abline(h=0.95)
abline(h=alpha)
abline(v=delsig90)
text(x=-2,y=0.972,labels="Power=95%")
text(x=-1.1,y=alpha+0.02,labels=low.95)
text(x=1.1,y=alpha+0.02,labels=high.95)

df3 <- cbind(n,alpha,delsig, power)
write.csv(df3,"20220512 Power for Two-Sample Equivalence T test.csv")
#####
```

Fig. 11.7 (continued)

11.5 Comparing Two Standard Deviations

Generally, one-sided, "noninferiority" tests apply to measures of variability, such as standard deviations. Often, the standard deviation from an "evaluation" product or method is said to be noninferior to a comparator, or conventional product or method, if:

$$\sigma_e \le k_0 \sigma_c$$

$$k_0 \ge 1.$$

If S represents a sample standard deviation, then:

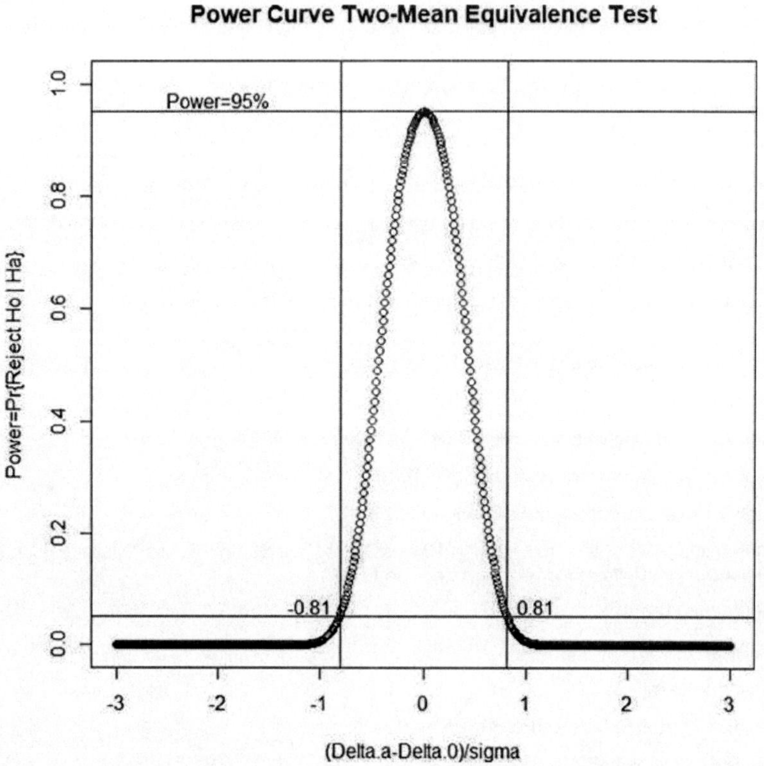

Fig. 11.8 Power curve of equivalence test for two means

$$F = \frac{S_e^2}{k_0^2 S_c^2}$$

would have an F distribution with degrees of freedom ($n_e - 1$, $n_c - 1$) under the condition:

$$\sigma_e = k_0 \sigma_c$$

As k_1 gets large, the ratio $F = \frac{S_e^2}{k_1^2 S_c^2}$ decreases. Thus:

$$\Pr\left\{ \frac{S_e^2}{k_1^2 S_c^2} \geq F_\alpha \middle| n_e, n_c \right\}$$

also decreases (F_α is the 100(α) percentile of an g distribution with ($n_e - 1$, $n_c - 1$) degrees of freedom). The noninferiority hypotheses are:

```
setwd("<your path here>")
n <- 40 #assume same sample size per group
alpha<- 0.05
power0.text <- as.character(round(1-alpha,2))
del <- seq(from=0.1,to=3.3,by=0.1)
k0 <- 1.2
k0.text <- as.character(round(k0,1))
F.crit <- qf(p=alpha,df1=n-1,df2=n-1,lower.tail=TRUE)
power <- 1-pf(q=del*F.crit,df1=n-1,df2=n-1,lower.tail=TRUE)
plot(x=del,y=power,main="Power for Two SD Non-Inferiority Test",xlab="delta;
k1=delta*k0",ylab="Pr{Reject H0}",type="b")
abline(h=1-alpha)
abline(v=1.0)
text(x=0.50,y=0.6,labels="k0 = ")
text(x=0.70,y=0.6,labels=k0.text)
text(x=1.5,y=(1-alpha)+0.023,labels="Pr{Reject H0} = ")
text(x=2.0,y=(1-alpha)+0.023,labels=power0.text)
df2 <- cbind(k0,del,power)
write.csv(df2,"20220513 Power for Two-Sample Equivalence of Two SDs.csv")
#####
```

Fig. 11.9 R code for power curve: comparing two standard deviations

$$H_0 : \sigma_e > k_0\sigma_c$$

and:

$$H_1 : \sigma_e \leq k_0\sigma_c$$

If $k_1 = \delta k_0$ then:

$$\Pr\left\{\frac{S_e^2}{k_1^2 S_c^2} \geq F_\alpha | n_e, n_c\right\} = \Pr\left\{\frac{S_e^2}{k_0^2 S_c^2} \geq \delta F_\alpha | n_e, n_c\right\}$$

When $\delta \leq 1$, the probability of rejecting the null hypothesis is no less than $1 - \alpha$. The probability of rejecting the null increases as δ increases. Figure 11.9 shows R code for computing the power curve. Figure 11.10 shows a plot of the curve with $n_e = n_c = 40$.

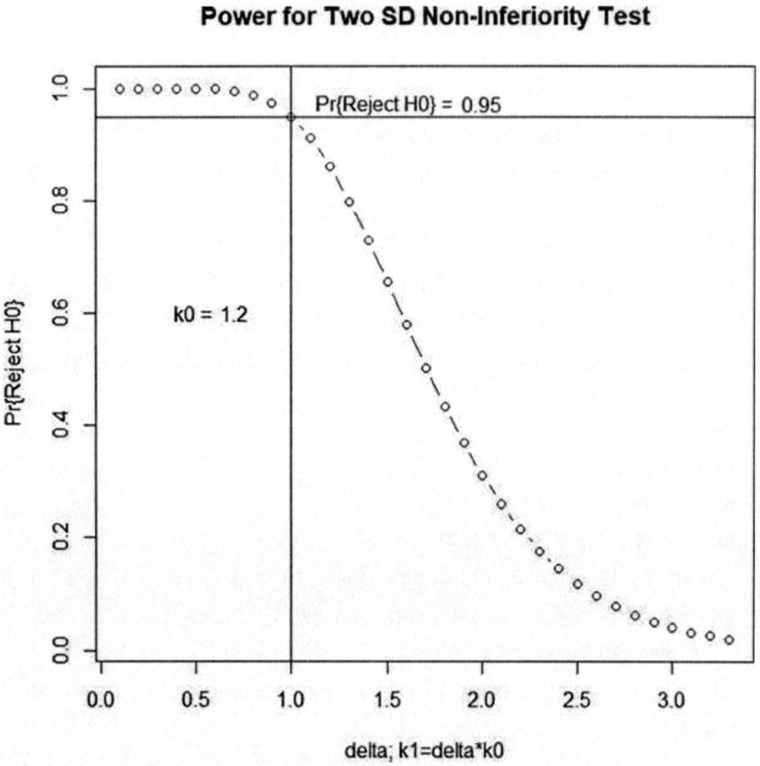

Fig. 11.10 Plot of power curve: comparing two standard deviations

11.6 Summary of Equivalence and Noninferiority

Both equivalence and noninferiority tests have wide and varied applicability in the world of medical devices. Even the submission to the FDA for clearance to market medical devices, under section 510(k) of the Food, Drug, and Cosmetics Act (Title 21, Ch. V, 2022), calls for "substantial equivalence" between the evaluation device and a predicate, or comparator device. Although the term "substantial equivalence" has no formal definition, providing the results of an equivalence or noninferiority test is often sufficient evidence of "substantial equivalence." The two main points of this chapter are:

1. Limits of equivalence/noninferiority are determined based on functionality and not statistically. Statistics are used to determine whether or not equivalence/noninferiority can be claimed.
2. It is crucial to formulate the equivalence/noninferiority hypotheses, in order to determine the adequacy of the sample size(s).

References

Efron, B. (1982). *The Jackknife, the bootstrap, and other resampling plans*. SIAM.

Fleiss, J. L., Levin, B., & Paik, M. C. (2003). *Statistical Methods for Rates and Proportions* (3rd ed.). John Wiley and Sons, New York.

Good, P. (1994). *Permutation tests*. Springer.

Pardo, S. A. (2014). *Equivalence and Noninferiority Tests for Quality, Manufacturing, and Test Engineers*. Chapman & Hall/CRC, Boca Raton.

Schuirmann, D. (1987). A comparison of the two one-sided tests procedure and the power approach to assessing the equivalence of average bioavailability. *Pharmacometrics, 15*, 657.

Welch, B. L. (1947). The generalization of 'Student's' problem when several different population variances are involved. *Biometrika, 34*, 28–35.

Chapter 12
Nonparametrics

Abstract Traditional nonparametric tests, mostly involving rank transformations, are described. In addition, the bootstrap and permutation tests are used when parametric methods are intractable.

12.1 Introduction

Many conventional statistical methods rely on an assumption about the distributions from which the data are drawn. Notably, the t-test, the F-test (ANOVA), and even tests of hypotheses about a single standard deviation all require that the data are drawn from normal distributions. Not only are the distributional assumptions potentially violated, but in many cases it is not possible to assess the degree to which those assumptions are violated.

There are techniques that are similar in goals to the more conventional methods but do not require the restrictive distributional assumptions. While they may seem very appealing, they are, in general, not as powerful when those assumptions are either not violated at all or are at least reasonably approximate. Nevertheless, so-called nonparametric methods are very useful. There are two classes of nonparametric methods that will be focused on in this chapter:

1. Methods based on the rank transformation
2. Resampling methods, such as bootstrap and permutation tests

12.2 Rank-Based Methods

The basic idea of rank transformation is to first order the data, regardless of any groupings, and number, or rank, them from smallest (1) to largest (N), replace the data with their respective ranks, and then form a statistic using the ranks instead of the original values. Conover (1999) describes most of the common rank-based methods. Only two will be described here.

S. A. Pardo, *Statistical Methods and Analyses for Medical Devices*,
https://doi.org/10.1007/978-3-031-26139-8_12

12.2.1 The Mann-Whitney-Wilcoxon Test: Analogous to the T-Test

A rank-based method analogous to a two-sample t-test is the Mann-Whitney-Wilcoxon (MWW) test. The procedure is simple:

1. Obtain two groups of data. The sample sizes are not necessarily identical. Say that for group 1, you obtain n values and the group 2 m values. $N = n + m$.
2. Take all the N values and sort them from smallest to largest. The rank is the order in the sorted list.
3. Let $R(x_i)$ = the rank of value x_i (this is regardless of the group from which it came).
4. Just for group 1, compute:

$$U = \sum_{i=1}^{n} R(x_i)$$

5. The critical values for U will result in p-values that can be computed using the R function wilcox.test() (in base R). It allows the user to specify whether the test is lower tail, upper tail, or two-sided (default).

The degree to which the assumptions of the Student's t-test (or Welch's variant that does not require equal variances between the groups) will determine the amount of discrepancy between the conventional test and the rank-based test.

Figure 12.1 shows code for computing the traditional (Welch) t-test, the t-test with the data replaced by the ranks, and finally the MWW test. These tests are performed on data drawn from normal populations (variable X in the dataset) and from gamma populations (variable ZORK). For both variables, the mean for group 1 is 10 and 15 for group 2.

Figure 12.2 shows all the output.

For variable X, there was no change in the p-values reported, regardless of whether the t-test was performed on the data directly (X); the t-test was performed in the ranks of the data (Rank.X) or the MWW test was used.

For variable ZORK, which was gamma distributed, the p-value was considerably larger for the t-test on the original data, but the p-values for both the MWW test and the t-test on the ranks were much closer. The implication is that perhaps substituting ranks for the raw response data and using a more conventional "parametric" method might yield results very close to those obtained with a "nonparametric" method.

```
setwd("<your path here>")
df1 <- read.csv("20220516 Ch 12 Data for Mann Whitney Test.csv")
#
#
# VARIABLES:
#

# Group
# X
# R.X
# ZORK
#
Rank.X <- rank(df1$X)
Rank.Z <- rank(df1$ZORK)

t.test(df1$X ~ df1$Group,var.equal=FALSE)
t.test(Rank.X ~ df1$Group,var.equal=FALSE)
wilcox.test(df1$X ~ df1$Group)
#
t.test(df1$ZORK ~ df1$Group,var.equal=FALSE)
t.test(Rank.Z ~ df1$Group,var.equal=FALSE)
wilcox.test(df1$ZORK ~ df1$Group)
#####
```

Fig. 12.1 R code for the Mann-Whitney-Wilcoxon test

12.2.2 One-Way Nonparametric ANOVA: The Kruskal-Wallis Test

When there are more than two groups to compare, as in the case of the one-way layout, the MWW test is not adequate. Another rank-based test, called the Kruskal-Wallis test (Conover, 1999), can be used. As in the case of MWW, the ranks are computed regardless of the group or treatment/condition under which the data were observed. So if x_{ij} is the jth observation in group i, then the ranks are $R(x_{ij})$. For each of k treatment groups, compute the sum of the ranks:

$$R_i = \sum_{j=1}^{n_i} R\left(x_{ij}\right)$$

The treatment groups are not required to have the same number of observations.

```
data:  df1$X by df1$Group
t = -14.847, df = 77.872, p-value < 2.2e-16
alternative hypothesis: true difference in means is not equal to 0
95 percent confidence interval:
 -5.125055 -3.913108
sample estimates:
mean in group 1 mean in group 2
       10.24831        14.76739

> t.test(Rank.X ~ df1$Group,var.equal=FALSE)

        Welch Two Sample t-test

data:  Rank.X by df1$Group
t = -14.716, df = 77.998, p-value < 2.2e-16
alternative hypothesis: true difference in means is not equal to 0
95 percent confidence interval:
 -44.95739 -34.24261
sample estimates:
mean in group 1 mean in group 2
           20.7            60.3

> wilcox.test(df1$X ~ df1$Group)

        Wilcoxon rank sum exact test

data:  df1$X by df1$Group
W = 8, p-value < 2.2e-16
alternative hypothesis: true location shift is not equal to 0

> #
```

Fig. 12.2 Output for two-sample tests: Welch's and Mann-Whitney-Wilcoxon

```
> t.test(df1$ZORK ~ df1$Group,var.equal=FALSE)

        Welch Two Sample t-test

data:  df1$ZORK by df1$Group
t = -2.2148, df = 76.497, p-value = 0.02975
alternative hypothesis: true difference in means is not equal to 0
95 percent confidence interval:
 -8.3190220 -0.4417034
sample estimates:
mean in group 1 mean in group 2
       11.09800        15.47836
> t.test(Rank.Z ~ df1$Group,var.equal=FALSE)

        Welch Two Sample t-test

data:  Rank.Z by df1$Group
t = -2.6995, df = 75.807, p-value = 0.00856
alternative hypothesis: true difference in means is not equal to 0
95 percent confidence interval:
 -23.460688  -3.539312
sample estimates:
mean in group 1 mean in group 2
          33.75           47.25


> wilcox.test(df1$ZORK ~ df1$Group)

        Wilcoxon rank sum exact test

data:  df1$ZORK by df1$Group
W = 530, p-value = 0.008991
alternative hypothesis: true location shift is not equal to 0
```

Fig. 12.2 (continued)

Then compute the total sample size:

$$N = \sum_{i=1}^{k} n_i$$

and a variance estimate:

$$S^2 = \frac{1}{N-1}\left(\sum_{i,j} R(x_{ij})^2 - \frac{N(N+1)^2}{4}\right)$$

The test statistic is:

$$\tau^2 = S^{-2}\left(\sum_{i=1}^{k} \frac{R_i^2}{n_i} - \frac{N(N+1)^2}{4}\right)$$

The test statistic approximately has, under the null hypothesis of no difference (on the average) between groups, a Chi-squared distribution with $k - 1$ degrees of freedom. If it is larger than the $100(1 - \alpha)$ percentile of the Chi-squared distribution having $k - 1$ degrees of freedom, then reject the null hypothesis of no difference.

Figure 12.3 shows R code for computing the Kruskal-Wallis test. It also has a parametric ANOVA on the ranks for comparison. Figure 12.4 shows output for a dataset having three treatment groups ($k = 3$). The sample data were all generated from different gamma distributions. Table 12.1 shows the parameter values used to generate the data.

The Kruskal-Wallis function uses the Chi-squared approximation to compute p-values.

The p-values from the Kruskal-Wallis test and the conventional ANOVA on ranks only differed by 0.000632.

12.3 Resampling Methods: Bootstrap

The bootstrap was introduced in Chap. 11 for computing confidence intervals on differences of proportions or odds ratios, neither of which have any simple closed-form expressions for confidence intervals. The bootstrap approach can be used to test hypotheses and compute confidence intervals when parametric assumptions are either violated or their status is unknown.

Consider the data from the MWW example (variable ZORK). Although not strictly necessary, a bootstrap approach can be used to compute a confidence interval for the difference between the two means.

Figure 12.5 shows R code for creating the bootstrap confidence interval for the difference in the means. The conventional confidence interval is also computed. The

```
setwd("<your path here>")
df1 <- read.csv("20220517 Ch 12 Complete Blocks.csv")
#
# VARIABLES:
# Block
# Treatment
# Response
#
# the car package contains the Anova() function, which allows the use of Type III
sums of squares
library(car)
#
k <- 4
f.Block <- factor(df1$Block)
f.Treatment <- factor(df1$Treatment)
Rank.Y <- rank(df1$Response)
kruskal.test(df1$Response ~ f.Treatment)
kw.test <- kruskal.test(df1$Response ~ f.Treatment)
tau.squared <- as.numeric(kw.test$statistic)
kw.p.approx <- 1-pchisq(q=tau.squared,df=k-1)
kw.p.approx
rank.model <- lm(Rank.Y ~ f.Treatment)
Anova(mod=rank.model,type="III")
#####
```

Fig. 12.3 R code for the Kruskal-Wallis test

two intervals are very close, despite the fact that the data were not sampled from normal distributions and did not have equal variances. Table 12.2 shows the two intervals.

The two intervals appear to be fairly close; the conventional interval is wider than the bootstrap. In both cases, the null hypothesis:

$$H_0 : \mu_1 = \mu_2$$

would have been rejected in favor of the two-sided alternative:

$$H_1 : \mu_1 \neq \mu_2$$

The rejection is based on the fact that the interval does not contain 0.

Fig. 12.4 Output for Kruskal-Wallis test

```
Kruskal-Wallis rank sum test

data:  df1$Response by f.Treatment
Kruskal-Wallis chi-squared = 11.11, df = 2, p-value = 0.003867

> kw.test <- kruskal.test(df1$Response ~ f.Treatment)
> tau.squared <- as.numeric(kw.test$statistic)
> kw.p.approx <- 1-pchisq(q=tau.squared,df=k-1)
> kw.p.approx
[1] 0.00386742
> rank.model <- lm(Rank.Y ~ f.Treatment)
> Anova(mod=rank.model,type="III")
Anova Table (Type III tests)
Response: Rank.Y
            Sum Sq  Df  F value  Pr(>F)
(Intercept) 101204   1  90.7026  2.889e-16 ***
f.Treatment 13443    2  6.0243   0.003235 **
Residuals   130546 117
---
Signif. codes:  0 '***' 0.001 '**' 0.01 '*' 0.05 '.' 0.1 ' ' 1
> #
```

Table 12.1 Gamma distribution parameters for generating example data

Treatment	Shape	Scale
1	5	2
2	5	3
3	3	4

Figure 12.6 is a histogram of the bootstrap distribution.

12.4 Resampling Methods: Permutation Test

As described in Chap. 11, permutation tests (Good, 1994) are useful, especially when the test statistic's sampling distribution is either unknown or intractable. The procedure is as follows:

1. Compute the test statistic with the "unpermuted" data.
2. For *N* repetitions:

```
setwd("<your path here>")
df1 <- read.csv("20220516 Ch 12 Data for Mann Whitney Test.csv")
#
#
# VARIABLES:
#
# Group
# X
# R.X
# ZORK
#
t.stat <- c()
CL <- c()
diff <- c()
mean.1 <- c()
mean.2 <- c()
sd.1 <- c()
sd.2 <- c()
#t.stat <- c()
#SE <- c()
#
Rank.X <- rank(df1$X)
Rank.Z <- rank(df1$ZORK)
group1.ZORK <- df1$ZORK[df1$Group==1]
n.1 <- length(group1.ZORK)
group2.ZORK <- df1$X[df1$Group==2]
n.2 <- length(group2.ZORK)
mean.obs.1 <- mean(group1.ZORK)
mean.obs.2 <- mean(group2.ZORK)
diff.obs <- mean.obs.1 - mean.obs.2
```

Fig. 12.5 R code: bootstrap confidence interval for $\mu_1 - \mu_2$

(a) Permute the data by randomly sampling (with replacement) and randomly assigning each value to a treatment group.
(b) Use the permuted data to compute the test statistic, and store its value.

3. Find the tail that the original test statistic falls in the distribution of the permuted data statistics.

```
sd.obs.1 <- sd(group1.ZORK)
sd.obs.2 <- sd(group2.ZORK)
SE.diff <- sqrt((sd.obs.1**2)/n.1 + (sd.obs.2**2)/n.2)
LCL.conv <- diff.obs - qt(p=0.975,df=n.1+n.2-2)*SE.diff
UCL.conv <- diff.obs + qt(p=0.975,df=n.1+n.2-2)*SE.diff
#
diff.obs <- mean.obs.1 - mean.obs.2
set.seed(26)
bootn <- 2000
for (i in 1:bootn) {
  samp1 <- sample(group1.ZORK,size=n.1,replace=TRUE)
  samp2 <- sample(group2.ZORK,size=n.2,replace=TRUE)
  mean.1[i] <- mean(samp1)
  mean.2[i] <- mean(samp2)
  sd.1[i] <- sd(samp1)
  sd.2[i] <- sd(samp2)
#  SE[i] <- sqrt((sd.1[i]**2)/n.1 + (sd.2[i]**2)/n.2)
  diff[i] <- mean.1[i] - mean.2[i]
#  t.stat[i] <- (mean.1[i] - mean.2[i]) / SE
  }
CL <- quantile(x=diff,probs=c(0.025,0.975),type=8)
LCL <- CL[1]
UCL <- CL[2]
LCL
UCL
LCL.conv
UCL.conv
hist.diff <- hist(diff,plot=FALSE)
```

Fig. 12.5 (continued)

```
#
hist.diff$density <- 100*hist.diff$counts / sum(hist.diff$counts)
plot(hist.diff,freq=FALSE,col="light grey",main="Histogram of Bootstrap mean.1 -
mean.2",xaxt="n",yaxt="n",xlim=c(-10,13),xlab="mean.1-mean.2",ylab="Frequency
(%)")
axis(side=1,at=seq(from=-10,to=10,by=1)) #controls x-axis
axis(side=2,at=seq(from=0,to=50,by=1)) #controls y-axis
abline(v=diff.obs,lty=1)
abline(v=LCL,lty=2)
abline(v=UCL,lty=2)
abline(v=0,lty=3)
legend(x=5,y=10,legend=c("mean.1-mean.2","Conf.Lim","0"),lty=c(1,2,3))

df2 <- cbind(mean.1,mean.2,sd.1,sd.2,diff)
write.csv(df2,"20220518 Ch 12 Bootstrap Diff of Means.csv")
#####
```

Fig. 12.5 (continued)

Table 12.2 95% confidence intervals for $\mu_1 - \mu_2$: conventional and bootstrap

Type	LCL	UCL	UCL-LCL
Conventional	−6.2850	−1.0538	5.2312
Bootstrap	−6.0436	−1.1444	4.8992

The tail probability gives the *p*-value. The permuted data distribution is the "null" distribution, inasmuch as the assignment of data to treatment groups was random and therefore meaningless.

Figure 12.7 shows R code for computing a permutation test of the hypothesis:

$$H_0 : \mu_1 = \mu_2$$

versus:

$$H_1 : \mu_1 \neq \mu_2$$

Figure 12.8 shows the histogram of the permutation sample results of the MWW test, with the original statistic illustrated.

The original (unpermuted) test statistic is in the 0.5th percentile of the permutation distribution. Therefore, the null hypothesis is rejected in favor of the alternate.

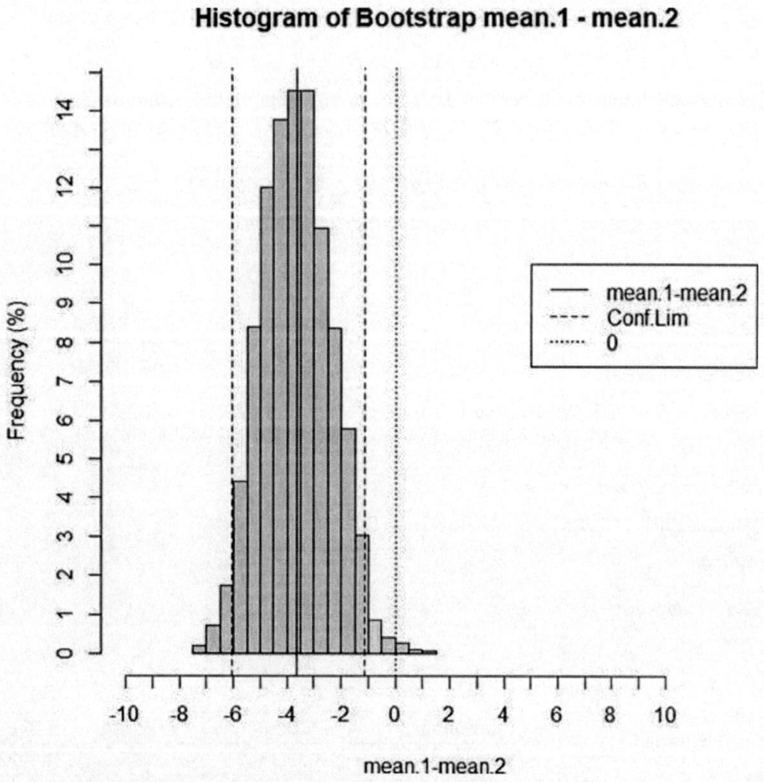

Fig. 12.6 Histogram of bootstrap distribution: mean.1–mean.2

12.5 Summary

Nonparametric methods (in particular, rank-based tests) are useful when assumptions required for parametric tests are violated, and methods such as bootstrapping and permutation tests are useful when the sampling distributions of test statistics are either unknown or intractable. In the case of rank tests, it may be reasonable to replace the original data with their ranks and perform more conventional tests on the ranks. One of the uses of the rank transformation is that it reduces the degree to which highly disparate response values affect the result. This is particularly true of ANOVA, which relies on having equal variances for all treatment groups. Bootstrap and permutation tests are very useful for situations where assumptions are violated or where the test statistic of interest has an unknown or mathematically intractable sampling distribution.

```
setwd("<your path here>")
df1 <- read.csv("20220516 Ch 12 Data for Mann Whitney Test.csv")
#
#
# VARIABLES:
#
#  Group
#  X
#  R.X
#  ZORK
#
MWW.statistic <- c()
MWW.p.value <- c()
#
Rank.X <- rank(df1$X)
Rank.Z <- rank(df1$ZORK)
group1.ZORK <- df1$ZORK[df1$Group==1]
n.1 <- length(group1.ZORK)
group2.ZORK <- df1$X[df1$Group==2]
n.2 <- length(group2.ZORK)
mean.obs.1 <- mean(group1.ZORK)
mean.obs.2 <- mean(group2.ZORK)
diff.obs <- mean.obs.1 - mean.obs.2
sd.obs.1 <- sd(group1.ZORK)
sd.obs.2 <- sd(group2.ZORK)
SE.diff <- sqrt((sd.obs.1**2)/n.1 + (sd.obs.2**2)/n.2)
LCL.conv <- diff.obs - qt(p=0.975,df=n.1+n.2-2)*SE.diff
UCL.conv <- diff.obs + qt(p=0.975,df=n.1+n.2-2)*SE.diff
#
```

Fig. 12.7 R code for permutation test

```
diff.obs <- mean.obs.1 - mean.obs.2
MWW.test <- wilcox.test(df1$ZORK ~ df1$Group)
MWW.original.statistic <- as.numeric(MWW.test$statistic)
MWW.original.p.value <- as.numeric(MWW.test$p.value)
set.seed(26)
bootn <- 2000
treatments <- df1$Group #this is the list of Group ID's - used with the permuted data
for (i in 1:bootn) {
  samp1 <- sample(group1.ZORK,size=n.1,replace=TRUE)
  samp2 <- sample(group2.ZORK,size=n.2,replace=TRUE)
  big.samp <- c(samp1,samp2)
  rand.order <- runif(length(big.samp),0,1)
  big.sample.shuffle <- big.samp[order(rand.order)]
  big.permute <- as.data.frame(cbind(treatments,big.sample.shuffle))
  permute.test <-
wilcox.test(big.permute$big.sample.shuffle~big.permute$treatments)
  MWW.statistic[i] <- as.numeric(permute.test$statistic)
  MWW.p.value[i] <- as.numeric(permute.test$p.value)
  }
F.statistic <- ecdf(MWW.statistic)
F.p.value <- ecdf(MWW.p.value)
permute.p <- F.statistic(MWW.original.statistic)
permute.p.alt <- F.p.value(MWW.original.p.value)
permute.p
permute.p.alt
hist.stat <- hist(MWW.statistic,plot=FALSE)
#
hist.stat$density <- 100*hist.stat$counts / sum(hist.stat$counts)
stat.text <- as.character(round(MWW.original.statistic,1))
plot(hist.stat,freq=FALSE,col="light grey",main="Histogram of Bootstrap MWW
Statistics",xaxt="n",yaxt="n",xlim=c(50,1200),xlab="MWW",ylab="Frequency (%)")
axis(side=1,at=seq(from=50,to=1200,by=25)) #controls x-axis
axis(side=2,at=seq(from=0,to=50,by=1)) #controls y-axis
abline(v=MWW.original.statistic,lty=2)
text(x=175,y=11,labels="MWW Statistic = ")
text(x=350,y=11,labels=stat.text)
#####
```

Fig. 12.7 (continued)

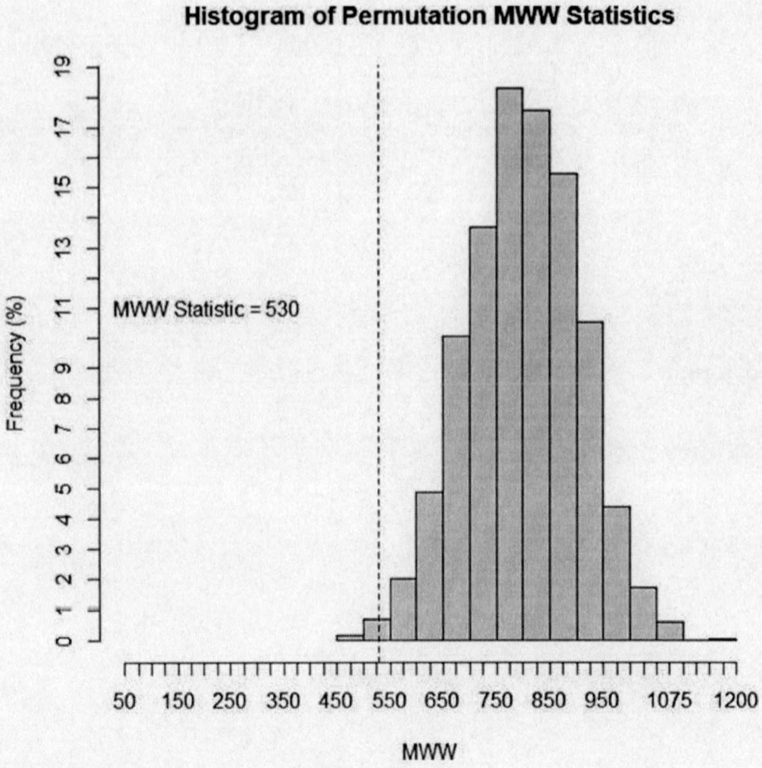

Fig. 12.8 Histogram of permutation MWW statistics

References

Conover, W. J. (1999). *Practical nonparametric statistics* (3rd ed.). Wiley.
Good, P. (1994). *Permutation tests*. Springer.

Chapter 13
Bayesian Methods

Abstract Earlier, Bayes' theorem was introduced. Now Bayesian methods are described for inference and information, especially using Markov Chain Monte Carlo (MCMC) methods.

13.1 The Basic Idea of Bayesian Inference

The primary notion of Bayesian analyses is that the analyst has some information about the parameters of interest prior to gathering data. The nature of this information is probabilistic. That is to say, the analyst has some idea of the range of possible values for the (unknown) parameters, and the information is contained in a probability distribution. This is not to necessarily say that parameters do not have "true" values; rather, we are quantifying our knowledge of the parameter values in terms of uncertainty. Thus, the analyst gathers data, then using the likelihood function (which is admittedly a model) the analyst will "update" the information contained in the prior distribution using Bayes' theorem, to create the "posterior" distribution, as described in Chap. 1. The formula for computing the posterior density function, given the prior density and the likelihood function, is:

$$f_{\text{posterior}}(\tilde{\mu}|x_1, x_2, \ldots, x_n) = \frac{f_{\text{prior}}(\tilde{\mu})L[x_1, x_2, \ldots, x_n|\tilde{\mu}]}{\int_{-\infty}^{+\infty} f_{\text{prior}}(\xi)L[x_1, x_2, \ldots, x_n|\xi]d\xi}$$

The function $L[x_1, x_2, \ldots, x_n|\tilde{\mu}]$ is the likelihood of obtaining the data $x_1, x_2, \ldots, x_n$ given that the parameter (or vector of parameters) is $\tilde{\mu}$. The likelihood function for a random sample $x_1, x_2, \ldots, x_n$, is:

$$L[x_1, x_2, \ldots, x_n|\tilde{\mu}] = \prod_{i=1}^{n} f_x(x_i|\tilde{\mu})$$

S. A. Pardo, *Statistical Methods and Analyses for Medical Devices*,
https://doi.org/10.1007/978-3-031-26139-8_13

That is to say, $f_x(x_i|\tilde{\mu})$ is a density function conditioned on $\tilde{\mu}$ and is the same function for all the x_i.

So, Bayesian analyses, i.e., updating the prior, depends on the ability to compute the integral:

$$\int_{-\infty}^{+\infty} f_{\text{prior}}(\xi) L[x_1, x_2, \ldots, x_n|\xi] d\xi$$

There are certain combinations of prior densities and likelihood functions for which that integral can be evaluated in a closed form and in fact result in a posterior density of the same family as the prior. Such combinations are referred to as "conjugate pairs" (Gelman et al., 1997). Such pairs were at one time the only practical models for which Bayes' theorem could be applied.

13.2 The Markov Chain Monte Carlo (MCMC) Method

While various numerical methods might be applied to the evaluation of the integral for non-conjugate pairs, such methods are not necessarily practical, due to limitations in computing. In more recent times, a methodology referred to as the Markov Chain Monte Carlo (MCMC) approach simplified the computations. An algorithm called the Metropolis-Hastings algorithm made implementing MCMC fairly straightforward. This is how it works:

1. Initially and arbitrarily choose a value for $\tilde{\mu}$, call it $\tilde{\mu}_0$.
2. Randomly choose another value for $\tilde{\mu}$, call it $\tilde{\mu}_1$. You can use any method you like for randomly generating a value (or a vector) $\tilde{\mu}_1$. One way would be to randomly generate a value (or a vector, if there are multiple parameters) from the prior distribution $g(\tilde{\mu})$.
3. Compute the ratio:

$$R = \frac{g(\tilde{\mu}_1) f_X(x|\tilde{\mu}_1)}{g(\tilde{\mu}_0) f_X(x|\tilde{\mu}_0)}$$

   $f_X(x|\tilde{\mu})$ is the likelihood function.
4. Randomly generate a number between 0 and 1, and call it u. If $R \geq u$, then the new value for˜is˜$_1$. Store it in the˜list. Otherwise, store˜$_0$.
5. Assign to $\tilde{\mu}_0$ the newly stored value.
6. Repeat the process n times from step #2.

The idea is that the ratio, R, is actually the ratio of two posterior density valuations. Since both the numerator and denominator are of the form:

$$\frac{g(\tilde{\mu})f[x|\tilde{\mu}]}{\int_{-\infty}^{+\infty} g(\xi)f[f|\xi]d\xi}$$

The integral cancels, so evaluating it is unnecessary. Thus, as long at the product:

$$g(\gamma)f(x|\gamma)$$

is easily computed, the forms of $g(.)$ and $f(.)$ are far less restricted, i.e., they do not have to be a conjugate pair.

Once the posterior distribution is found, a lower and upper percentile of the posterior can be used as a sort of interval estimate, called a credible interval, for the parameter of interest. Typically, the 2.5th and 97.5th percentiles are used for a 95% credible interval. The interpretation of such intervals is that we believe parameter value falls in the interval with probability determined by the tail probabilities of the interval limits (e.g., if the lower limit is the 2.5th percentile and the upper limit is the 97.5th percentile, then the probability content is 97.5 – 2.5% = 95%).

As an example, consider a binomial experiment, where x "successes" are observed out of n independent and identically distributed Bernoulli trials (a Bernoulli trial is an experiment where only one of two possible outcomes can occur, where one type of outcome is considered a "success"). The probability of a success on any given trial, p, is the unknown parameter. Suppose that we postulate a prior for p and that it is of the beta distribution family. The beta-binomial distributions form a conjugate pair.

The prior density can be described as:

$$g(p|\alpha,\beta) = \frac{\Gamma(\alpha+\beta)}{\Gamma(\alpha)\Gamma(\beta)}p^{\alpha-1}(1-p)^{\beta-1}$$

The "meta-parameters," α and β, are values that define the exact shape of the prior density for p.

The expected value of the beta is:

$$E(p) = \frac{\alpha}{\alpha+\beta}$$

The likelihood function is the binomial mass function:

$$f(y|p) = \binom{n}{y}p^{y}(1-p)^{n-y}$$

So, applying Bayes' theorem would yield a posterior density, call it g', that would have the form:

$$g'(p|\alpha',\beta') = \frac{\Gamma(\alpha'+\beta')}{\Gamma(\alpha')\Gamma(\beta')} p^{\alpha'-1}(1-p)^{\beta'-1}$$

Suppose that an experiment was performed in which $n = 50$ individuals were determined to be either "good" or "not good." Furthermore, suppose that of those 50, $y = 40$ were "good"; therefore, a point estimate of the binomial parameter, p, is:

$$\widehat{p} = \frac{y}{n} = \frac{40}{50} = 0.80$$

The analyst chooses to use a Bayesian analysis, with a beta prior having "meta" parameters:

$$\alpha = 1.0 \quad \text{(shape)}$$
$$\beta = 0.25 \ \text{(scale)}$$

As to why those prior "meta" parameters, in theory they provide a prior distribution that captures the analyst's guess about the range of possible values for p. In this case, the prior expected value is:

$$E(p) = \frac{\alpha}{\alpha+\beta} = \frac{1}{1+0.25} = 0.80$$

This is the value of the point estimate.

Figure 13.1 shows the prior density for p.

Based on the conjugate beta-binomial pair, the posterior meta parameters are:

$$\alpha' = \alpha + y$$

and:

$$\beta' = \beta + n - y$$

Figure 13.2 shows the R code for implementing the MCMC (Metropolis-Hastings) algorithm. Figure 13.3 shows the posterior histogram.

Table 13.1 shows the prior meta-parameters, the prior expected value, and the data (n, y). Table 13.2 shows the posterior median and a 95% credible interval (LCL, UCL) for the parameter, based on the MCMC results and using the conjugate beta-binomial pair.

It is apparent that the MCMC approach yielded posterior results close to that which was expected from the conjugate pair's posterior distribution.

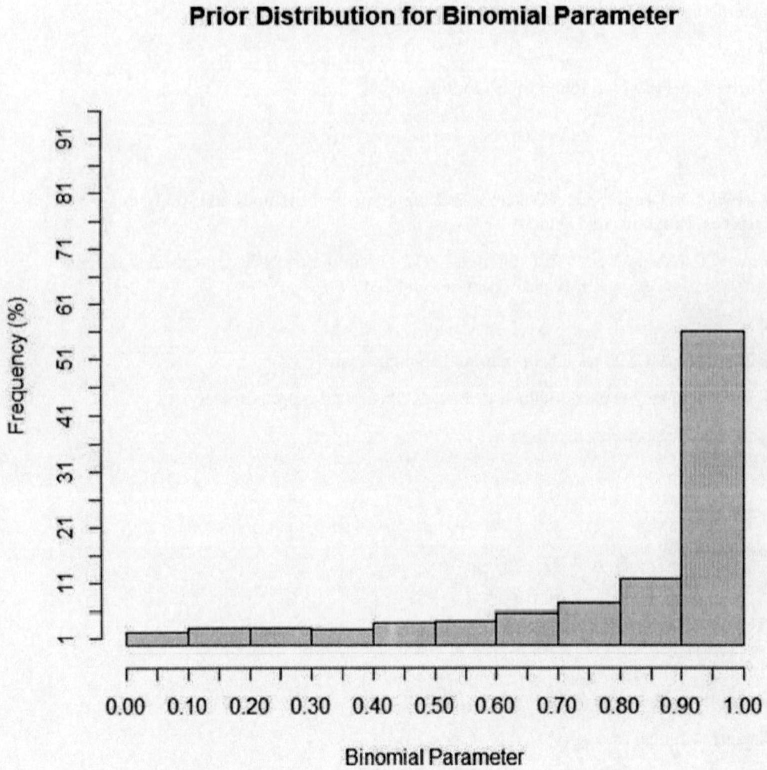

Fig. 13.1 Prior beta (1,0.25) density for binomial parameter p

13.3 Summary

Bayesian methods provide a means of incorporating prior information about parameters in the data analysis. Prior to the past 20 years or so, Bayesian methods were virtually never used for clinical data analyses. However, today it is not true. Nevertheless, considerable thought must be put into choosing priors and likelihoods. It was thought that Bayesian methods would shorten studies, by using them in a sequential fashion, continually updating the posterior distributions as more data are obtained. However, it is not strictly true that Bayesian techniques will yield shorter studies. Furthermore, defining the stopping rules is more involved than those used for non-Bayesian, or "frequentist," approaches. The frequentist approach requires some statement of critical values, sample size, and power. A Bayesian stopping rule would require some criteria concerning the posterior distribution. Whereas sample size and critical values are determined before any data are ever gathered, achieving the stopping criteria for a Bayesian study involves some uncertainty. It is easier to

```
options(na.action=na.exclude)
#
# setwd tells R which folder to use for this script
#

#setwd("C:\\Users\\minvr\\Desktop\\Failsafe\\Statistical Methods and Analyses for Medical
Devices\\Programs and Data")
setwd("C:\\Users\\MINVR\\OneDrive - PHC Group\\Failsafe\\Statistical Methods and
Analyses for Medical Devices\\Programs and Data")
#
# This is an MCMC for a beta-binomial conjugate pair
# The conjugacy is used to illustrate how MCMC yields approximately
# the correct posterior distribution
#
set.seed(26)
pstore <- c()
#
a.beta <- 1
b.beta <- .25
alphapri <- a.beta
betapri <- b.beta
#
# Values of the parameter will be randomly generated
# from the Beta prior, to illustrate the shape of
# the prior density.
#
beta.prior <- rbeta(n=2000,shape1=a.beta,shape2=b.beta)
beta.hist <- hist(beta.prior,plot=FALSE)
beta.hist$density <- 100*beta.hist$counts / sum(beta.hist$counts)
plot(beta.hist,freq=FALSE,col="light grey",main="Prior Distribution for Binomial
Parameter",xaxt="n",yaxt="n",xlim=c(0,1),ylim=c(0,max(beta.hist$density)+1),xlab="Binom
ial Parameter",ylab="Frequency (%)")
axis(side=1,at=seq(from=0,to=1,by=0.05))
axis(side=2,at=seq(frpom=0,to=100,by=5))
#
# Now create the observed data:
#
```

Fig. 13.2 R code for MCMC algorithm: beta-binomial

```
sampsize <- 50
y <- 40
phat <- y / sampsize
p0 <- a.beta / (a.beta + b.beta)
a.post <- a.beta + y
b.post <- b.beta + sampsize - y
nmcmc <- 2000
for (i in 1:nmcmc) {
  p1 <- min(abs(rnorm(n=1,mean=p0,sd=0.05)),1)#use folded & truncated normal to generate
p1
#                                          insures p1 will be in interval (0,1)
  num <- dbeta(x=p1,shape1=alphapri,shape2=betapri)*dbinom(x=y,size=sampsize,prob=p1)
  den <- dbeta(x=p0,shape1=alphapri,shape2=betapri)*dbinom(x=y,size=sampsize,prob=p0)
  rat <- num / den
  uvar <- runif(n=1,min=0,max=1)
  if (rat >= uvar) {
   pstore[i] <- p1
  }
   else {
    pstore[i] <- p0
    }
   p0 <- pstore[i] #change p0 to current value
   }
  #
  #
  #
  alphapos <- alphapri + sampsize
  betapos <- betapri + sampsize - y
  meanpos <- alphapos / (alphapos + betapos)
  dev.new()
  prob.hist <- hist(pstore,plot=FALSE)
  prob.hist$density <- 100*prob.hist$counts / sum(prob.hist$counts)
  #
  plot(prob.hist,freq=FALSE,main="Histogram of Posterior Sample",col="light grey",
```

Fig. 13.2 (continued)

```
xaxt="n",yaxt="n",xlim=c(0.60,1.00),ylim=c(0,max(prob.hist$density)+1),xlab="Binomial
Parameter",ylab="Frequency (%)")
axis(side=1,at=seq(from=0.60,to=1.00,by=0.05))#x-axis
axis(side=2,at=seq(from=0,to=max(prob.hist$density)+1,by=1))#y-axis
lowlim <- quantile(x=pstore,probs=c(0.025),type=8)
medprob <- quantile(x=pstore,probs=c(0.50),type=8)
upplim <- quantile(x=pstore,probs=c(0.975),type=8)
lowlim
medprob
upplim
conjugate.low <- qbeta(p=0.025,shape1=a.post,shape2=b.post)
conjugate.med <- qbeta(p=0.500,shape1=a.post,shape2=b.post)
conjugate.upp <- qbeta(p=0.975,shape1=a.post,shape2=b.post)
conjugate.low
conjugate.med
conjugate.upp

df2 <- cbind(alphapri,betapri,alphapos,betapos,sampsize,y,lowlim,upplim,pstore)
#write.csv(df2,"20220519 beta binomial posterior.csv")   #####
```

Fig. 13.2 (continued)

Table 13.1 Prior values

Parameter	Value
n	50
y	40
Prior α	1.00
Prior β	0.25
Prior $E(p)$	0.80

Table 13.2 Posterior values: two methods

		LCL	UCL	
Method	Median	2.5th percentile	97.5th percentile	$E(p\|y,n)$
MCMC	0.8016	0.6976	0.8862	0.8057
Conjugate	0.8039	0.6813	0.8965	0.8000

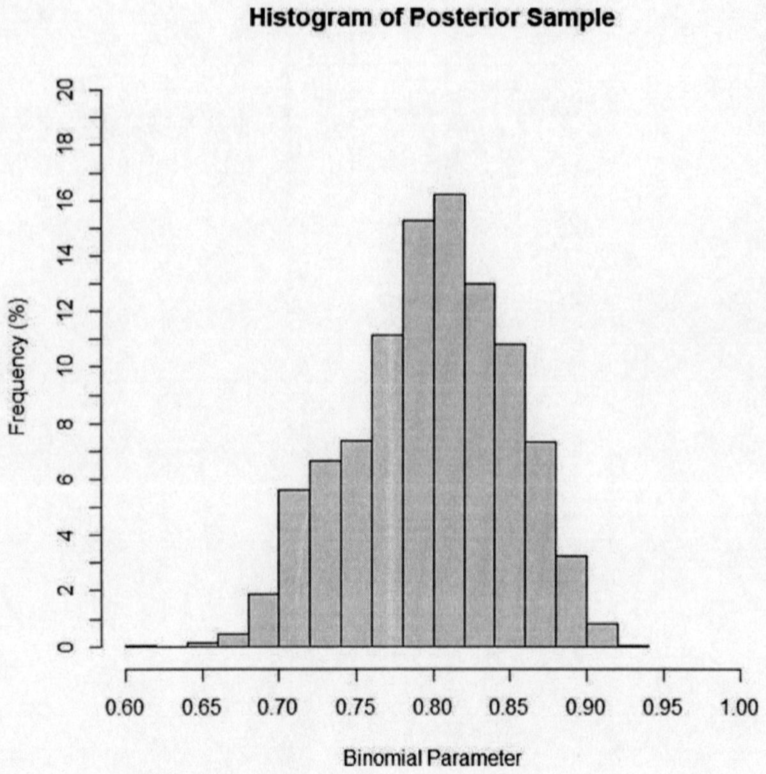

Fig. 13.3 Posterior histogram from MCMC

apply Bayesian methods in situations that are more organically sequential, such as the occurrence of monthly complaints about a marketed product. In theory, there could be no complaints for the entire life of the product. Every month, a value of 0 would be entered into the updating computations for the posterior distribution. There is no need for a stopping rule. However, initially an "action limit" could be specified, such as if the posterior probability that percentage of complaints exceeds x_c in a single month at least 80%, then corrective action must be taken.

Reference

Gelman, A., Carlin, J. B., Stern, H. S., & Rubin, D. B. (1997). *Bayesian data analysis*. Chapman & Hall/CRC.

Chapter 14
Prediction, Classification, and Nonlinear Modeling

Abstract Various multivariate methods for building predictive and classificatory models are explained. Most of these methods are often described as "machine learning."

14.1 Introduction

In some studies, perhaps especially observational studies, numerous variables are recorded for each experiment unit (e.g., subjects). Generally, there is some interest in relating different variables to one another and in finding a mapping that allows one to predict a response variable in terms of some subset of the potentially important explanatory, or "regressor," variables. For medical devices, it is possibly true that most of these sorts of studies are "post-market," designed to enhance the marketability or even claims to be made about a product. There was also a move to obtain "real world data" on the use of medical devices. Such studies will include concomitant variables for users, such as age, years since diagnosis, length of time the device had been used, etc. Without any information indicating which regressor/concomitant variables are more important or better indicators for some response variable(s), the model builder is faced with the choice of either including all the regressors, possibly inducing some collinearity, or somehow choosing to remove some of those regressors from the model. Some of the methods described in this chapter are referred to as "machine learning" algorithms. The notion is that the analyst has limited intervention in specifying a model. Rather, she or he may select a large number of potential input variables (regressors) and then the algorithm chooses the model. In some cases, the algorithm will select a subset of the inputs as being "important" and ultimately only include the "important" inputs in the model. Other methods will include all inputs presented to the algorithm, regardless of whether those inputs contribute a lot, a little, or not at all to the predictive accuracy. In the language of machine learning, there are two basic categories of algorithm. One is called "supervised," and the other, oddly enough, is called "unsupervised." Supervised algorithms require a set or sample of predictor variable values and a set of response variable values and then attempt to find a set of rules that allows one to make a prediction of the responses as a

S. A. Pardo, *Statistical Methods and Analyses for Medical Devices*,
https://doi.org/10.1007/978-3-031-26139-8_14

function of the predictors. Unsupervised algorithms dispense with the dichotomy of predictor and response; rather, the unsupervised algorithm will simply take the data and then find features that could be predictable for new individuals. This chapter will deal with a few methods for choosing and selecting models, and all of them would be considered "supervised."

14.2 Stepwise Regression

Suppose there are p-1 regressors (with an intercept, there would be p parameters) available for use in a predictive model; for a single response, we will call y. One method of model selection is called backward stepwise regression (Draper & Smith, 1981). In this approach, a model is fit first with all regressors and then removed one at a time, and new models are fit. After each fit, a measure of goodness is computed. At the end of the procedure, after variables have been deleted, or replaced back, through several iterations, a final model is selected, using those regressors that yielded the "optimal" value of the measure of goodness. Possibly the most widely used measure is called Akaike information criterion (AIC) (not to be confused with the measure of blood glucose control, hbA1c):

$$\mathrm{AIC}_p = \frac{\mathrm{SSE}_p}{\widehat{\sigma}^2} + \left(\ln(2\pi) + \ln(\widehat{\sigma}^2)\right)N + 2p$$

$p =$ number of parameters in the model
$N =$ sample size
$\widehat{\sigma}^2 =$ root mean squared error term from the model (an estimate of model error variance)

The reason that AIC is used as opposed to adjusted R^2 is that as the number of parameters increases, adjusted R^2 will get closer and closer to 1, regardless of how good the model is as a predictive equation. For AIC, the lower it is, the "better" the model. Its cousin, C_p, has an ideal value equal to p, the number of parameters:

$$C_p = \frac{\mathrm{SSE}_p}{\widehat{\sigma}^2} - N + 2p$$

SSE_p is the sum of squared residuals with a model having p parameters. So if $p = m$:

$$\widehat{\sigma}^2 = \frac{\mathrm{SSE}_m}{N - m}$$

The "best" model could be considered to be that which yields the lowest value of AIC (of those models fit) or that model whose C_p is closest to the number of parameters in that model.

As an example, consider a situation in which there are five regressors and a single response. Suppose the "full" model is given by:

$$\text{Response} = \beta_0 + \sum_{i=1}^{5} \beta_i X_i + \sum_{i \neq j} \beta_{ij} X_i X_j + \varepsilon$$

The data were simulated using the following equation:

$$\text{Response} = -10 - 1.2\text{X.01} - 2.3\text{X.03} + 5\text{X0.06} - 1.9\text{X0.07} - 3.2\text{X0.06X0.07} + 7\text{X.10} + \varepsilon$$

The noise variable, ε, was randomly generated from a normal distribution with mean 0 and standard deviation 1.2.

The question is which of the main effects (X_i) or two-way interactions (X_iX_j) actually have an effect on the response. In this example, we would expect that only those terms actually used in the simulation would have significant parameter estimates.

The *step*() function in R will take a linear model object from function *lm*() and perform a stepwise model selection process. The default method uses AIC as a criterion. The R code is given in Fig. 14.1. The output is shown in Fig. 14.2. Finally, Fig. 14.3 shows a plot of the predicted values from the final model versus the actual observed response values. The algorithm went through eight iterations, or steps, before finding the final model.

The *summary*(model_step) statement only shows results for the final model. Notice that the interaction term X.06:X.10 was included in the final model, although its corresponding parameter estimate (0.006507) was not significantly different from 0. This interaction was the only term in the final model that was not actually used in generating the simulated data. Table 14.1 shows the terms from the final model, their parameter estimates, and the actual parameter values used to simulate the data.

14.3 Bayesian Model Averaging

Like stepwise regression, Bayesian model averaging (BMA) takes an initial "full" model and iteratively finds new models. BMA uses prior distributions for model parameters to yield a Bayesian posterior model that is an "average" of all the iterations. The final posterior model has all the same terms as the initial model, but the parameter estimates are based on a Bayesian approach rather than ordinary least squares. So, the BMA approach ultimately does not eliminate parameters but provides a different estimate of those parameter values. The posterior values of the parameters are actually an average from the posterior distribution and depend on the choice of the prior. The R function *bic.glm*() in package BMA implements BMA.

```
setwd("<your path here>")
df1 <- read.csv("20220524 Ch 14 Stepwise Regression.csv")
library(car) #needed to use function Anova() - not the same as anova()
# Variables
#
# X.01
# X.02
# X.03
# X.04
# X.05
# X.06
# X.07
# X.08
# X.09
# X.10
# Response
attach(df1)

# For illustration, only include X.01,X.03,X.06,X.07,X.10 and their two-way interactions
model_big <- lm(Response ~ X.01 + X.03 + X.06 + X.07 + X.10 + X.01:X.03 + X.01:X.06 +
X.01:X.07 + X.01:X.10 + X.03:X.06 + X.03:X.07 + X.03:X.10 + X.06:X.07 + X.06:X.10 +
X.07:X.10)

Anova(model_big,test.statistic="F",type=3)

#
model_step <- step(model_big)
summary(model_step)
lin.mod <- lm(model_step$fitted.values ~ Response)
int <- lin.mod$coefficients[1]
slo <- lin.mod$coefficients[2]
plot(x=Response,y=model_step$fitted.values,main="Observed Response vs.
Predicted",xlab="Observed",ylab="Predicted")
abline(a=int,b=slo)
#####
```

Fig. 14.1 Stepwise regression code

```
Anova Table (Type III tests)

Response: Response
           Sum Sq  Df   F value   Pr(>F)
(Intercept)     92  1 5.5177e+01 3.990e-12 ***
X.01           275  1 1.6554e+02 < 2.2e-16 ***
X.03           133  1 7.9960e+01 4.026e-16 ***
X.06          9697  1 5.8469e+03 < 2.2e-16 ***
X.07           351  1 2.1140e+02 < 2.2e-16 ***
X.10          6984  1 4.2112e+03 < 2.2e-16 ***
X.01:X.03        1  1 3.3250e-01   0.5649
X.01:X.06        0  1 6.0000e-04   0.9809
X.01:X.07        0  1 9.0700e-02   0.7636
X.01:X.10        0  1 3.0000e-03   0.9562
X.03:X.06        0  1 7.9900e-02   0.7777
X.03:X.07        1  1 3.4450e-01   0.5579
X.03:X.10        0  1 1.5690e-01   0.6925
X.06:X.07   454607  1 2.7412e+05 < 2.2e-16 ***
X.06:X.10        3  1 2.1027e+00   0.1487
X.07:X.10        0  1 5.6200e-02   0.8128
Residuals      305 184
---
Signif. codes:  0 '***' 0.001 '**' 0.01 '*' 0.05 '.' 0.1 ' ' 1
>
> #
> model_step <- step(model_big)
Start:  AIC=116.5
```

Fig. 14.2 Stepwise regression output

```
Response ~ X.01 + X.03 + X.06 + X.07 + X.10 + X.01:X.03 + X.01:X.06 +
   X.01:X.07 + X.01:X.10 + X.03:X.06 + X.03:X.07 + X.03:X.10 +
   X.06:X.07 + X.06:X.10 + X.07:X.10

             Df  Sum of Sq  RSS    AIC
- X.01:X.06  1       0      305   114.50
- X.01:X.10  1       0      305   114.50
- X.07:X.10  1       0      305   114.56
- X.03:X.06  1       0      305   114.58
- X.01:X.07  1       0      305   114.60
- X.03:X.10  1       0      305   114.67
- X.01:X.03  1       1      306   114.86
- X.03:X.07  1       1      306   114.87
<none>                      305   116.50
- X.06:X.10  1       3      309   116.77
- X.06:X.07  1  454607 454912 1575.91

Step:  AIC=114.5
Response ~ X.01 + X.03 + X.06 + X.07 + X.10 + X.01:X.03 + X.01:X.07 +
   X.01:X.10 + X.03:X.06 + X.03:X.07 + X.03:X.10 + X.06:X.07 +
   X.06:X.10 + X.07:X.10

             Df Sum of Sq  RSS   AIC
- X.01:X.10  1       0   305  112.50
- X.07:X.10  1       0   305  112.56
- X.03:X.06  1       0   305  112.59
- X.01:X.07  1       0   305  112.60
- X.03:X.10  1       0   305  112.67
```

Fig. 14.2 (continued)

```
- X.01:X.03  1       1    306  112.87
- X.03:X.07  1       1    306  112.87
<none>                    305  114.50
- X.06:X.10  1       4    309  114.83
- X.06:X.07  1  455431 455736 1574.27

Step:  AIC=112.5
Response ~ X.01 + X.03 + X.06 + X.07 + X.10 + X.01:X.03 + X.01:X.07 +
    X.03:X.06 + X.03:X.07 + X.03:X.10 + X.06:X.07 + X.06:X.10 +
    X.07:X.10

            Df Sum of Sq    RSS     AIC
- X.07:X.10  1       0    305  110.56
- X.03:X.06  1       0    305  110.60
- X.01:X.07  1       0    305  110.60
- X.03:X.10  1       0    305  110.67
- X.01:X.03  1       1    306  110.88
- X.03:X.07  1       1    306  110.88
<none>                    305  112.50
- X.06:X.10  1       4    309  112.88
- X.06:X.07  1  456227 456532 1572.62

Step:  AIC=110.56
Response ~ X.01 + X.03 + X.06 + X.07 + X.10 + X.01:X.03 + X.01:X.07 +
    X.03:X.06 + X.03:X.07 + X.03:X.10 + X.06:X.07 + X.06:X.10

            Df Sum of Sq    RSS     AIC
- X.01:X.07  1       0    305  108.65
```

Fig. 14.2 (continued)

```
- X.03:X.06  1       0    305  108.65
- X.03:X.10  1       0    306  108.73
- X.01:X.03  1       1    306  108.93
- X.03:X.07  1       1    306  108.96
<none>                    305  110.56
- X.06:X.10  1       4    309  110.99
- X.06:X.07  1  462398 462703 1573.30

Step:  AIC=108.65
Response ~ X.01 + X.03 + X.06 + X.07 + X.10 + X.01:X.03 + X.03:X.06 +
   X.03:X.07 + X.03:X.10 + X.06:X.07 + X.06:X.10

            Df Sum of Sq  RSS    AIC
- X.03:X.06  1       0    306  106.75
- X.03:X.10  1       0    306  106.82
- X.03:X.07  1       1    306  107.02
- X.01:X.03  1       1    306  107.03
<none>                    305  108.65
- X.06:X.10  1       4    309  109.08
- X.06:X.07  1  465789 466094 1572.77

Step:  AIC=106.75
Response ~ X.01 + X.03 + X.06 + X.07 + X.10 + X.01:X.03 + X.03:X.07 +
   X.03:X.10 + X.06:X.07 + X.06:X.10

            Df Sum of Sq  RSS    AIC
- X.03:X.10  1       0    306  104.94
- X.03:X.07  1       1    306  105.16
```

Fig. 14.2 (continued)

```
- X.01:X.03  1       1    306  105.19
<none>                    306  106.75
- X.06:X.10  1       4    309  107.08
- X.06:X.07  1  474776 475081 1574.58

Step:  AIC=104.94
Response ~ X.01 + X.03 + X.06 + X.07 + X.10 + X.01:X.03 + X.03:X.07 +
    X.06:X.07 + X.06:X.10

            Df Sum of Sq    RSS     AIC
- X.03:X.07  1       1    306  103.34
- X.01:X.03  1       1    307  103.48
<none>                    306  104.94
- X.06:X.10  1       4    309  105.30
- X.06:X.07  1  475278 475584 1572.80

Step:  AIC=103.34
Response ~ X.01 + X.03 + X.06 + X.07 + X.10 + X.01:X.03 + X.06:X.07 +
    X.06:X.10

            Df Sum of Sq    RSS     AIC
- X.01:X.03  1       1    307  101.99
<none>                    306  103.34
- X.06:X.10  1       4    310  103.70
- X.06:X.07  1  498017 498323 1580.14

Step:  AIC=101.99
Response ~ X.01 + X.03 + X.06 + X.07 + X.10 + X.06:X.07 + X.06:X.10
```

Fig. 14.2 (continued)

```
            Df Sum of Sq    RSS     AIC
<none>                      307  101.99
- X.06:X.10  1        4     312  102.69
- X.03       1     4142    4450  634.45
- X.01       1     7204    7512  739.18
- X.06:X.07  1   498097  498404 1578.17
> summary(model_step)

Call:
lm(formula = Response ~ X.01 + X.03 + X.06 + X.07 + X.10 + X.06:X.07 +
   X.06:X.10)

Residuals:
    Min      1Q  Median      3Q     Max
-3.5779 -0.7930 -0.0357  0.8336  3.6010

Coefficients:
             Estimate Std. Error  t value Pr(>|t|)
(Intercept) -9.681519   0.898690  -10.773   <2e-16 ***
X.01        -1.212309   0.018073  -67.077   <2e-16 ***
X.03        -2.305816   0.045335  -50.862   <2e-16 ***
X.06         4.944910   0.055181   89.613   <2e-16 ***
X.07        -1.912936   0.095023  -20.131   <2e-16 ***
X.10         6.939454   0.066546  104.280   <2e-16 ***
X.06:X.07   -3.200941   0.005739 -557.743   <2e-16 ***
X.06:X.10    0.006507   0.004028    1.616    0.108
---
```

Fig. 14.2 (continued)

```
Signif. codes:  0 '***' 0.001 '**' 0.01 '*' 0.05 '.' 0.1 ' ' 1

Residual standard error: 1.265 on 192 degrees of freedom
Multiple R-squared:      1,     Adjusted R-squared:      1
F-statistic: 9.379e+05 on 7 and 192 DF,  p-value: < 2.2e-16

> summary(model_step)

Call:
lm(formula = Response ~ X.01 + X.03 + X.06 + X.07 + X.10 + X.06:X.07 +
    X.06:X.10)

Residuals:
    Min      1Q  Median      3Q     Max
-3.5779 -0.7930 -0.0357  0.8336  3.6010

Coefficients:
             Estimate Std. Error t value  Pr(>|t|)
(Intercept) -9.681519  0.898690  -10.773  <2e-16 ***
X.01        -1.212309  0.018073  -67.077  <2e-16 ***
X.03        -2.305816  0.045335  -50.862  <2e-16 ***
X.06         4.944910  0.055181   89.613  <2e-16 ***
X.07        -1.912936  0.095023  -20.131  <2e-16 ***
X.10         6.939454  0.066546  104.280  <2e-16 ***
X.06:X.07   -3.200941  0.005739 -557.743  <2e-16 ***
X.06:X.10    0.006507  0.004028    1.616   0.108
---
Signif. codes:  0 '***' 0.001 '**' 0.01 '*' 0.05 '.' 0.1 ' ' 1

Residual standard error: 1.265 on 192 degrees of freedom
Multiple R-squared:      1,     Adjusted R-squared:      1
F-statistic: 9.379e+05 on 7 and 192 DF,  p-value: < 2.2e-16
```

Fig. 14.2 (continued)

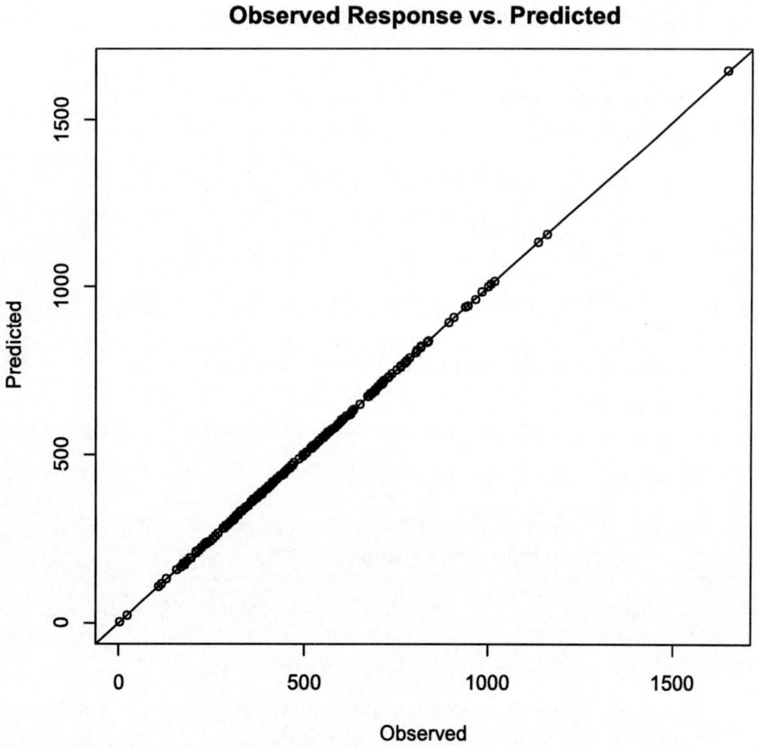

Fig. 14.3 Predicted versus observed response: final model from stepwise procedure

Table 14.1 Final stepwise model parameter estimates

Term	Estimate	Actual
(Intercept)	−9.682	−10.0
X.01	−1.212	−1.2
X.03	−2.306	−2.3
X.06	4.945	5.0
X.07	−1.913	−1.9
X.10	6.939	7.0
X.06:X.07	−3.201	3.2
X.06:X.10	0.007	0.0

The parameter *prior.param* allows the user to specify a prior probability "weight" for each variable in the initial "full" model. The function will then use some internal rules to decide which subset of models (different combinations of the variables in the initial model) should be used for averaging. The choice of models to include depends in part on the *prior.param* input. If that input is not specified, the program will assume a uniform prior probability, which is based on the number of variables in the full model. The values in *prior.param* are not required to sum to 1.

As an example, suppose there was a set of 10 regressors, X.01, X0.02, . . ., X.10, and a single response, called "Response." A sample of $n = 200$ observations were simulated using the equation:

```
setwd("<your path here>")
df1 <- read.csv("20180701 Example 11.2 BMA v.2.csv")
#
#
# need to execute library(BMA) once during R session
library(BMA)
#
# VARIABLES:
# X.01
# X.02
# X.03
# X.04
# X.05
# X.06
# X.07
# X.08
# X.09
# X.10
# Response
#
nbin <- nrow(df1)
testind <- rbinom(n=nbin,size=1,prob=0.1)

df2 <- data.frame(cbind(df1,testind))

attach(df2)
```

Fig. 14.4 BMA R code

```
dftest <- subset(df2,testind==1)

dftrain <- subset(df2,testind==0)

#

detach(df2)

attach(dftrain)

#prior.p <- c(0.9,0.001,0.9,0.001,0.001,0.9,0.9,0.9,0.9,0.001)

prior.p <- c(0.1,0.1,0.1,0.1,0.1,0.1,0.1,0.1,0.1,0.1)

bma.model <- bic.glm(f=Response ~ X.01 + X.02 + X.03 + X.04 + X.05 + X.06 + X.07 +
X.08 + X.09 + X.10,data=dftrain,glm.family=gaussian(),prior.param=prior.p)

linreg <- lm(Response ~ X.01 + X.02 + X.03 + X.04 + X.05 + X.06 + X.07 + X.08 + X.09 +
X.10,data=dftrain)

postcoeff <- bma.model$postmean

Response.pred <- postcoeff[1] + postcoeff[2]*X.01 + postcoeff[3]*X.02 + postcoeff[4]*X.03
+ postcoeff[5]*X.04 + postcoeff[6]*X.05 + postcoeff[7]*X.06 + postcoeff[8]*X.07 +
postcoeff[9]*X.08 + postcoeff[10]*X.09 + postcoeff[11]*X.10;

# Make some plots:

plot(x=Response,y=Response.pred,pch=1,xlab="Observed",ylab="Predicted",main="Predicte
d Response vs. Observed")

points(x=Response,y=linreg$fitted.values,pch=2)

points(x=Response,y=Response,pch=3)

legend(x=0.0,y=150,legend=c("BMA","OLS","Observed Response"),pch=c(1,2,3))

#

# dev.new() allows more plots to be made without overwriting previous plots

#

summary(linreg)

lm.coeffs <- linreg$coefficients


detach(dftrain)

attach(dftest)

BMA.OOS <- postcoeff[1] + postcoeff[2]*X.01 + postcoeff[3]*X.02 + postcoeff[4]*X.03 +
postcoeff[5]*X.04 + postcoeff[6]*X.05 + postcoeff[7]*X.06 + postcoeff[8]*X.07 +
postcoeff[9]*X.08 + postcoeff[10]*X.09 + postcoeff[11]*X.10;
```

Fig. 14.4 (continued)

```
lm.OOS <- lm.coeffs[1] + lm.coeffs[2]*X.01 + lm.coeffs[3]*X.02 + lm.coeffs[4]*X.03 +
lm.coeffs[5]*X.04 + lm.coeffs[6]*X.05 + lm.coeffs[7]*X.06 + lm.coeffs[8]*X.07 +
lm.coeffs[9]*X.08 + lm.coeffs[10]*X.09 + lm.coeffs[11]*X.10;

dev.new()

plot(x=Response,y=BMA.OOS,pch=1,xlab="Observed",ylab="Predicted",main="OOS
Predicted Response vs. Observed")

points(x=Response,y=lm.OOS,pch=2)

points(x=Response,y=Response,pch=3)

legend(x=0.0,y=100,legend=c("BMA OOS","OLS OOS","Observed
Response"),pch=c(1,2,3))

#####
```

Fig. 14.4 (continued)

Table 14.2 BMA example parameter estimates

Term	Estimate-uniform	Estimate-weighted	OLS	Actual
Intercept	−9.71	−9.61	−9.85	−10.0
X.01	−1.21	1.22	−1.21	−1.2
X.02	0.00	0.00	0.06	0.0
X.03	−2.34	−2.34	−2.34	−2.3
X.04	0.00	0.00	0.00	0.0
X.05	0.00	0.00	0.02	0.0
X.06	5.00	5.01	5.00	5.0
X.07	−1.98	−1.95	−1.99	−1.9
X.08	−3.22	−3.23	−3.23	−3.2
X.09	4.72	4.77	4.66	7.0
X.10	0.00	0.00	−0.01	0.0

$$\text{Response} = 10 + 1.2\text{X.01} + 2.3\text{X.03} - 5\text{X.06} + 1.9\text{X.07} + 3.2\text{X.08} - 7\text{X.09} + \varepsilon$$

where $\varepsilon \sim N(\mu = 0, \sigma = 1.2)$

Figure 14.4 shows code for executing *bic.bma*(). Table 14.2 shows the estimates of the coefficients using the uniform and "weighted" priors, the ordinary least squares (OLS) estimates, and the actual coefficients used to simulate the data. Note that the code saves 10% of the data for potential checking (a test set).

Figure 14.5 shows the predicted response values plotted against the "observed" values, using the uniform priors for BMA. Figure 14.6 shows the same plot, but using the "weighted" priors.

As in the case of the stepwise procedure, the parameter estimates' signs have been reversed relative to the values actually used for generating the simulated data. Figure 14.7 shows a plot of out-of-sample (OOS) predictions, i.e., predictions of

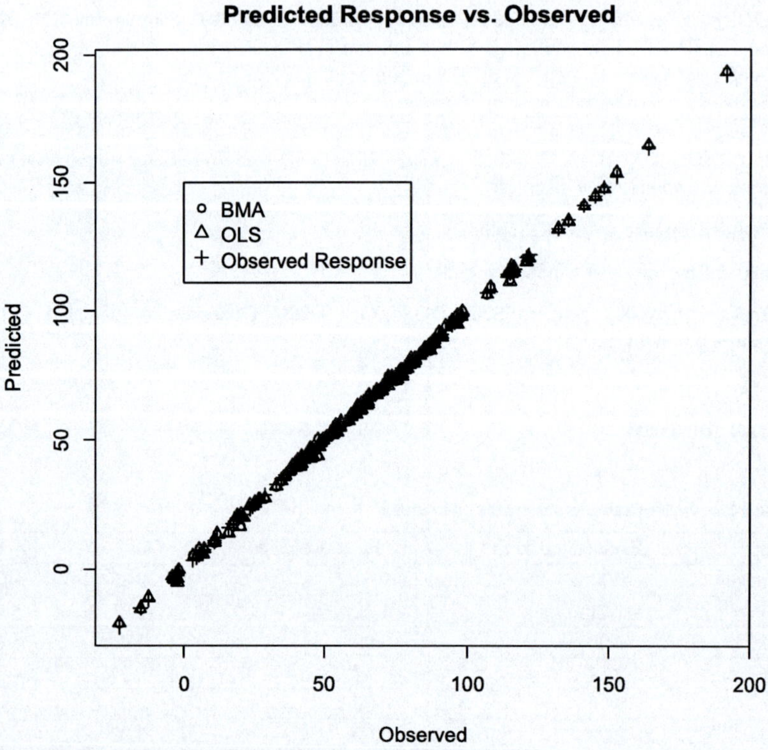

Fig. 14.5 Predicted vs. observed: uniform priors for BMA

the test data, with uniform priors for the BMA predictions. Figure 14.8 shows the OOS predictions with weighted priors for BMA.

Keep in mind that in this example, the weights were chosen based on knowledge of the actual equation used to generate the data. Of course, this is very unlikely to be the case with real data. However, it may be true that, based on previous analyses with the same regressors and response, the experimenter may have a better idea as to which regressors deserve higher weights.

14.4 GLMULTI: An Automated Model Selection Procedure

In R package glmulti, there is a function called *glmulti*(), which selects models in a fashion similar to stepwise regression, with some distinct differences. First, *glmulti*() does not depend on an "initial" model. The response and regressors are input in a formula that looks like a simple main effects model. However, *glmulti*() will

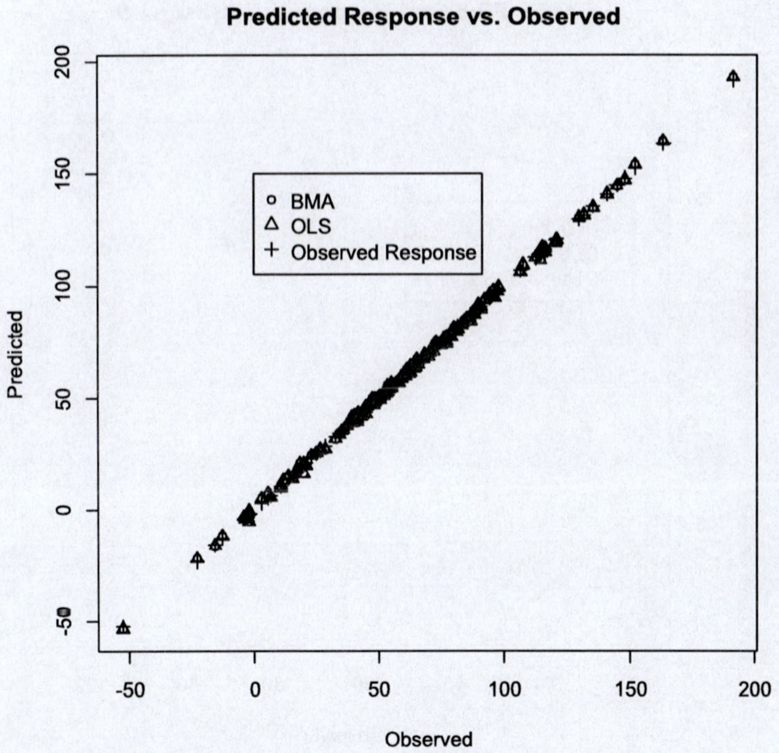

Fig. 14.6 Predicted vs. observed: weighted priors for BMA

incorporate two-way interactions in its algorithm for selecting a final model. Secondly, the function can use several different algorithms for trying to find a good model. One algorithm is called "genetic," because it will change the current model by random "mutation," i.e., randomly selecting to add or subtract terms, and by "crossover," where a whole set of terms is replaced with a different set. Code for using *glmulti*() is given in Fig. 14.9. This code uses the same data that were used for the BMA example. In addition to the *glmulti*() approach, an OLS multiple regression (including all main effects) and a stepwise approach were applied. Approximately 10% of the data (~20 observations) were reserved as a test set, to compare out-of-sample (OOS) predictions. Table 14.3 shows the parameter estimates for each of the methods, together with the actual parameter values used to generate the data. Figure 14.10 shows the predictions for the "training" set, and Fig. 14.11 shows the OOS predictions. Due to the fact that a random 10% of the total data set was extracted for use as a test set, the parameter estimates for *step*() and *lm*() in the BMA and GLMULTI examples are not exactly identical between those examples.

From the plots in Figs. 14.10 and 14.11, it does not appear that any particular modeling method had substantial improvement over any other.

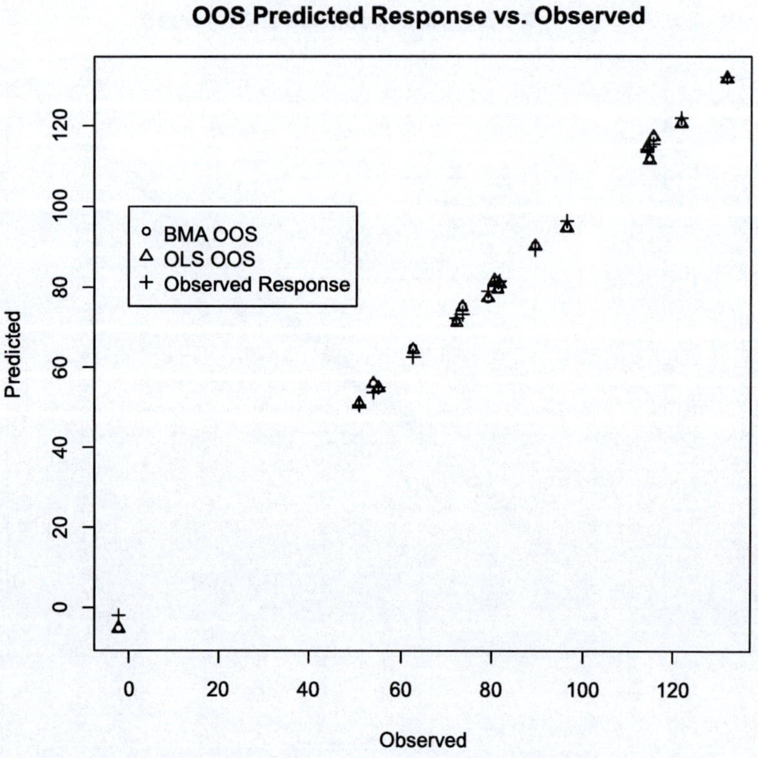

Fig. 14.7 Out-of-sample predictions from BMA (uniform priors) and OLS

14.5 Neural Networks

Neural networks (Kuhn & Johnson, 2016) are based on a model of how a natural brain makes connections between observations and concepts. Unfortunately, the model was largely incorrect. Nevertheless, neural networks, or more correctly, artificial neural networks, provide a convenient means of predicting responses given a "training" set of regressors and the associated response values.

Neural networks are organized into "layers" which comprise of transformations with unknown parameters whose values are estimated via a fitting algorithm, such as least squares. The predictor is called a neural network, because there are layers of "neurons" through which the information moves. The outmost layers are the inputs and the responses. The layers are composed of an array of neurons. Each neuron will get "stimulated" by each input and then produce some sort of "output" that can then feed into the next layer, etc., until the final neural layer produces a "response." The internal layers of neurons are called "hidden" layers. Most applications seem to do best, in terms of computing speed and quality of predictions, with a single hidden layer. Each neuron consists of a little function, h_k, with some parameter that is

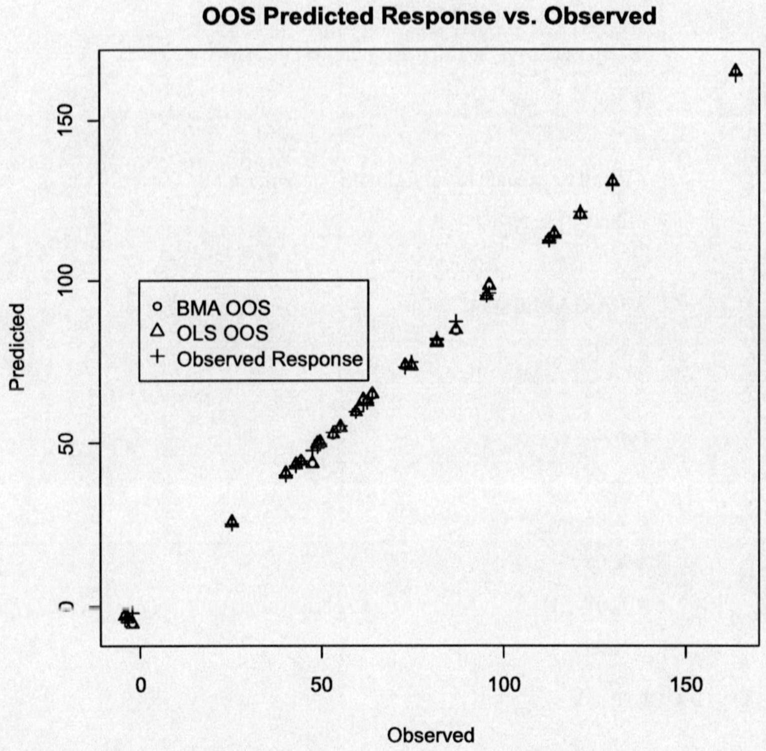

Fig. 14.8 Out-of-sample predictions from BMA (weighted priors) and OLS

estimated using the training data. In other words, the algorithm finds the "best" values of the parameters so that the predicted responses most closely match the observed responses. Figure 14.12 illustrates the concept of the artificial neural network (referred to from here on as neural networks).

Neural networks can be used in two ways. First, they can be used to classify unknown individuals into one of several a priori defined classes, as in the case of linear discriminant analyses. They can also be used as multiple regression models, to compute a predictive equation for continuous value responses. Here we will only use them as predictive models.

The little functions within each neuron are often the logistic "S" curve function (usually most appropriate when the neural net is used for classification):

$$h(\boldsymbol{x}) = \frac{1}{1+e^{-\boldsymbol{\beta}^T\boldsymbol{x}}}$$

However, the functions in the hidden layer(s) can be simply linear:

```
setwd("<your path here>")
dfl <- read.csv("20180701 Example 11.2 BMA v.2.csv")
#
#
# need to execute library(glmulti) once during R session
#library(glmulti)
#
# VARIABLES:
# X.01
# X.02
# X.03
# X.04
# X.05
# X.06
# X.07
# X.08
```

Fig. 14.9 GLMULTI code

$$h(\boldsymbol{x}) = \boldsymbol{\beta}^T \boldsymbol{x}$$

There is no actual restriction on the forms of the $h(\boldsymbol{x})$ functions. However, the S-shaped curves and linear curves seem to be the most common choices. The variable $\boldsymbol{x}$ is a vector of inputs from the previous layer and $\boldsymbol{\beta}$ a vector of unknown "weights." The idea is that the input and response data are used to estimate the weights to minimize the error in associating the known inputs with their corresponding response values. In fact, a numerical least squares approach is often used to estimate the weights.

Figure 14.13 shows R code for implementing the function *nnet*() in package nnet.

Figure 14.14 shows the predicted response values plotted against the actual observed values. Figure 14.15 shows the out-of-sample predicted values from ordinary least squares (OLS), neural network, stepwise, BMA, and GLMULTI models for the test set (i.e., data reserved for testing; not used in the fitting).

There does not appear to be a large discrepancy between the predictions. A measure of average error between model predictions and actual observations is root mean square error (RMSE):

```
# X.09
# X.10
# Response
#
nbin <- nrow(df1)
testind <- rbinom(n=nbin,size=1,prob=0.1)

df2 <- data.frame(cbind(df1,testind))

attach(df2)
dftest <- subset(df2,testind==1)
dftrain <- subset(df2,testind==0)
#
detach(df2)
attach(dftrain)
reg.names <- c("X.01","X.02","X.03","X.04","X.05","X.06","X.07","X.08","X.09","X.10")
multi.model <-
glmulti(Response~X.01+X.02+X.03+X.04+X.05+X.06+X.07+X.08+X.09+X.10,data=dftrain
,plotty=FALSE,report=FALSE,level=1,maxsize=10,method="g")
lm.model <-
lm(Response~X.01+X.02+X.03+X.04+X.05+X.06+X.07+X.08+X.09+X.10,data=dftrain)
step.model <- step(lm.model)
# Make some plots:
multi.predict <- predict(multi.model)
step.predict <- step.model$fitted.values
lm.predict <- lm.model$fitted.values
prediction <- multi.predict$averages
plot(x=dftrain$Response,y=prediction,pch=1,xlab="Observed",ylab="Predicted",main="Pred
icted Response vs. Observed")
points(x=dftrain$Response,y=step.predict,pch=2)
points(x=dftrain$Response,y=lm.predict,pch=3)
```

Fig. 14.9 (continued)

```
points(x=dftrain$Response,y=dftrain$Response,pch=4,type="b")

legend(x=0.0,y=150,legend=c("GLMULTI","STEPWISE","OLS LM","Observed
Response"),pch=c(1,2,3,4))

#

# dev.new() allows more plots to be made without overwriting previous plots

#

#summary(lm.model)

lm.coeffs <- lm.model$coefficients

step.coeffs <- step.model$coefficients

multi.params <- coefficients(multi.model)

multi.coeffs <- multi.params[,1]

detach(dftrain)

attach(dftest)

OOS.multi <- multi.coeffs[6] + multi.coeffs[7]*X.01 + multi.coeffs[4]*X.02 +
multi.coeffs[8]*X.03 + multi.coeffs[3]*X.04 + multi.coeffs[1]*X.05 + multi.coeffs[9]*X.06
+ multi.coeffs[10]*X.08 + multi.coeffs[5]*X.09 + multi.coeffs[2]*X.10

OOS.lm <- lm.coeffs[1] + lm.coeffs[2]*X.01 + lm.coeffs[3]*X.02 + lm.coeffs[4]*X.03 +
lm.coeffs[5]*X.04 + lm.coeffs[6]*X.05 + lm.coeffs[7]*X.06 + lm.coeffs[8]*X.07 +
lm.coeffs[9]*X.08 + lm.coeffs[10]*X.09 + lm.coeffs[11]*X.10

OOS.step <- step.coeffs[1] + step.coeffs[2]*X.01 + step.coeffs[3]*X.02 +
step.coeffs[4]*X.03 + step.coeffs[5]*X.04 + step.coeffs[6]*X.05 + step.coeffs[7]*X.06 +
step.coeffs[8]*X.07 + step.coeffs[9]*X.08 + step.coeffs[10]*X.09 + step.coeffs[11]*X.10

dev.new()

plot(x=dftest$Response,y=OOS.multi,pch=1,xlab="Observed",ylab="Predicted",main="OOS
Predicted Response vs. Observed")

points(x=dftest$Response,y=OOS.step,pch=2)

points(x=dftest$Response,y=OOS.lm,pch=3)

points(x=dftest$Response,y=dftest$Response,pch=4,type="b")

legend(x=5,y=110,legend=c("GLMULTI","STEPWISE","OLS LM","Observed
Response"),pch=c(1,2,3,4))

#####
```

Fig. 14.9 (continued)

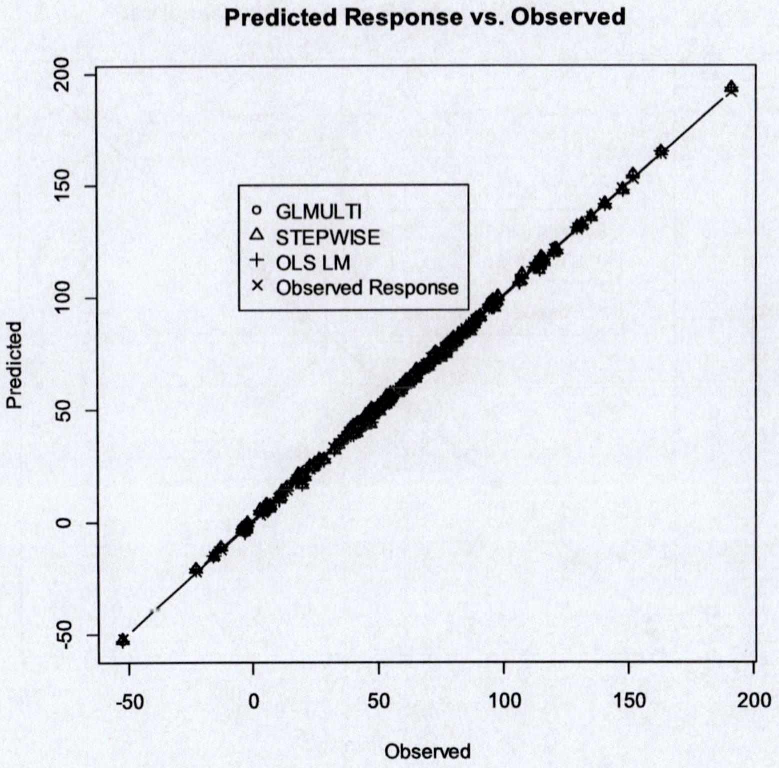

Fig. 14.10 In-sample (training set) predictions

$$\mathrm{RMSE} = \left(\frac{1}{n} \sum_{i=1}^{n} (y_i - \widehat{y}_i)^2 \right)^{½}$$

The y_i represents the observed (or measured) values of the response variable, and $\widehat{y}_i$ represents the corresponding model predictions. Table 14.4 shows the RMSE values for each of the models described above. The RMSE values were computed using out-of-sample (OOS) data.

The magnitude of any of these RMSE values is relatively small compared to the mean and standard deviation of the OOS response data. So, although the NNET model produced the smallest RMSE (entry in bold face), and GLMULTI the largest, the differences in these "average errors" may be relatively meaningless.

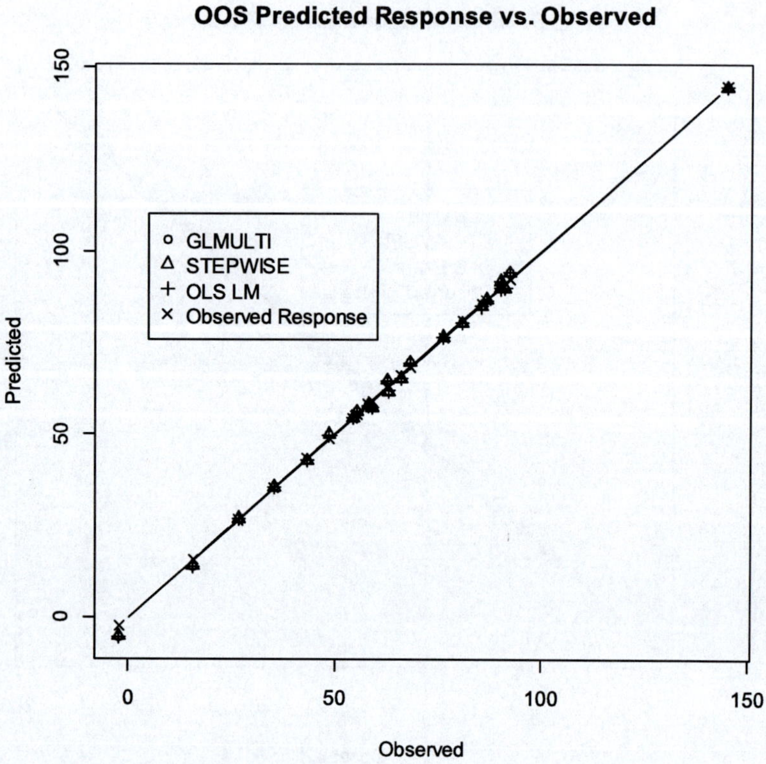

Fig. 14.11 Out-of-sample (test set) predictions

Fig. 14.12 A conceptual neural network

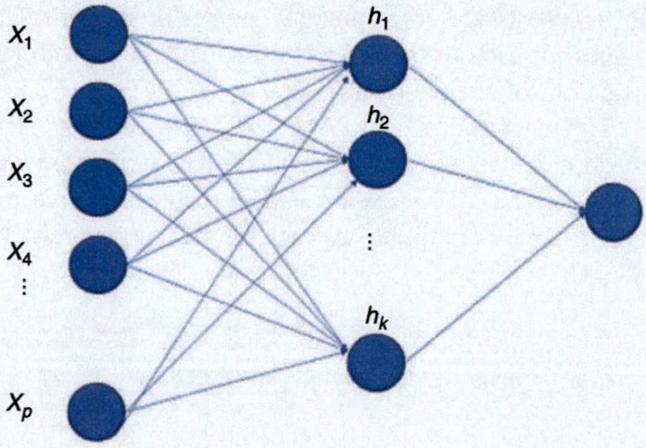

```
setwd("<your path here>")
df1 <- read.csv("20180701 Example 11.2 BMA v.2.csv")
#
#
# need to execute library(nnet) once during R session
#library(nnet)
#
# VARIABLES:
# X.01
# X.02
# X.03
# X.04
# X.05
# X.06
# X.07
# X.08
# X.09
# X.10
# Response
#
nbin <- nrow(df1)
testind <- rbinom(n=nbin,size=1,prob=0.1)

df2 <- data.frame(cbind(df1,testind))

attach(df2)
```

Fig. 14.13 Neural network predictor code

```
dftest <- subset(df2,testind==1)

dftrain <- subset(df2,testind==0)



#

detach(df2)

attach(dftrain)

reg.names <- c("X.01","X.02","X.03","X.04","X.05","X.06","X.07","X.08","X.09","X.10")

nnet.model <-
nnet(Response~X.01+X.02+X.03+X.04+X.05+X.06+X.07+X.08+X.09+X.10,data=dftrain,si
ze=5,skip=TRUE,linout=TRUE)

lm.model <-
lm(Response~X.01+X.02+X.03+X.04+X.05+X.06+X.07+X.08+X.09+X.10,data=dftrain)

step.model <- step(lm.model)

# Make some plots:

nnet.predict <- nnet.model$fitted.values

step.predict <- step.model$fitted.values

lm.predict <- lm.model$fitted.values

#prediction <- multi.predict$averages

plot(x=dftrain$Response,y=nnet.predict,pch=1,xlab="Observed",ylab="Predicted",main="Pr
edicted Response vs. Observed")

points(x=dftrain$Response,y=step.predict,pch=2)

points(x=dftrain$Response,y=lm.predict,pch=3)

points(x=dftrain$Response,y=dftrain$Response,pch=4,type="b")

legend(x=-5,y=150,legend=c("NNET","STEPWISE","OLS LM","Observed
Response"),pch=c(1,2,3,4))

#

# dev.new() allows more plots to be made without overwriting previous plots

#

#summary(lm.model)

lm.coeffs <- lm.model$coefficients
```

Fig. 14.13 (continued)

```
step.coeffs <- step.model$coefficients

#multi.params <- coefficients(multi.model)

#multi.coeffs <- multi.params[,1]

detach(dftrain)

attach(dftest)

OOS.nnet <- predict(nnet.model,dftest)

OOS.lm <- lm.coeffs[1] + lm.coeffs[2]*X.01 + lm.coeffs[3]*X.02 + lm.coeffs[4]*X.03 +
lm.coeffs[5]*X.04 + lm.coeffs[6]*X.05 + lm.coeffs[7]*X.06 + lm.coeffs[8]*X.07 +
lm.coeffs[9]*X.08 + lm.coeffs[10]*X.09 + lm.coeffs[11]*X.10

OOS.step <- step.coeffs[1] + step.coeffs[2]*X.01 + 0.0*X.02 + step.coeffs[3]*X.03 +
0.0*X.04 + 0.0*X.05 + step.coeffs[4]*X.06 + step.coeffs[5]*X.07 + step.coeffs[6]*X.08 +
step.coeffs[7]*X.09 + 0.0*X.10

dev.new()

plot(x=dftest$Response,y=OOS.nnet,pch=1,xlab="Observed",ylab="Predicted",main="OOS
Predicted Response vs. Observed")

points(x=dftest$Response,y=OOS.step,pch=2)

points(x=dftest$Response,y=OOS.lm,pch=3)

points(x=dftest$Response,y=dftest$Response,pch=4,type="b")

legend(x=5,y=110,legend=c("NNET","STEPWISE","OLS LM","Observed
Response"),pch=c(1,2,3,4))

#####
```

Fig. 14.13 (continued)

14.6 Classification and Regression Trees (CART)

CART (Breiman et al., 1984) is a method for fitting models, especially linear models with many regressors. It employs an algorithm that continually partitions the data by splitting on the value of one regressor and then continually splitting until there is a prediction of the response for each "branch." CART will continue to split the data until either the sample size for a branch is "too" small, or Mallow's C_p begins to increase. CART has been implemented in the R library rpart, function *rpart*().

The data used for the previous methods will be used to demonstrate how CART works in R. The code in Fig. 14.16 shows the implementation with ten regressors. Figure 14.17 shows the tree output. In the tree, the label for the branch indicates the left side. So, for example, the first split was using X.06. The left side is for predictions with X.06 < 17.82. The next branch (split) on the left is for X.08 ≥ 13.1 (left side of this branch). The last split on the left is for X.08 ≥ 18.5. The predicted value of response when X.06 < 17.82, X.08 ≥ 18.5 is −7.844. There were $n = 10$ observations used to make this prediction.

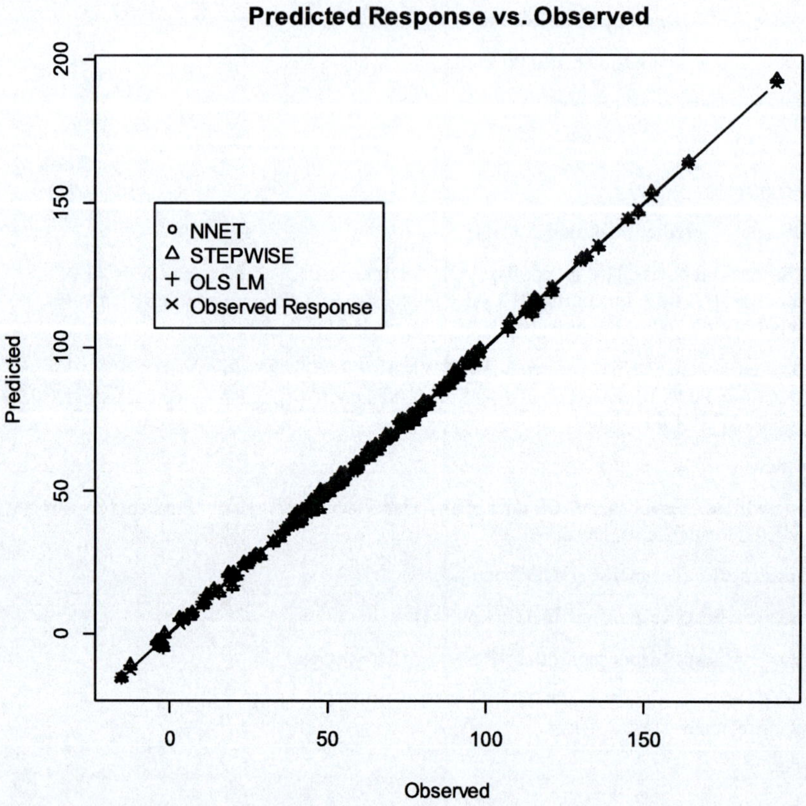

Fig. 14.14 In-sample predictions from a neural network versus observed responses

Figure 14.18 shows a plot of the predicted versus actual observed values of the response for a "test" set of data (not used for the model fitting).

Finally, Fig. 14.19 shows a progression of Mallow's C_p that were obtained after each split.

14.7 Random Forests

Random forests (Breiman, 2001) are a bootstrapped CART models. That is to say, data are resampled, a regression tree is computed, and then the same thing is done many times. The predictions are sort of averaged over all the randomly resampled data. Random forests are implemented in R with function *randomForest*(), in package randomForest.

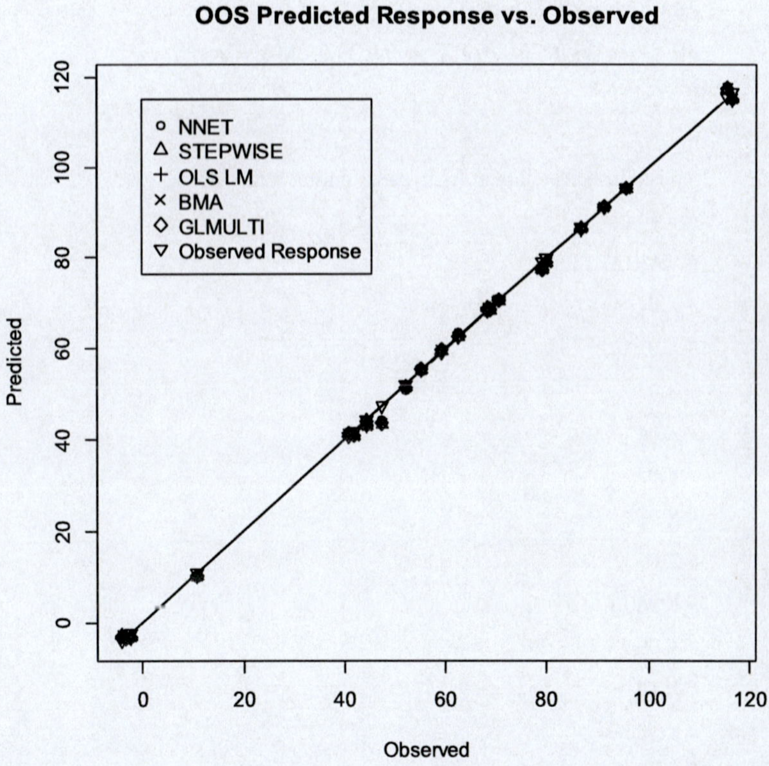

Fig. 14.15 Out-of-sample predictions from several modeling methods versus observed responses

Table 14.4 RMSE values: continuous response

Model	RMSE
NNET	**0.3350**
Stepwise	0.3535
OLS	0.3840
BMA	0.3535
GLMULTI	0.3647
Mean response:	57.4211
Std.dev. response:	34.3371

The output of *randomForest*() is obtained through the use of several other functions in the same package. Figure 14.20 shows R code for implementing a random forest model, together with a CART model and a stepwise regression model.

Random forest models offer a means of judging the relative importance of each regressor. There are two measures provided by the R function: IncMSE and IncNodePurity. The IncMSE measure is based on the mean squared error between predicted values and observed values (residuals). The IncNodePurity measure is based on the sums of squares of residuals. In both cases, larger values indicate

```
setwd("<your path here>")
dfl <- read.csv("20180701 Example 11.2 BMA v.2.csv")
#
#
# need to execute library(rpart) once during R session
#
# VARIABLES:
# X.01
# X.02
# X.03
# X.04
# X.05
# X.06
# X.07
# X.08
# X.09
# X.10
# Response
#
library(rpart)
nbin <- nrow(dfl)
testind <- rbinom(n=nbin,size=1,prob=0.1) #randomly selected 10%
df2 <- cbind(dfl,testind)
detach(dfl)
attach(df2)
dftest <- subset(df2,testind==1)
dftrain <- subset(df2,testind==0)
```

Fig. 14.16 CART R code

```
detach(df2)

attach(dftrain)

linreg <- lm(Response ~ X.01 + X.02 + X.03 + X.04 + X.05 + X.06 + X.07 + X.08 + X.09 +
X.10,data=dftrain)

# Make some plots:

#

# dev.new() allows more plots to be made without overwriting previous plots

#


summary(linreg)

#

#

#

cart.model <- rpart(Response ~ X.01 + X.02 + X.03 + X.04 + X.05 + X.06 + X.07 + X.08 +
X.09 + X.10,data=dftrain)

#plot(cart.model,xlim=c(0.75,5.33),ylim=c(-0.01,1.10))

plot(cart.model,ylim=c(0.1,1.10),main="Classification and Regression Tree")


text(cart.model,use.n=TRUE)

dev.new()

rsq.rpart(cart.model)

dev.new()

predcart <- predict(cart.model,dftest)

plot(dftest$Response,predcart,xlab=c("Actual Response"),ylab=c("Predicted
Response"),main="CART Predictions for Test Data")

#####
```

Fig. 14.16 (continued)

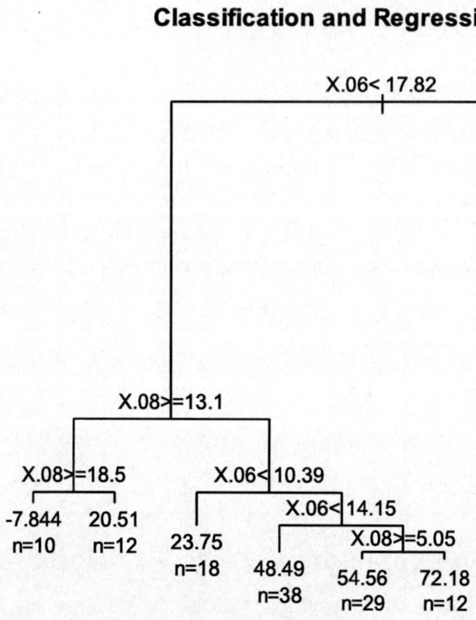

Fig. 14.17 A classification and regression tree

greater importance (notice the importance = TRUE input parameter to the *randomForest*() function). The attribute "importance" provides the IncMSE and IncNodePurity values for each regressor in the random forest model.

Figure 14.21 shows the importance measure calculations.

Regressor X.06 has the largest values for both IncMSE and IncNodePurity. The stepwise regression procedure included regressors X.01, X.03, X.06, X.07, X.08, and X.09 in the final model. Figure 14.22 shows the ANOVA output for the stepwise model.

The t-statistic for the X.06 coefficient was largest. Table 14.5 shows the regressors, sorted by IncMSE, with the IncNodePurity and the t-statistics. Regressor X.06 had the largest IncMSE, IncNodePurity, and largest t-statistic (in absolute value). There were, however, some differences in the order of importance (magnitude of measures) for the three ranking methods in the table. While IncMSE and t-statistics agreed on the order of regressor importance (at least for those regressors included in the final stepwise model), there was some disagreement with IncNodePurity.

One advantage that random forest models have over other machine learning methods is the assessment of regressor importance, which can possibly provide the modeler with some insight into the nature of relationships between regressors and response.

Figure 14.23 shows the predicted values plotted against the actuals. In this case, which was contrived to be a linear function of regressors, the random forest and

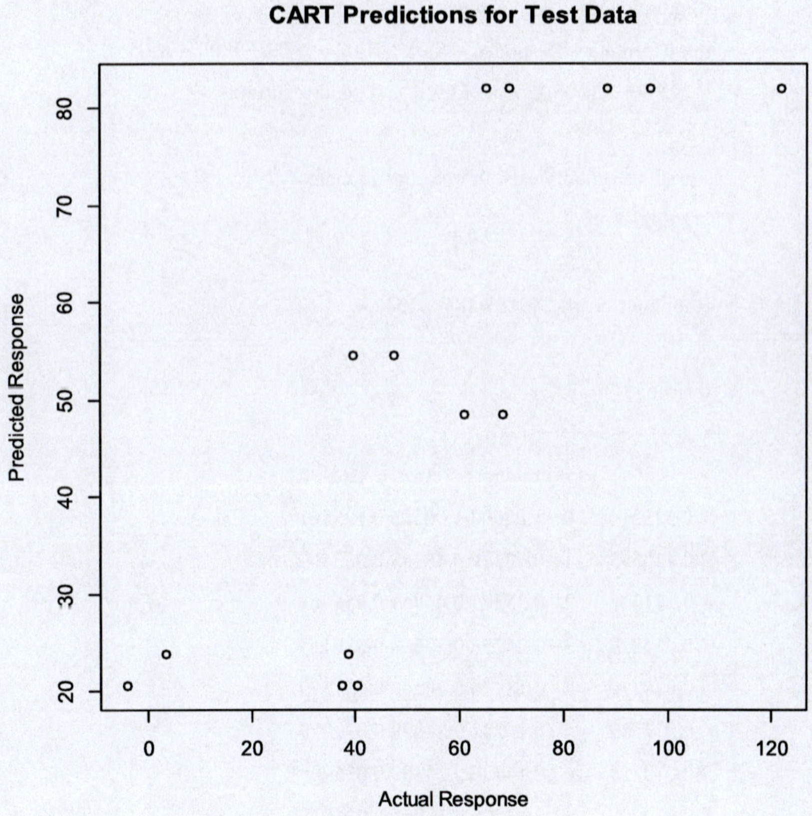

Fig. 14.18 CART predictions and actual observations

CART predictions did not appear to be as accurate as those of the linear model or the final stepwise model.

14.8 Logistic Regression and Model Selection

So far model selection and machine/algorithm-based model building have been applied to linear regression models. The same methods can be applied to generalized linear models. As an example, model selection methods will be applied to a logistic regression model.

Consider a ten-factor logistic regression model. The code in Fig. 14.24 shows how stepwise regression, neural networks, *glmulti*(), and Bayesian model averaging can all be applied to the logistic regression model.

Figure 14.25 shows the predicted probabilities, Pr{Ybin = 1}, compared to the single logistic regression model predicted values.

```
Regression tree:
rpart(formula = Response ~ X.01 + X.02 + X.03 + X.04 + X.05 +
    X.06 + X.07 + X.08 + X.09 + X.10, data = dftrain)

Variables actually used in tree construction:
[1] X.06 X.08

Root node error: 249685/186 = 1342.4

n= 186

        CP  nsplit rel error  xerror     xstd
1 0.525214      0   1.00000 1.00585 0.125801
2 0.120893      1   0.47479 0.48753 0.063460
3 0.111318      2   0.35389 0.44969 0.056463
4 0.054879      3   0.24257 0.29608 0.036113
5 0.025279      4   0.18770 0.24549 0.031879
6 0.017559      5   0.16242 0.24879 0.031979
7 0.010255      6   0.14486 0.24309 0.035906
8 0.010000      8   0.12435 0.23901 0.035918
```

Fig. 14.19 CART model output and C_p sequence

Figure 14.26 shows the out-of-sample (OOS) predictions using a "test set" of data (not used for fitting).

Inasmuch as the predictions are probabilities, and the response data are binary results (1s and 0s), RMSE as defined in the previous section cannot be computed. However, a predicted value of each model can be computed at a particular set of values for the regressors. In particular, suppose the average value of the binary response is computed for the OOS dataset, and the average values of the regressors are then input to the models. The prediction of probabilities can be compared to the average binary result (which is the proportion of 1s). Table 14.6 shows the results. In this case, the GLMULTI prediction was closest to the observed proportion.

```
setwd("<your path here>")
df1 <- read.csv("20180701 Example 11.2 BMA v.2.csv")
#
#
# need to execute library(randomForest) once during R session
#
# VARIABLES:
# X.01
# X.02
# X.03
# X.04
# X.05
# X.06
# X.07
# X.08
# X.09
# X.10
# Response
#
library(rpart) #package for CART
library(randomForest) #package for the randomForest() function
nbin <- nrow(df1)
```

Fig. 14.20 Random forest model code

14.9 Summary

Model selection in the face of many regressors whose influence on the response is not only unknown, but for which there is not a strong "hypothesis," is at best difficult. Modelers sometimes resort to an algorithmic approach, where the objective is generally defined in terms of prediction or classification errors. Such methods do not generally offer much in the way of insight into the nature of relationships between regressors and responses, but they can be useful in generating a predictive or classification machine.

```
set.seed(13) #fixes random seed for repeatability

testind <- rbinom(n=nbin,size=1,prob=0.1) #randomly selected 10%


df2 <- cbind(df1,testind)

#detach(df1)

attach(df2)

dftest <- subset(df2,testind==1)

dftrain <- subset(df2,testind==0)


detach(df2)

attach(dftrain)

linreg <- lm(Response ~ X.01 + X.02 + X.03 + X.04 + X.05 + X.06 + X.07 + X.08 + X.09 +
X.10,data=dftrain)

step.model <- step(linreg)

# Make some plots:

#

# dev.new() allows more plots to be made without overwriting previous plots

#


forest.model <- randomForest(Response ~ X.01 + X.02 + X.03 + X.04 + X.05 + X.06 + X.07
+ X.08 + X.09 + X.10,importance=TRUE,data=dftrain)

forest.pred <- predict(forest.model,newdata=dftest,type="response")


# Make some plots:


summary(linreg)

summary(step.model)

#

#

#
```

Fig. 14.20 (continued)

```
cart.model <- rpart(Response ~ X.01 + X.02 + X.03 + X.04 + X.05 + X.06 + X.07 + X.08 +
X.09 + X.10,data=dftrain)


#plot(cart.model,xlim=c(0.75,5.33),ylim=c(-0.01,1.10))

plot(cart.model,ylim=c(0.1,1.10),main="Classification and Regression Tree")


text(cart.model,use.n=TRUE)

dev.new()

rsq.rpart(cart.model)

dev.new()

pred.lin <- predict(linreg,newdata=dftest)

pred.step <- predict(step.model,newdata=dftest)

predcart <- predict(cart.model,dftest)

plot(dftest$Response,predcart,pch=1,xlab=c("Actual Response"),ylab=c("Predicted
Response"),main="Random Forest and Other Predictions for Test Data")

points(x=dftest$Response,forest.pred,pch=2)

points(x=dftest$Response,y=pred.lin,pch=3)

points(x=dftest$Response,y=pred.step,pch=4)

legend(x=0,y=100,legend=c("CART","Random Forest","Linear
Model","Stepwise"),pch=c(1,2,3,4))

forest.model$importance

#####
```

Fig. 14.20 (continued)

Fig. 14.21 Importance measures for random forest model

> forest.model$importance

	%IncMSE	IncNodePurity
X.01	13.253005	8068.391
X.02	-12.018055	6594.380
X.03	12.234598	12174.001
X.04	13.263183	7412.899
X.05	-9.246903	8045.997
X.06	**1300.450647**	**117270.818**
X.07	31.359970	15222.417
X.08	299.547329	41462.192
X.09	-3.140281	7010.163
X.10	-2.174465	4791.224

Harrell (2018) has defined two concepts, statistical modeling (SM) and machine learning (ML). Briefly summarized, SMs are useful when the dataset is not huge, the numbers of regressors are small, and the relationships between regressors and responses van be understood at a more molecular level. MLs are more useful with very large (Big Data) datasets, with many regressors whose relationships to the responses are not only completely unknown, but not of any particular interest. Rather, the ML approach is useful when the modeler simply needs a device for making predictions or classifying individuals based on a fairly large number of regressor variables. Harrell provided a "road map" which can help the data analyst which methodological direction to choose.

Finally, a word of caution from the fourteenth century. William Ockham wrote "Non est ponenda sine necessitate," which means do not add (complexity) without necessity. It is roughly what people call "Ockham's razor." The point is to be parsimonious with adding predictor variables and take care when considering non-linearities in a model.

```
Call:
lm(formula = Response ~ X.01 + X.03 + X.06 + X.07 + X.08 + X.09,
   data = dftrain)

Residuals:
    Min      1Q  Median      3Q     Max
-2.7525 -0.7372 -0.0872  0.7130  3.8250

Coefficients:
            Estimate Std. Error t value Pr(>|t|)
(Intercept) -9.66992    0.59185  -16.34  < 2e-16 ***
X.01        -1.22945    0.01848  -66.52  < 2e-16 ***
X.03        -2.34681    0.04651  -50.46  < 2e-16 ***
X.06         5.01173    0.01531  327.38  < 2e-16 ***
X.07        -1.94908    0.03244  -60.08  < 2e-16 ***
X.08        -3.22088    0.01883 -171.09  < 2e-16 ***
X.09         5.00976    1.18723    4.22 3.97e-05 ***
---
Signif. codes:  0 '***' 0.001 '**' 0.01 '*' 0.05 '.' 0.1 ' ' 1

Residual standard error: 1.207 on 170 degrees of freedom
Multiple R-squared:  0.999,     Adjusted R-squared:  0.9989
F-statistic: 2.713e+04 on 6 and 170 DF,  p-value: < 2.2e-16
```

Fig. 14.22 ANOVA for stepwise model

Table 14.5 Regressor importance measures

Regressor	IncMSE	IncNodePurity	t-Statistic from stepwise[a]
X.02	−12.0181	6594.38	N/A
X.05	−9.2469	8045.997	N/A
X.09	−3.14028	7010.163	4.22
X.10	−2.17447	4791.224	N/A
X.03	*12.2346*	*12174.001*	*−50.46*
X.01	*13.25301*	*8068.391*	*−66.52*
X.04	*13.26318*	*7412.899*	*N/A*
X.07	31.35997	15222.417	−60.08
X.08	299.5473	41462.192	−171.09
X.06	**1300.451**	**117270.818**	**327.38**

[a]Order based on absolute value

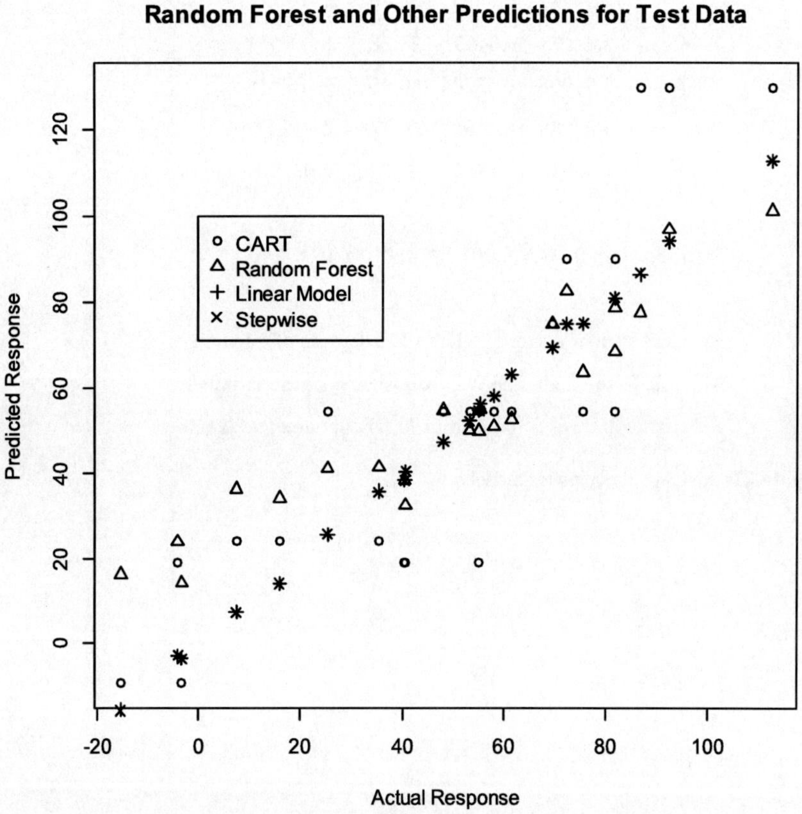

Fig. 14.23 Predictions from random forest, CART, linear, and stepwise models

```
setwd("<your path here>")
df1 <- read.csv("20220527 CH 15 Logistic Regression.csv")
#
#
#
# VARIABLES:
# X.01
# X.02
# X.03
# X.04
# X.05
# X.06
# X.07
# X.08
# X.09
# X.10
# Binary
#
library(BMA)
library(glmulti)
library(nnet)
nbin <- nrow(df1)
testind <- rbinom(n=nbin,size=1,prob=0.1)

df2 <- data.frame(cbind(df1,testind))

attach(df2)
dftest <- subset(df2,testind==1)
dftrain <- subset(df2,testind==0)
```

Fig. 14.24 Code for comparing model selection methods with logistic regression

```
#
detach(df2)
attach(dftrain)

reg.names <- c("X.01","X.02","X.03","X.04","X.05","X.06","X.07","X.08","X.09","X.10")
prior.p <- c(0.9,0.001,0.9,0.001,0.001,0.9,0.9,0.9,0.9,0.001)
bma.model <- bic.glm(f=Binary ~ X.01 + X.02 + X.03 + X.04 + X.05 + X.06 + X.07 + X.08
+ X.09 + X.10,data=dftrain,glm.family=binomial(),prior.param=prior.p)
nnet.model <-
nnet(Binary~X.01+X.02+X.03+X.04+X.05+X.06+X.07+X.08+X.09+X.10,data=dftrain,size
=5,skip=TRUE,linout=TRUE)
glm.model <-
glm(Binary~X.01+X.02+X.03+X.04+X.05+X.06+X.07+X.08+X.09+X.10,family=binomial(l
ink="logit"),data=dftrain)
step.model <- step(glm.model)
multi.model <-
glmulti(glm.model,plotty=FALSE,report=FALSE,level=1,maxsize=10,method="g")
# Make some plots:
bma.predict <- predict(bma.model,dftrain)
nnet.predict <- predict(nnet.model,dftrain)
#nnet.predict <- nnet.model$fitted.values
step.predict <- 1 / (1 + exp(-predict(step.model,dftrain)))
#step.predict <- step.model$fitted.values
multi.big <- predict(multi.model)
multi.predict <- multi.big$averages
#multi.predict <- 1 / (1 + exp(-(predict(multi.model,dftrain))))
glm.predict <- 1 / (1 + exp(-predict(glm.model,dftrain)))
#glm.predict <- glm.model$fitted.values
#prediction <- multi.predict$averages
plot(x=glm.predict,y=nnet.predict,pch=1,xlab="Logistic
Model",ylab="Predicted",main="Predicted Response vs. Logistic")
points(x=glm.predict,y=step.predict,pch=2)
points(x=glm.predict,y=multi.predict,pch=3)
```

Fig. 14.24 (continued)

```
points(x=glm.predict,y=bma.predict,pch=4)

#points(x=dftrain$Response,y=dftrain$Response,pch=4,type="b")

legend(x=0.1,y=1.05,legend=c("NNET","STEPWISE",
"GLMULTI","BMA"),pch=c(1,2,3,4))

#

# dev.new() allows more plots to be made without overwriting previous plots

#

#summary(lm.model)

#lm.coeffs <- lm.model$coefficients

#step.coeffs <- step.model$coefficients

#multi.params <- coefficients(multi.model)

#multi.coeffs <- multi.params[,1]

detach(dftrain)

attach(dftest)

OOS.bma <- predict(bma.model,dftest)

OOS.nnet <- predict(nnet.model,dftest)

OOS.glm <- 1 / (1 + exp(-predict(glm.model,dftest)))

#OOS.lm <- lm.coeffs[1] + lm.coeffs[2]*X.01 + lm.coeffs[3]*X.02 + lm.coeffs[4]*X.03 +
lm.coeffs[5]*X.04 + lm.coeffs[6]*X.05 + lm.coeffs[7]*X.06 + lm.coeffs[8]*X.07 +
lm.coeffs[9]*X.08 + lm.coeffs[10]*X.09 + lm.coeffs[11]*X.10

OOS.step <- 1 / (1 + exp(-predict(step.model,dftest)))

multi.test <- predict(multi.model,,dftest)

OOS.multi <- multi.test$averages

dev.new()

plot(x=OOS.glm,y=OOS.nnet,pch=1,xlab="Logistic OOS",ylab="Predicted
OOS",main="OOS Predicted Response vs. Logistic")

points(x=OOS.glm,y=OOS.step,pch=2)

points(x=OOS.glm,y=OOS.multi,pch=3)

points(x=OOS.glm,y=OOS.bma,pch=4)

#points(x=dftest$Response,y=dftest$Response,pch=4,type="b")

legend(x=0.1,y=0.85,legend=c("NNET","STEPWISE","GLMULTI","BMA"),pch=c(1,2,3,4)
)
```

Fig. 14.24 (continued)

```
row.X <- 1

test.X <- 1

meanX1.df <-
data.frame(cbind(row.X,mean(X.01),mean(X.02),mean(X.03),mean(X.04),mean(X.05),mean
(X.06),mean(X.07),mean(X.08),mean(X.09),mean(X.10),mean(Binary),test.X))

row.X <- 2

test.X <- 2

meanX2.df <-
data.frame(cbind(row.X,mean(X.01),mean(X.02),mean(X.03),mean(X.04),mean(X.05),mean
(X.06),mean(X.07),mean(X.08),mean(X.09),mean(X.10),mean(Binary),test.X))

meanX.df <- rbind(meanX1.df,meanX2.df)

names(meanX.df) <-
c("row.names","X.01","X.02","X.03","X.04","X.05","X.06","X.07","X.08","X.09","X.10","
Binary","testind")


center.bma <- predict(bma.model,newdata=meanX.df)

center.nnet <- predict(nnet.model,newdata=meanX.df)

center.step <- 1 / (1 + exp(-predict(step.model,newdata=meanX.df)))

center.multi <- predict(multi.model,newdata=meanX.df)

center.glm <- predict(glm.model,newdata=meanX.df)

center.glm <- 1 / (1 + exp(-predict(glm.model,newdata=meanX.df)))


diff.bma <- meanX.df$Binary[1] - center.bma[1]

diff.nnet <- meanX.df$Binary[1] - center.nnet[1]

diff.step <- meanX.df$Binary[1] - center.step[1]

diff.multi <- meanX.df$Binary[1] - center.multi$averages[1]

diff.glm <- meanX.df$Binary[1] - center.glm[1]

#####
```

Fig. 14.24 (continued)

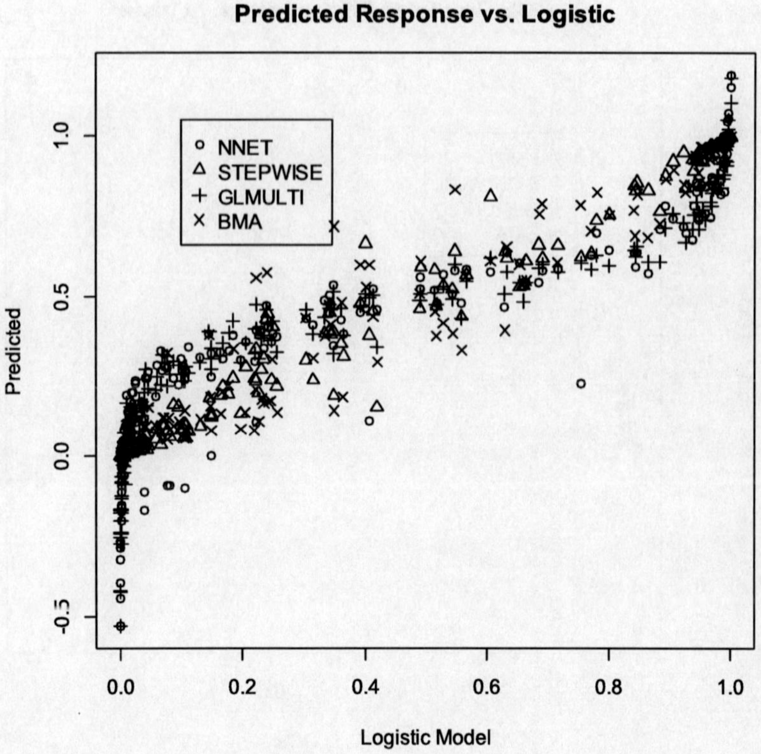

Fig. 14.25 In-sample predicted probabilities

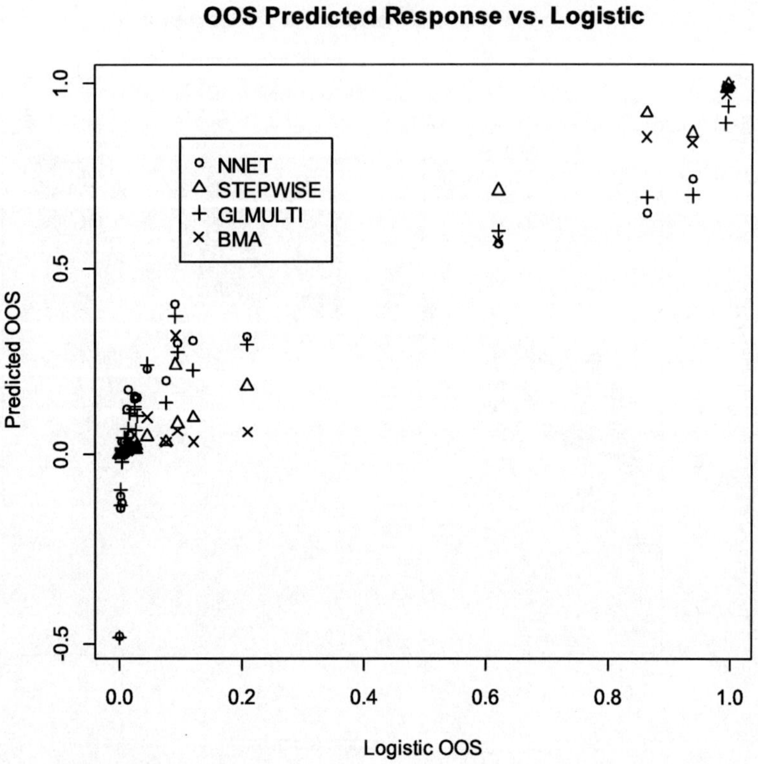

Fig. 14.26 OOS predicted probabilities

Table 14.6 RMSE values: binary response

Model	RMSE
NNET	−0.0731
Stepwise	0.1827
GLM	0.2025
BMA	0.1845
GLMULTI	**0.0179**
Mean response:	0.2667
Std.dev. response:	0.4577

References

Breiman, L. (2001). Random forests. *Machine Learning, 45*, 5–32.

Breiman, L., Friedman, J., Stone, C. J., & Olshen, R. A. (1984). *Classification and regression trees*. Chapman and Hall/CRC.

Draper, N. R., & Smith, H. (1981). *Applied regression analysis* (2nd ed.). Wiley.

Harrell, F. (2018). *Road map for choosing between statistical modeling and machine learning*. http://www.fharrell.com/post/stat-ml/

Kuhn, M., & Johnson, K. (2016). *Applied predictive modeling*. Springer.

Chapter 15
Variance Components and Precision

Abstract Often the precision of performance is as important as accuracy. Methods for estimating measures of precision are described, along with associated confidence in intervals.

15.1 Introduction

When response variables are observed or measured under different conditions, it is natural to want to know how those conditions affect the value of the response. Earlier, we focused on the average effects of those conditions and in fact relied on the assumption that the variability of the response was not affected by varying conditions. Variance components analysis (VCA) is an attempt to quantify the amount of variation in the response caused by different factors that may vary in their values or levels.

Factors (regressors with discrete values, or levels) will be categorized as either fixed (all the levels of the factor are specifically interesting) or random (the levels of the factor represent a random sample of all possible levels and as such are not specifically interesting). The VCA will attempt to estimate how much the random effects contribute to the variance of the response. Often, both fixed and random factors (also referred to as "effects") are included in a single analysis or model. Such models are referred to as "mixed."

15.2 The Mixed Model: Randomized Complete Block Design

Suppose an adhesive requires heat curing. Furthermore, the adhesive comes in "units," or packages, that have enough material for two applications. The engineer is considering two possible heat treatment protocols. The cure time, in seconds, can actually be measured, as the material changes color once it is fully cured. The questions are:

S. A. Pardo, *Statistical Methods and Analyses for Medical Devices*,
https://doi.org/10.1007/978-3-031-26139-8_15

1. How much of a difference in mean cure time is there for the two treatments?
2. How much variability do units impart on the cure time (response)?

The model is:

$$\text{Response} \sim \text{Unit} + \text{Treatment} + \text{Noise}$$

In this case, we will be concerned with the differences in average response between the two treatments and in the portion of the variance of the response attributed to the units.

In R, to specify that the factor "unit" is random, use the following:

```
fUnit <- factor(Unit)
model_rcb <- lmer(Response ~ (1|fUnit) + Treatment)
```

Figure 15.1 shows R code for analyzing this model.
The console output is shown in Fig. 15.2.
The total variance for response is:

$$104.149 + 9.259 = 113.408$$

The total estimated standard deviation is:

$$\widehat{\sigma} = \sqrt{113.408} \approx 10.649$$

Notice that this estimate is close to the average standard deviation for the data under each treatment:

$$\widehat{\sigma} = \sqrt{\frac{12.04^2 + 9.05^2}{2}} \approx 10.649$$

The "component" of variance due to unit variation is $\frac{104.149}{113.408} \approx 0.9184$, or about 91.84% of the total variance.

The method for estimating the variance components was restricted maximum likelihood (REML), which is described by Searle et al. (1992). REML depends on the assumption that noise (residuals) are normally distributed, with mean 0 and a variance that is constant, and independent of the factor levels.

Aside from the fact that a variance component for unit was much larger than the residual component, it might be of interest to test the hypothesis that the variance component for the random effect is actually larger than the noise. To do this, compute:

```
setwd("<your path here>")
df1 <- read.csv("20220526 Ch 15 Mixed Model.csv")
#library(car) #needed to use function Anova() - not the same as anova()
library(lmerTest)
# Variables
#
# Unit
# Group
# Order
# Treatment
# Response
attach(df1)

fGroup <- factor(Group)
fUnit <- factor(Unit)
fOrder <- factor(Order)
#
# fUnit is nested in Group
# There are two ways to represent this in the model statement:
# fGroup/fUnit
# fUnit%in%fGroup
#
# fGroup/fUnit = fGroup + fUnit%in%fGroup
#
#  Note the different order in which factors are listed in these two specifications
#
model_rcb <- lmer(Response ~ (1|fUnit) + Treatment)
anova(model_rcb)
summary(model_rcb)
#Anova(model_rcb,test.statistic="F",type=3)
tapply(Response,INDEX=Treatment,FUN=mean)
tapply(Response,INDEX=Treatment,FUN=sd)
#####
```

Fig. 15.1 Variance components and the mixed model

```
Type III Analysis of Variance Table with Satterthwaite's method
           Sum Sq Mean Sq NumDF DenDF F value    Pr(>F)
Treatment 31.104   31.104    1     29  3.3593   0.07711 .
---
Signif. codes:  0 '***' 0.001 '**' 0.01 '*' 0.05 '.' 0.1 ' ' 1
> summary(model_rcb)
Linear mixed model fit by REML. t-tests use Satterthwaite's method [
lmerModLmerTest]

Formula: Response ~ (1 | fUnit) + Treatment
REML criterion at convergence: 392
Scaled residuals:
     Min       1Q   Median       3Q      Max
-1.22266 -0.58168  0.03333  0.54907  1.35228

Random effects:
 Groups   Name        Variance Std.Dev.
 fUnit    (Intercept) 104.149  10.205
 Residual               9.259   3.043
Number of obs: 60, groups:  fUnit, 30

Fixed effects:
            Estimate  Std. Error      df  t value   Pr(>|t|)
(Intercept)  25.3533     1.9443  31.4640  13.040   3.14e-14 ***
TreatmentB    1.4400     0.7857  29.0000   1.833   0.0771 .
---
Signif. codes:  0 '***' 0.001 '**' 0.01 '*' 0.05 '.' 0.1 ' ' 1
```

Fig. 15.2 Output for mixed model

$$\widehat{F} = \frac{\widehat{\sigma}_{\text{Unit}}^2}{\widehat{\sigma}_{\varepsilon}^2}$$

$\widehat{\sigma}_{\text{Unit}}^2 =$ REML estimate of variance component for unit
$\widehat{\sigma}_{\varepsilon}^2 =$ REML estimate of variance component for residuals

If dfU is the degrees of freedom for unit (number of units minus 1), and dfe is the degrees of freedom for residuals, then if $\widehat{F} \geq F(\alpha|\text{dfU}, \text{dfe})$, reject the hypothesis:

$$H_0 : \sigma^2_{\text{Unit}} = \sigma^2_\varepsilon$$

in favor of the alternative:

$$H_1 : \sigma^2_{\text{Unit}} > \sigma^2_\varepsilon$$

$F(\alpha|\,\text{dfU}, \text{dfe})$ = the 100(1 − α) percentile of an F distribution with dfU and dfe degrees of freedom in the numerator and denominator, respectively.

For more details on VCA, see Searle et al. (1992)

15.3 Measures of Precision: Standard Deviation and Coefficient of Variation

Probably the two most recognized measures of precision (really, imprecision) are the standard deviation (sd) and the coefficient of variation (cv). The sample estimator formulas commonly employed are:

$$\text{sd} = \sqrt{\frac{1}{n-1}\sum_{i=1}^{n}(x_i - \overline{x})^2}$$

and:

$$\text{cv} = \frac{\text{sd}}{\overline{x}}$$

The sample mean is:

$$\overline{x} = \frac{1}{n}\sum_{i=1}^{n} x_i$$

and x_i = ith observation in a random sample of values, $x_1, x_2, \ldots, x_n$.

Both the quantities sd and cv are sample estimates and as such benefit from some confidence bounds. For sd, the 100(1 − α)% (upper) confidence limit is for the population parameter σ, and the formula is:

$$\widehat{\sigma}_{1-\alpha} = \sqrt{\frac{(n-1)\text{sd}^2}{\chi^2_{1-\alpha}}}$$

$\chi^2_{1-\alpha}$ = the 100(1 − α) percentile of a Chi-squared distribution having n − 1 degrees of freedom.

For cv, there is no "exact": formula for a confidence limit, but there are several approximations. The approximate confidence limits are for the parameter:

$$c = \frac{\sigma}{\mu}$$

The first approximation presented here is based on Kang et al. (2007)
The limit is computed as:

$$\widehat{\text{cv}}_{1-\alpha} = \frac{\sqrt{n}}{T'_{1-\alpha}}$$

$T'_{1-\alpha}$ is the 100(1 – α) percentile of a non-central t distribution with degrees of freedom (df) = n – 1 and (estimated) non-centrality parameter:

$$\text{nct} = \frac{\sqrt{n}}{\text{cv}}$$

The value of cv is the sample coefficient of variation.
Another formula for an approximate confidence limit is given by Vangel (1996):

$$\widehat{\text{cv}}_{1-\alpha} = \text{cv}\left[\left(\frac{\chi^2_{1-\alpha}+1}{n-1} - 1\right)\text{cv}^2 + \frac{\chi^2_{1-\alpha}+1}{n-1}\right]^{-\frac{1}{2}}$$

Figure 15.3 shows an R script for computing confidence limits with both sd and cv.

15.4 Imprecision and Quality

Dimensions and performance parameters may have an allowable range of variation, or tolerance interval, associated with them. As such, the manufacturers and designers of a product might want to have a measure of variation that would allow a "large" proportion of the product to have those measures fall within acceptable limits.

If (L, U) represents the lower and upper specification limit for a quality variable X (either a performance measure or a physical dimension on a part), then a measure of compliance in variability is (Kushler & Hurley, 1992):

$$\text{Cp} = \frac{U - L}{6\sigma_X}$$

The notion is that if the expected value, μ, of X is actually at the center of the specification range:

```
setwd("<your path here>")
#
#
# Kang,C.W.,  Lee, M.S., Seong, Y.J., Hawkins, D.M. (2007)
# "A Control Chart for the  Coefficient of Variation",
# JQT, April 2007, Vol. 39, No. 2
#
# Mark Vangel (1996), "Confidence Intervals for a Normal Coefficient of Variation",
# American Statistician, Vol. 15, No. 1, pp. 21-26.
#
#
conf <- 0.99
#conf <- 0.999975
#conf <- 0.99
alpha <- 1 - conf
k <- 7
alpha.prime <- 1-conf**(1/k)
n <- 36
C0 <- 1.7 #CV in %
C <- C0 / 100
C0.text <- as.character(round(C0,4))
NCT <-sqrt(n) / C #This is the noncentrality parameter
T.quant <- qt(p=alpha.prime,df=n-1,ncp=NCT)
C.crit <- 100*sqrt(n) / T.quant
C.crit.text <- as.character(round(C.crit,2))
C.alt <- seq(from=0.01,to=0.05,by=0.0001)
C.alt.pct <- C.alt*100
NCT.alt <- sqrt(n) / C.alt
Power <- pt(T.quant,df=n-1,ncp=NCT.alt)
plot(x=C.alt.pct,y=Power,main="Power Curve for Testing CV Hypothesis",xlab="Pop.
CV (%)")
```

Fig. 15.3 Confidence limits for measures of imprecision

```
text(x=4.0,y=0.50,labels="Crit. CV(%) =")
text(x=4.85,y=0.50,labels=C.crit.text)
text(x=4.0,y=0.40,labels="Target Pop. CV(%) =")
text(x=4.85,y=0.40,labels=C0.text)
abline(v=3.27)
abline(v=1.7)
abline(h=0.95)
abline(h=0.05)
#
#Kang et al limit:
#
CV.UCL <- 100*sqrt(n) / T.quant
CV.UCL
#
# Vangel Modified McKay Limit (for lower limit, use qchisq(1-alpha.prime,df=n-1)
#
CV.UCL.V <- 100*C / sqrt(((qchisq(alpha.prime,df=n-1)/n) - 1)*C**2 +
qchisq(alpha.prime,df=n-1)/(n-1))
CV.UCL.V
#
# This computes an upper confidence limit for standard deviation:
#
S <- 1.3
Sig.UCL <- S*sqrt((n-1)/qchisq(p=alpha.prime,df=n-1))
Sig.UCL
df2 <- cbind(NCT.alt,C.alt,Power)
#####
```

Fig. 15.3 (continued)

$$\mu = \frac{L + U}{2}$$

and the standard deviation is such that:

$$L = \mu - 3\sigma_X$$
$$U = \mu + 3\sigma_X$$

```
setwd("<your path here>")
df1 <- read.csv("20220531 Ch 15 Confidence Limit Capability Cp.csv")
#
# Closed-Form and Bootstrap Confidence Limits for Cp
#
set.seed(26)
boot.X <- c()
sd.boot <- c()
Cp.boot <- c()
Cp.est <- c()
conf <- 0.95

alpha <- 1 - conf
k <- 1 #k set to > 1 for multiplicity adjustment if more than one confidence limit
needed
alpha.prime <- 1-conf**(1/k)

#alpha.prime <- alpha / k
n <- length(df1$X)
sd.X <- sd(df1$X)
U <- 45
L <- 35
point.Cp.cf <- (U - L) / (6*sd.X)
UCL.Cp.cf <- ((U - L)/6)*sqrt(qchisq(p=conf,df=n-1)/((n-1)*sd.X**2))
#
#
#
n.reps <- 2000
for (i in 1:n.reps){
```

Fig. 15.4 Confidence limits for Cp

```
  boot.X <- sample(df1$X,size=n,replace=TRUE)
  sd.boot[i] <- sd(boot.X)
  Cp.boot[i] <- (U-L) / (6*sd.boot[i])
  }
hist.Cp <- hist(Cp.boot,plot=FALSE)
hist.Cp$density <- 100*hist.Cp$counts / sum(hist.Cp$counts)
plot(hist.Cp,freq=FALSE,col="light gray",main="Histogram of Bootstrap Cp
Values",xlab="Cp Values",ylab="Percentage")
#
Cp.est <- quantile(x=Cp.boot,probs=c(0.50,conf))
point.Cp.boot <- Cp.est[1]
UCL.Cp.boot <- Cp.est[2]
#
point.Cp.cf
UCL.Cp.cf
point.Cp.boot
UCL.Cp.boot
UCL.Cp.boot
#####
```

Fig. 15.4 (continued)

then Cp = 1 and $\Pr\{L \leq X \leq U | \sigma_X\} \approx 0.9973$, assuming X is normally distributed. Do not confuse the capability index Cp with Mallow's C_p, the measure of overparameterization in linear models.

A sample estimate of Cp is:

$$\widehat{\text{C}}\text{p} = \frac{U - L}{6\text{sd}}$$

A confidence limit for Cp^{-1} is:

$$\widehat{\text{C}}^{-1}\text{p}(1-\alpha) = \frac{6}{U - L}\sqrt{\frac{(n-1)\text{sd}^2}{\chi^2_{1-\alpha}}}$$

To obtain an upper 100(1 − α)% confidence limit for Cp, use the reciprocal of $\widehat{\text{C}}^{-1}\text{p}(1-\alpha)$:

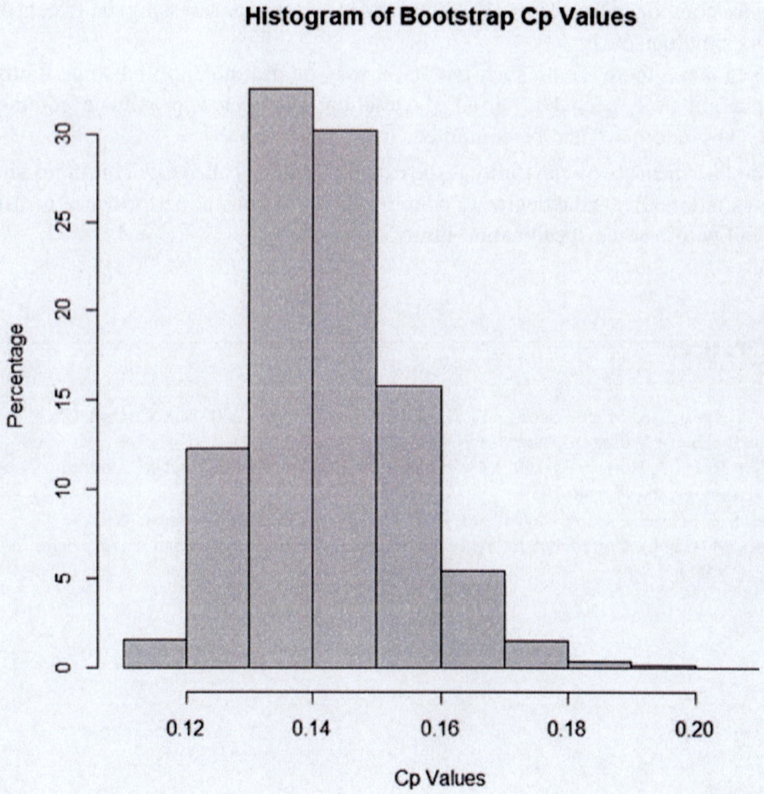

Fig. 15.5 Histogram of bootstrap Cp values

$$\widehat{\text{Cp}}(1-\alpha) = \frac{U-L}{6}\sqrt{\frac{\chi^2_{1-\alpha}}{(n-1)\text{sd}^2}}$$

Figure 15.4 has R code for computing both the closed form and bootstrap confidence limits for Cp.

Figure 15.5 shows a histogram of bootstrap Cp values.

15.5 Summary

It happens that variation in product performance is of greater concern than average performance. There are several ways of estimating measures of imprecision, or variation. When there are factors suspected of contributing to variation, and sometimes simultaneously some operating conditions or even competing product designs, variance components analysis (VCA) provides a means of assessing the effects of all factors simultaneously.

Even when there are no such factors, it may be desirable to find some limits on what might be expected in terms of variation. For that, approximate confidence limits on σ and $c = \frac{\sigma}{\mu}$ can be computed.

Another measure of variability is the capability index called Cp. This index shows how variation affects the degree to which part dimensions or performance measures will fall within some specification limits.

References

Kang, C. W., Lee, M. S., Seong, Y. J., & Hawkins, D. M. (2007). A Control Chart for the Coefficient of Variation. *Journal of Quality Technology, 39*(2).

Kushler, R. H., & Hurley, P. (1992). Confidence bounds for capability indices. *Journal of Quality Technology, 24*(4), 188–195.

Searle, S. R., Casella, G., & McCulloch, C. E. (1992). *Variance components*. Wiley.

Vangel, M. (1996). Confidence intervals for a normal coefficient of variation. *American Statistician, 15*(1), 21–26.

Chapter 16
Time Series and Dynamic Systems

Abstract When observations are not independent of each other, special methods are necessary for estimation and model-building. The Box-Jenkins approach is described for continuously valued response variables. The notion of Markov chains is also described for discretely valued variables.

16.1 Introduction

Most of the discussions in this text have been assuming that data form a random sample. That is to say, the set of observations $y_1, y_2, \ldots, y_n$ are all independent of each other, and they all have come from the same population. Sometimes, when data are gathered over time, the ith measurement, y_i, is correlated with the previous values. Such correlation is referred to as "autocorrelation" and requires special considerations.

16.2 Time Series

Time series will be defined here as a sequence of continuously valued random variables. For this text, we will assume that the variables are all normally distributed. The interesting part is that the values in the sequence are assumed to have non-zero correlations with each other. That is to say, the value observed at time t will have some correlation with previous values. These correlations will be exploited to help identify the underlying relationship that the values have with each other over time. Since the sequence is generally a set of values of the same measurement, but at different times, the correlations will be referred to as autocorrelations.

Suppose we observe a variable at uniformly spaced time intervals, for n intervals, and represent these values as:

S. A. Pardo, *Statistical Methods and Analyses for Medical Devices*,
https://doi.org/10.1007/978-3-031-26139-8_16

$$y_1, y_2, y_3, \ldots, y_t, y_{t+1}, \ldots, y_n$$

Mathematical statisticians would think of this as a realization of a sequence of random variables (Grimmett & Stirzaker, 2004). That is to say, suppose Y_1 was the first variable in the sequence. The value observed for Y_1 would be y_1. However, before Y_1 is observed, there are an infinite number of possible values it might have. Once Y_1 is observed, we think of it as being "realized." Suppose we further suspect that the value of y at time t is some unknown function not of time itself, but of the previous values of y, such as:

$$y_t = \beta_0 + \beta_1 y_{t-1} + \beta_2 y_{t-2} + a_t$$

The β_i are (unknown) constants, and a_t is a random "noise" variable with a mean value of 0 and a standard deviation, σ, that does not change with time. Furthermore, the a_t are assumed to be independent of each other as well as identically distributed. We will almost always assume that the noise variable is normally distributed but the fixed mean and standard deviation assumptions are the most critical. It is also possible that the value of y at the "current" time, t, is expressed as a function of noise values at previous times:

$$y_t = \alpha_0 + \alpha_1 a_{t-1} + \alpha_2 a_{t-2} + a_t$$

Again, a_t is a random "noise" variable, and the α_i are constants. Think of the values a_{t-1} and a_{t-2} as random variables whose values have (somehow) already been observed and recorded. The values of a_t and y_t have not yet been observed. We will restrict the discussion to those time series that are a linear function of previous values and noise values.

Both of the expressions described above are what we will call "second order," since they only reach two steps back in time. The first expression is referred to as an autoregressive model (symbolized as AR); the second is referred to as a moving average (symbolized as MA) model (Cryer, 1986). It is also possible that the future values of the series are related to both the series past values and the past values of the random "noise":

$$y_t = \delta + \beta_1 y_{t-1} + \beta_2 y_{t-2} + \alpha_{t-1} a_{t-1} + \alpha_{t-2} a_{t-2} + a_t$$

Such a "hybrid" expression, or model, is called an autoregressive moving average (ARMA). The order of the models (how far back they reach in time) is designated with integers, e.g., AR(2) is a second-order autoregressive model, MA(2) is a second-order moving average model, and ARMA(2,2) is an ARMA model of orders 2 and 2.

16.3 Identifying Time Series Model Types and Orders

Box and Jenkins (1976) discovered a methodology for identifying an appropriate model for time series. Their method involves computing two functions of the series data, called the autocorrelation function (acf) and partial autocorrelation function (pacf). First, some definitions are required. Recall the notion of expectation, or expected value. The covariance of two random variables, X and Y, was defined earlier to be:

$$\mathrm{Cov}(X,Y)=E[(X-\mu_X)(Y-\mu_Y)]=\int_{-\infty}^{+\infty}\int_{-\infty}^{+\infty}(x-\mu_X)(y-\mu_Y)f(x,y)dxdy$$

The correlation between X and Y is a sort of standardized covariance:

$$\mathrm{Corr}(X,Y)=\frac{\mathrm{Cov}(X,Y)}{\sigma_X\sigma_Y}$$

One could show that the correlation function is always a number between -1 and $+1$. In general, the correlation is positive if one variable increases when the other increases. It is negative if one variable decreases when the other increases. For time series, we are interested in how the values of the variable are related to values of the same variable, but observed at different times. So, for example, consider the values of a series at time t and at time $t-k$. The covariance would be:

$$\begin{aligned}\mathrm{Cov}(Y_t,Y_{t-k}) &= E\left[\left(Y_t-\mu_{Y_t}\right)\left(Y_{t-k}-\mu_{Y_{t-k}}\right)\right]\\ &= \int_{-\infty}^{+\infty}\int_{-\infty}^{+\infty}\left(y_t-\mu_{Y_t}\right)\left(y_{t-k}-\mu_{Y_{t-k}}\right)f(y_t,y_{t-k})dy_ydy_{t-k}\end{aligned}$$

Since this covariance is using the same variable, just observed at different times, it is called the autocovariance. Accordingly, the autocorrelation is:

$$\mathrm{Corr}(Y_t,Y_{t-k})=\frac{\mathrm{Cov}(Y_t,Y_{t-k})}{\sigma_{Y_k}\sigma_{Y_{t-k}}}$$

In this text, we will only concern ourselves with time series that have a particularly convenient property; namely, that the autocorrelation only depends on the length of time between the time points, k, and not the specific time itself, t. Such time series will be referred to as "stationary." This is not the strictest, most general definition of stationarity, but it is the one on which we will rely. Stationarity is a necessary condition for being able to identify the particular form and order of model to be fit to the data. The subscript k is called the "lag." If $k=0$, the autocorrelation is

1, and the covariance is the variance. For stationary time series, the covariance of lag k could be estimated as:

$$\widehat{\text{acov}}(k) = \frac{\sum_{t=k+1}^{n} (y_t - \bar{y})(y_{t-k} - \bar{y})}{n}$$

This expression is only a function of the lag, k. That is to say, the assumption of stationarity is required for this estimator to be meaningful. Also, the average, $\bar{y}$, is taken over all the observations, y_t, $t = 1, n$. Recall that this estimator is only a reasonable estimator if the time series is stationary.

The autocorrelation function (acf) can be estimated as:

$$r(k) = \widehat{\text{acf}}(k) = \frac{\widehat{\text{acov}}(k)}{\widehat{\sigma}^2}$$

where:

$$\widehat{\sigma}^2 = \frac{\sum_{t=1}^{n} (y_t - \bar{y})^2}{n}$$

A function of the lag that is closely related to the acf is called the partial autocorrelation function (pacf), mentioned earlier. The pacf is a sort of conditional acf and in general is the conditional expectation (Cryer, 1986):

$$\phi(k) = \text{pacf}(k) = \text{acf}\left(k | y_{t-1}, y_{t-2}, \ldots, y_{t-k+1}\right)$$

The estimated pacf values are obtained from the sample estimates of the acf, via a system of equations called the Yule-Walker equations (Box et al., 2008):

$$r(j) = \varphi_1 r(j-1) + \varphi_2 r(j-2) + \ldots + \phi(k) r(0), \ \ j = 1, k$$

That is to say, the coefficient for the 0th value of the acf is the estimate of the pacf at lag k. Of course, the 0th value of the acf is always 1. The unknown coefficients, φ_i, are actually the coefficients for an autoregressive model. The important point is that the acf and pacf are functions of the lag for stationary time series. The estimates of these functions will be used to identify a model to fit to the data.

16.4 The Box-Jenkins Approach

As mentioned earlier, Box and Jenkins discovered that different forms of ARMA times series had specific morphologies for their associated acf and pacf functions. Thus, if one can estimate the acf and pacf functions of a time series, one can then determine the appropriate time series model to fit to the data. Moving average models of order k (MA(k)) models have acf functions that "cut off" (i.e., drop to zero) after lag k, and pacf functions that "taper off" toward zero as the lag gets larger and larger. Autoregressive time series of order k (AR(k)) will have acf functions that taper off to zero as the lag gets large and pacf functions that will drop to zero after lag k. An autoregressive, moving average model of orders k and l (ARMA(k,l)) will have a pacf function that drops to zero after lag k and an acf function that drops to zero after lag l.

To illustrate the acf and pacf of various time series, we simulated four different time series. The first is called "white noise"; it is a non-autocorrelated sequence of random variables all having exactly the same distribution, generally with mean 0 and some standard deviation. If the variables are normally distributed, as in the case simulated here, the sequence is referred to as Gaussian white noise. In the case of white noise, we would expect the values of the acf and pacf to be patternless and relatively small. Figures 16.1 and 16.2 show the acf and pacf for the white noise sequence. The dashed horizontal lines are lower and upper limits of approximate 95% confidence intervals for the acf and pacf, under the hypothesis that the true values are 0. Thus, any values on the graph that extend beyond the dashed lines are suspected to be repeatably non-zero.

The next series is purely autoregressive and of order 2, referred to as AR(2). Figures 16.3 and 16.4 show the acf and pacf, respectively.

The acf for the AR(2) series ought to oscillate and be "damped," that is, gradually decrease in magnitude as lag increases. The pacf ought to be "large" for lags 1 and 2 and then become small for lags 3 and beyond. The lag 0 acf is always equal to 1.0.

There is no lag 0 pacf value, inasmuch as pacf must be conditioned on previous lags, and nothing comes before lag 0. Technically, negative lags are possible in some cases, but we will not consider them here.

The acf and pacf for a pure moving average process of order 2, MA(2), are shown in Figs. 16.5 and 16.6, respectively.

It is not necessarily clear that an acf or pacf "cuts off" at a particular lag, or is "damped."

Finally, Figs. 16.7 and 16.8 show the acf and pacf for a mixed autoregressive, moving average model or order 2,2, or ARMA(2,2). In theory both the acf and pacf should drop to 0 (i.e., fall within the 95% confidence limits) after lag 2.

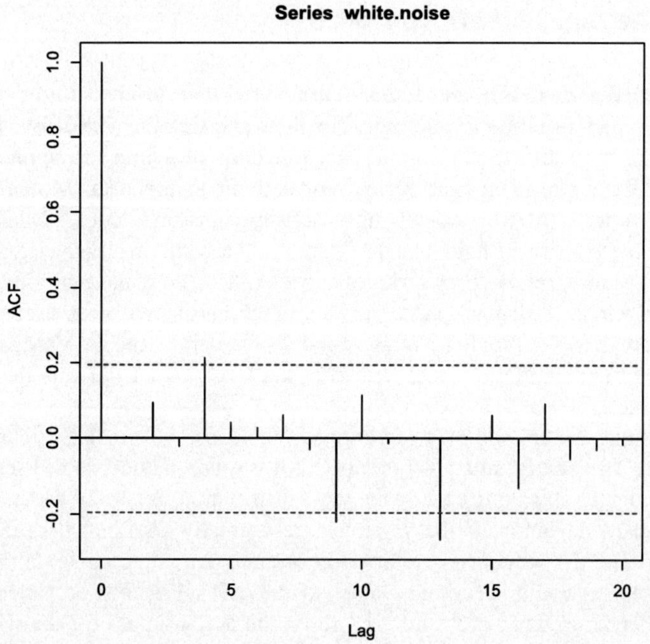

Fig. 16.1 ACF of white noise

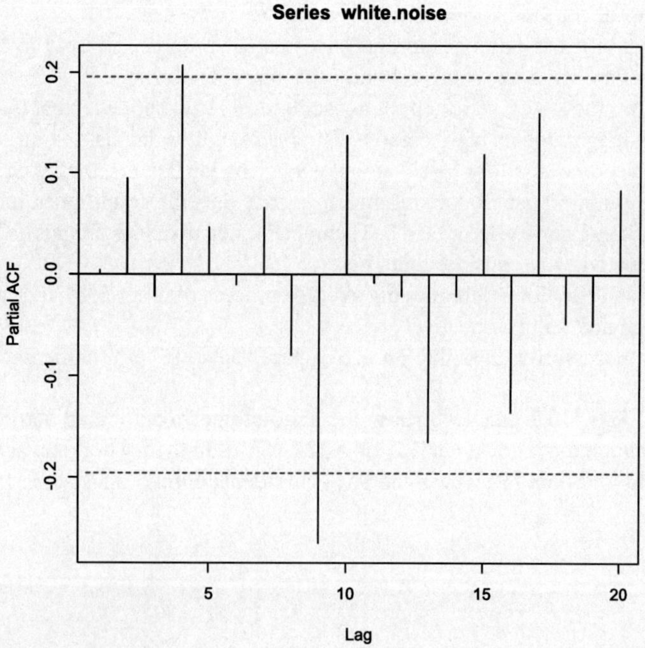

Fig. 16.2 PACF of white noise

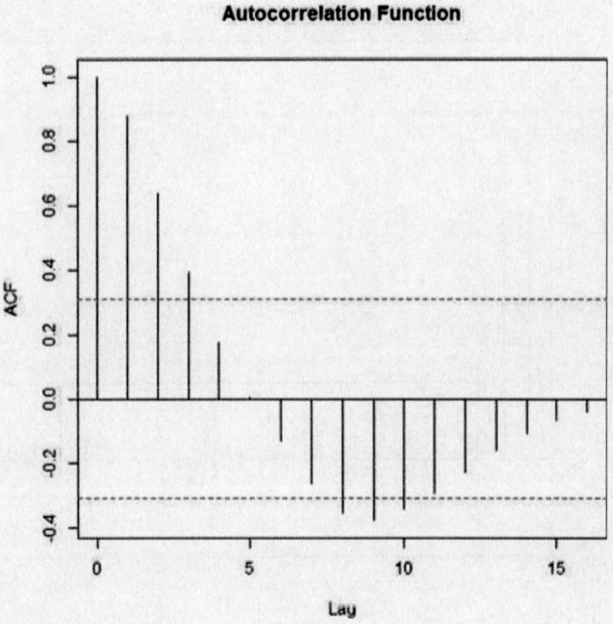

Fig. 16.3 ACF of AR(2)

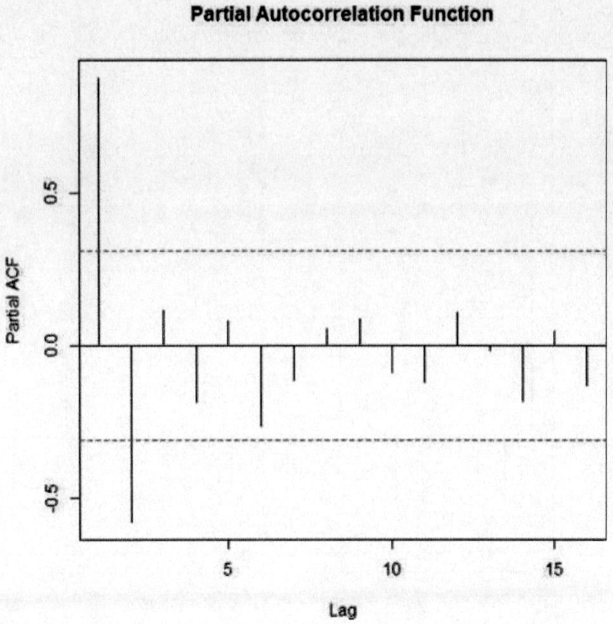

Fig. 16.4 PACF of AR(2)

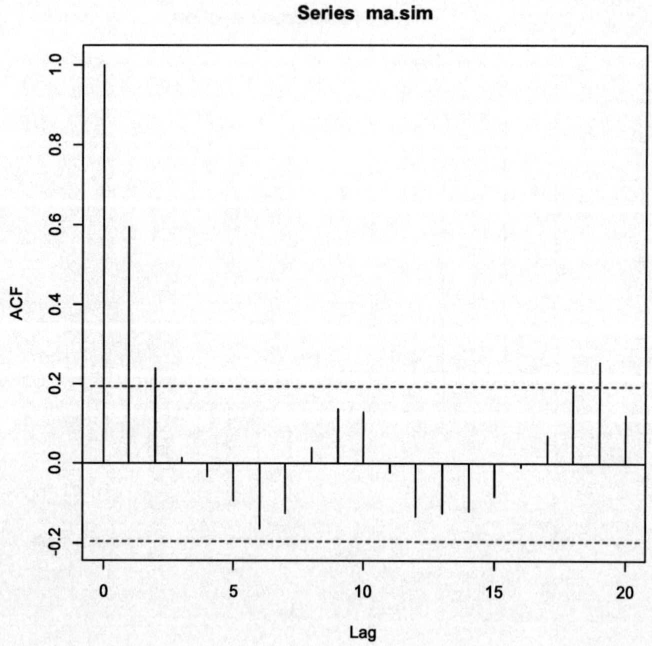

Fig. 16.5 ACF for MA(2)

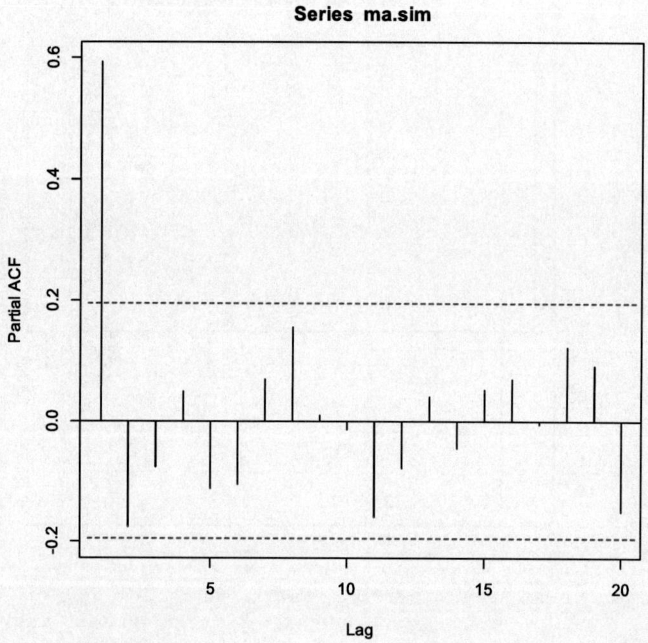

Fig. 16.6 PACF for MA(2)

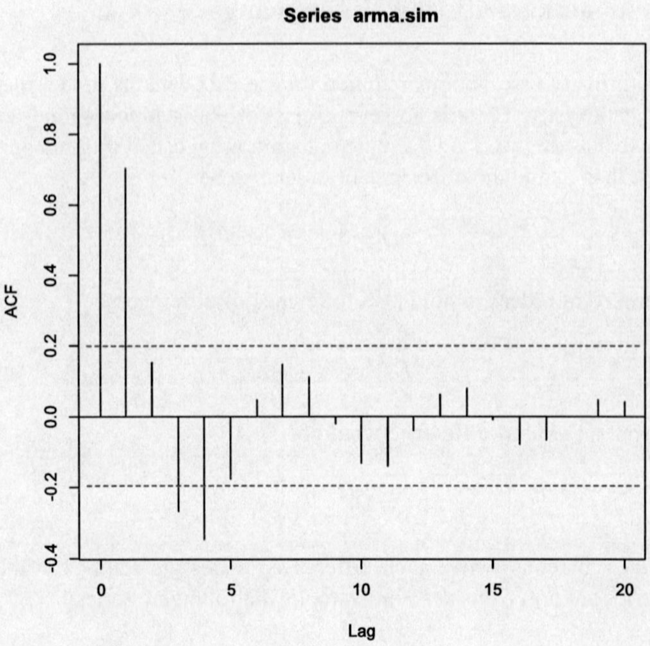

Fig. 16.7 ACF for ARMA(2,2)

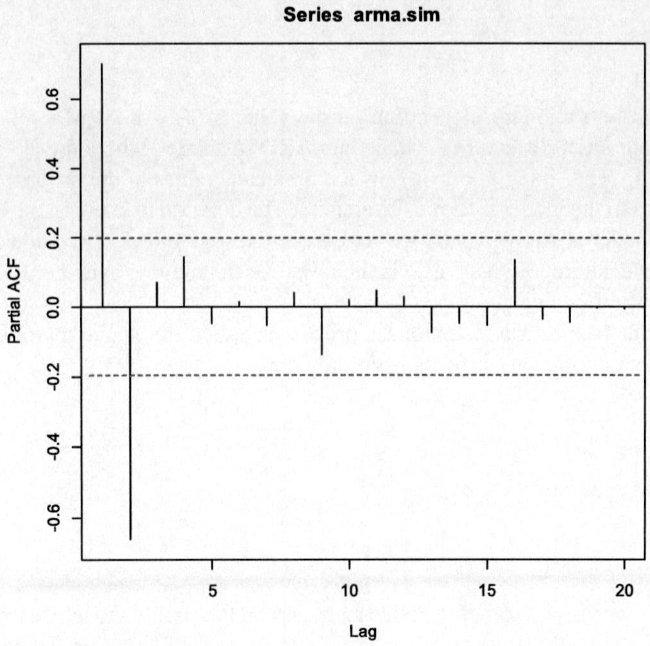

Fig. 16.8 PACF for ARMA(2,2)

16.5 Non-stationarity and Differencing

One of the primary assumptions required for the Box-Jenkins model identification process is stationarity. There is however a class of non-stationary time series where stationarity can be induced via a simple transformation called differencing. If Y_t is a time series, then define the difference of order 1 to be:

$$D_t = Y_t - Y_{t-1}$$

A difference of order 2 would be a difference of differences:

$$D_t^{(2)} = D_t - D_{t-1}$$

In general, an order d difference would be:

$$D_t^{(d)} = D_t^{(d-1)} - D_{t-1}^{(d-1)}$$

There are situations where such differenced series are stationary although the original series are not. Consider, for example, the following series:

$$y_t = y_{t-1} + \beta_2 y_{t-2} + a_t$$

This series is in fact non-stationary. However, the series:

$$d_t = y_t - y_{t-1} = \beta_2 y_{t-2} + a_t$$

may in fact be stationary, depending on the value of β_2. An ARMA model that is based on kth order differences is called an ARIMA model, where the "I" stands for "integrated." The "integration" is how the predicted values of the original series are obtained from the model fit to the differences. The notation to express the order of an ARIMA model is ARIMA(k,d,l) where, as in the case of ARMA models, k is the order for the autoregressive part, l is the order for the moving average part, and d is the order of differencing.

A special case of a non-stationary process is called the "random walk" (Cryer, 1986). The random walk process is described as:

$$y_t = y_{t-1} + a_t$$

The first-order difference is:

$$d_t = y_t - y_{t-1} = a_t$$

In other words, differencing yields a process that is simply noise. Hence, random walks are inherently unpredictable. In fact, the best prediction or forecast for the process value at time t is whatever its value was at time $t - 1$. The fact that the value at time t only depends on the immediately preceding time, $t - 1$, makes the random walk something called a Markov process (Grimmett & Stirzaker, 2004).

16.6 An Example of a Time Series Analysis

During any production day, nonconforming items are produced. The average hourly production of nonconforming items was recorded for 40 days. The autocorrelation and partial autocorrelation functions were computed. Figure 16.9 shows the autocorrelation function, and Fig. 16.10 shows the partial autocorrelation function for this series. Based on the autocorrelation and partial autocorrelation functions, it seems that an autoregressive model of order 2, AR(2), is suggested. Figure 16.11 shows the script for fitting the time series model. Figure 16.12 shows the time series plot of the data together with the fitted values from the model.

As another example, consider a random walk process. Figure 16.13 shows an R script for simulating a random walk process. Figures 16.14, 16.15, and 16.16 show the time plot of the data, ACF, and PACF, respectively. The PACF only has a lag

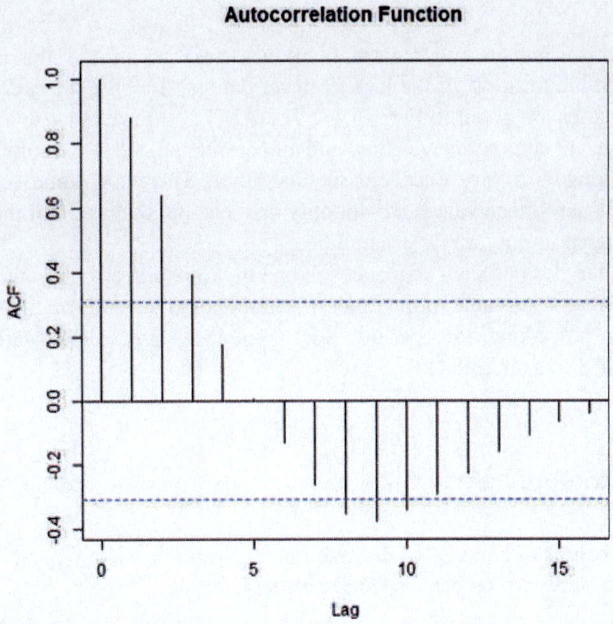

Fig. 16.9 Autocorrelation Function

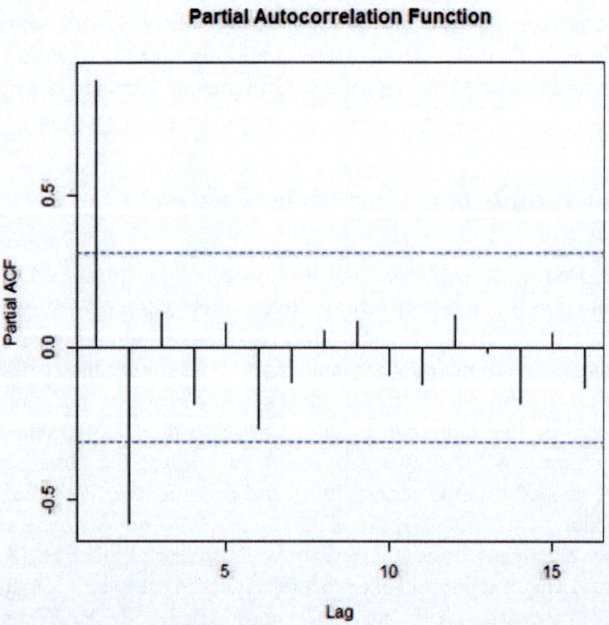

Fig. 16.10 Partial autocorrelation function

value of 1 (this is always true, regardless of the series) and an ACF that is very large in absolute value for most of the lags (it never damps). Finally, the estimates of the parameter values are given in Fig. 16.17. The AR(1) estimate is close to 1.

When data are continuously valued and autocorrelated, the Box-Jenkins approach to model fitting is a very excellent methodology. There are some sequences of random variables whose values are not only discrete, but the extent of their possible values is limited to only a few possibilities.

A particular class of such sequences is one in which the correlation/dependence on the current state or value of the random variable only depends on the value in the immediately preceding time period. Such sequences are called Markov chains (Grimmett & Stirzaker, 2004).

16.7 Markov Chains

A Markov chain is a sequence of discrete random variables, $X_1, X_2, \ldots$, changing in discrete time steps, which has a special property:

```
setwd("<your path here>")
df1 <- read.csv("20220601 Ch 16 Time Series.csv")
#
library(forecast) #Contains the function Arima()
#
#Variables:
#
# time.index
# Events.per.Hour
#
attach(df1)
acf(Events.per.Hour,main="Autocorrelation Function")
dev.new()
pacf(Events.per.Hour,main="Partial Autocorrelation Function")
dev.new()
arima(Events.per.Hour,order=c(2,0,3))
arma.model2 <- Arima(y=Events.per.Hour,order=c(2,0,3))
fitted.2 <- arma.model2$fitted
arma.model3 <- Arima(y=Events.per.Hour,order=c(2,0,0))#this is the model indicated
by the Box-Jenkins approach
fitted.3 <- arma.model3$fitted
plot(x=time.index,y=Events.per.Hour,main="Time Series Data - Ave. Events per
Hour",
    xlab="Day",pch=1,col="black",ylim=c(1,16),yaxt="n")
axis(side=2,at=seq(from=1,to=16,by=1))
#points(x=time.index,y=fitted.2,pch=17,col="red1")
points(x=time.index,y=fitted.3,pch=18,col="blue")
#####
```

Fig. 16.11 R code for fitting the AR(2) model

$$\Pr\{X_k = x_k | X_0 = x_0, X_1 = x_1, \ldots X_{k-1} = x_{k-1}\} = \Pr\{X_k = x_k | X_{k-1} = x_{k-1}\}$$

In other words, the probability that the current value in the sequence has any particular value only depends on the immediately preceding value and not on the entire "history" of the sequence. This property is called the "Markov property" (Grimmett & Stirzaker, 2004).

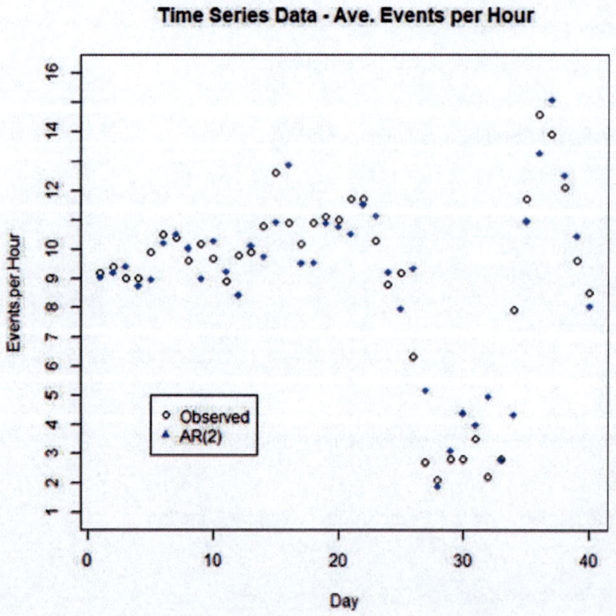

Fig. 16.12 Time series plot of observed and fitted values

As mentioned earlier, the random walk is a Markov process, which is more general than a Markov chain, except the random variables may be continuously valued.

For the sake of simplifying notation, we will refer to the values, or "states" of the X_k, as 1, 2, 3, ..., e.g., $X_k = 2$, although there is not necessarily any order to the states. For example, a traffic light can have the "states" of red, yellow, or green. We could arbitrarily assign to the "red" state the value 1, "yellow" could be 2, and "green" could be 3. For further simplification, we will use the notation for the conditional probabilities:

$$\Pr\{X_k = j | X_{k-1} = i\} = p_{ij}$$

For processes having the Markov property, p_{ij} is the conditional probability that the next state will be j given that the current state is i. These conditional probabilities are called transition probabilities. Most of the chains we will discuss have only a finite number of possible states, as in the case of traffic lights. Thus, the one-step transition probabilities for an n-state chain can be represented by a (finite-dimensional) $n \times n$ matrix:

```
setwd("<your path here>")
#
#
#
library(forecast)
Y.response <- c()
set.seed(52)
#
Y.response[1] <- 10
T.series <- 100
time <- seq(from=1,to=T.series,by=1)
for(i in 2:T.series) {
  Y.response[i] <- Y.response[i-1] + rnorm(n=1,mean=0,sd=1.5)
  }
#
plot(x=time,y=Y.response,main="Time Plot of Random Walk Process")
dev.new()
acf(Y.response,main="ACF of Random Walk")
dev.new()
pacf(Y.response,main="PACF of Random Walk")
#
arima(Y.response,order=c(1,0,0))
#####
```

Fig. 16.13 R script for simulating a random walk

$$\boldsymbol{P} = \begin{bmatrix} p_{11} & p_{12} & \cdots & p_{1n} \\ \vdots & \ddots & \ddots & \vdots \\ p_{n1} & p_{n2} & \cdots & p_{nn} \end{bmatrix}$$

As we have defined the transition probability matrix, each row represents a conditional distribution for the states $j = 1, n$, i.e.:

$$\sum_{j=1}^{n} p_{ij} = 1$$

Keep in mind that it is thoroughly possible that there are some states such that:

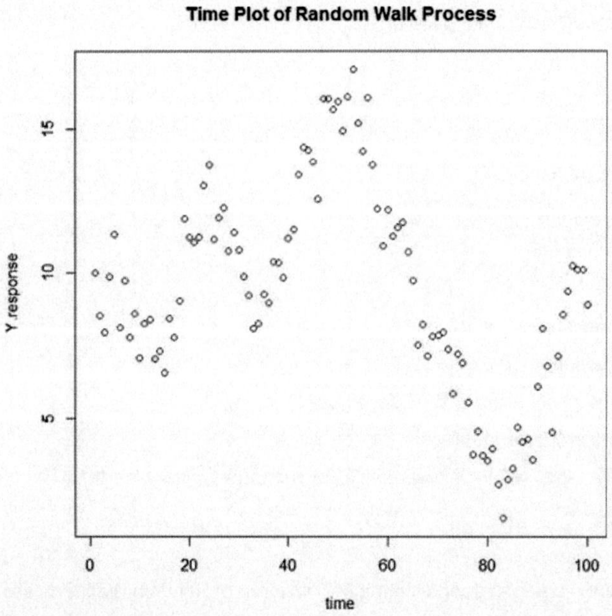

Fig. 16.14 Time plot of a random walk

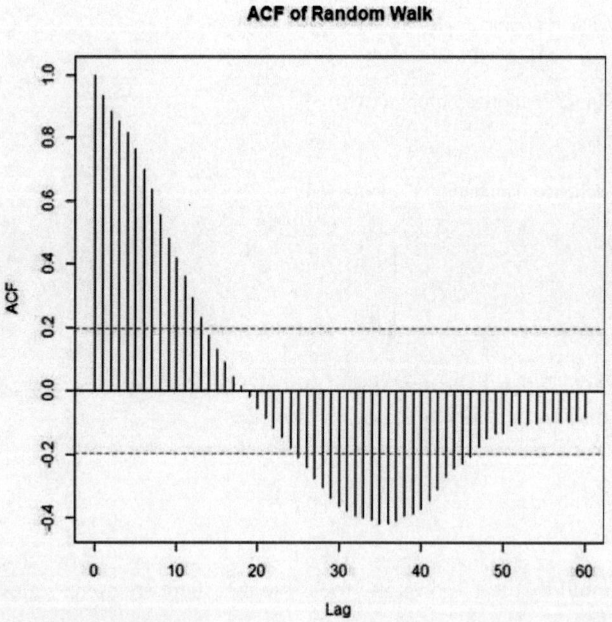

Fig. 16.15 ACF of a random walk

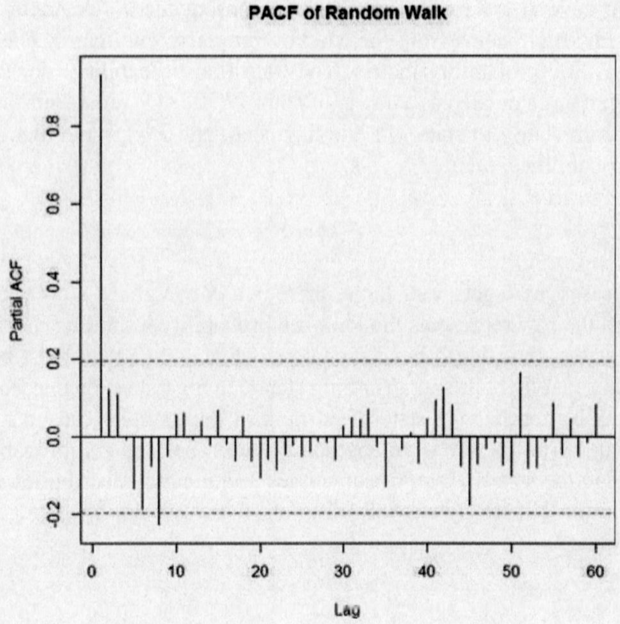

Fig. 16.16 PACF of a random walk

```
Call:
arima(x = Y.response, order = c(1, 0, 0))

Coefficients:
         ar1   intercept
      0.9246      9.4467
s.e.  0.0342      1.5117

sigma^2 estimated as 1.618:  log likelihood = -166.91,  aic = 339.83
```

Fig. 16.17 Parameter estimates for a random walk

1. $p_{ij} = 1$
2. $p_{ij} = 0$
3. $p_{ij} = 0$ for $j \neq i$ and $p_{ii} = 1$

The condition in (3) is referred to as "absorbing," so the state i is called an "absorbing state." That is to say, once the chain reaches state i, it can never transition to any other state.

There are several interesting results and consequences related to finite-state Markov chains. It all stems from the Markov property. The matrix $\boldsymbol{P}$ is called the one-step transition probability matrix. The transition probabilities for 2, 3, 4, etc. steps ahead can be calculated as a power of matrix $\boldsymbol{P}$. If $p_{ij}(1)$ represents the transition probability from state i to state j in one time step, and $\boldsymbol{P}(k)$ is the matrix of k-step transition probabilities, then:

$$\boldsymbol{P}(k) = \boldsymbol{P}^k$$

In many cases, as k gets very large, there is a convergence so that each row is identical, and the row represents the long-run probability of finding the "system" in each of the states. In order to have such convergence, the chain must be "irreducible," meaning that there are no absorbing states. In a gambling game, for example, once the gambler reaches the state of no money, the gambler can no longer play. Thus, the state of no money is an absorbing state. The long-run probability distribution is called the steady-state, or sometimes "stationary," distribution and can be calculated. From this steady-state distribution, the expected, or average, state can also be calculated.

16.8 Steady-State, or "Stationary," Distribution

To compute the stationary distribution, the one-step transition matrix must be known (or at least estimated from data). Table 16.1 shows the one-step transition matrix for a four-state Markov chain.

Figure 16.18 shows an R script for computing the stationary distribution.

See the Appendix A, Sect. A.14, for the formulas to compute powers of a matrix.

Figure 16.19 shows the output from this script.

Table 16.1 A one-step transition matrix: four-state Markov chain

	S1	S2	S3	S4
S1	0.879	0.060	0.060	0.001
S2	0.000	0.110	0.840	0.050
S3	0.000	0.840	0.110	0.050
S4	0.001	0.000	0.000	0.999

```
setwd("<your path here>")
df1 <- read.csv("20220602 Ch 16 Transition Probability Matrix.csv")
#
#
# VARIABLES:
# S1
# S2
# S3
# S4
#
attach(df1)
Amat <- as.matrix(df1)
Smat <- eigen(Amat)
Smat$values
Smat$vectors
Lamda <- diag(Smat$values)
#
# This will compute the stationary distribution (If It exists)
# of a finite, irreducible Markov Chain
# if the input matrix is a one-step transition matrix
# for that chain
#
n <- 600 #number of transitions to get to steady state
Lamdan <- diag(Smat$values**n)
Eig <- as.matrix(Smat$vectors)
EigInv <- solve(Eig)
Arecover <- Eig%*%Lamda%*%EigInv
An <- Eig%*%Lamdan%*%EigInv #The rows of An are identical, and are the
stationary #                                    distribution for the chain
Lamda
Eig
Amat
Arecover
n
An
#
# rows of An will be the stationary distribution if n is large enough
#####
```

Fig. 16.18 Computing the stationary distribution of a Markov chain

Fig. 16.19 Computing the stationary distribution

```
> Lamda
     [,1]    [,2]        [,3]     [,4]
[1,]    1 0.0000000 0.0000000  0.00
[2,]    0 0.9482929 0.0000000  0.00
[3,]    0 0.0000000 0.8797071  0.00
[4,]    0 0.0000000 0.0000000 -0.73
> Eig
     [,1]       [,2]         [,3]          [,4]
[1,]  0.5  0.77436976 -0.999929318  1.957048e-18
[2,]  0.5  0.44727970 -0.005962298 -7.071068e-01
[3,]  0.5  0.44727970 -0.005962298  7.071068e-01
[4,]  0.5 -0.01527141  0.008382139 -1.131896e-21
> Amat
        S1   S2   S3    S4
[1,] 0.879 0.06 0.06 0.001
[2,] 0.000 0.11 0.84 0.050
[3,] 0.000 0.84 0.11 0.050
[4,] 0.001 0.00 0.00 0.999
> Arecover
              [,1]          [,2]          [,3]  [,4]
[1,]  8.790000e-01  6.000000e-02  6.000000e-02 0.001
[2,] -5.194350e-18  1.100000e-01  8.400000e-01 0.050
[3,] -5.213991e-18  8.400000e-01  1.100000e-01 0.050
[4,]  1.000000e-03  9.367448e-17  3.556242e-17 0.999
> n
[1] 600
> An
            [,1]        [,2]        [,3]      [,4]
[1,] 0.008038585 0.009646302 0.009646302 0.9726688
[2,] 0.008038585 0.009646302 0.009646302 0.9726688
[3,] 0.008038585 0.009646302 0.009646302 0.9726688
[4,] 0.008038585 0.009646302 0.009646302 0.9726688
```

16.9 Extensions of Markov Chains

There are many extensions to the ideas of Markov chains. We will only mention two that have gained some popularity: higher-order chains and hidden Markov models (HMMs). As stated earlier, the Markov property for a stochastic process states that the probability of the chain being in a certain state at time k depends entirely on the state it was in at the immediately preceding time, $k - 1$. A higher-order chain is one in which that dependence is extended to $k - 1$, $k - 2$,..., $k - m$. So, for a second-order chain:

$$\Pr\{X_k = x_k | X_0 = x_0, X_1 = x_1, \ldots X_{k-1} = x_{k-1}\}$$
$$= \Pr\{X_k = x_k | X_{k-1} = x_{k-1}, X_{k-2} = x_{k-2}\}.$$

When such higher-order dependencies exist, the properties that make Markov chains so useful (e.g., finding stationary distributions) disappear. Thus, computing with higher-order processes becomes far more difficult. One possible way of finding limiting distributions of a higher-order chain's states is via simulation. This, however, is beyond the scope of this text.

Another extension is the notion of hidden Markov models. A hidden Markov model (HMM) is one whose states are not directly observable, but they depend on another chain whose states are observable. Suppose, for example, the change from season to season in the population size of some kind of grub cannot be assessed, but the change in the population size of a bird that feeds on those grubs can. If the state is population size, and the population size of the birds changes in a Markov fashion, then perhaps the change in the grub population size can be assessed indirectly.

While these extensions are beyond the scope of this book, the interested reader can find a straightforward tutorial by Fosler-Lussier (1998).

16.10 Summary

Time series are sequences of random variables whose values are potentially affected by previous values in the sequence. Such dependencies are referred to as autocorrelations. Box and Jenkins formulated a means of classifying time series in terms of those autocorrelations. The system is referred to as auto-regressive integrated moving average (ARIMA) models. Time series are a type of dynamic, stochastic processes. While ARIMA models involve continuously valued variables, there are stochastic processes with discrete variables. A particularly important class are called Markov processes, or Markov chains (with sequences in discrete steps). Markov processes have the convenient property that the current value is only dependent (at most) on the immediately proceeding time point or step. Dynamic processes arise in manufacturing, such as continuous production lines, and in biology, such as EKGs.

References

Box, G.E.P. & Jenkins, G.M., (1976) *Time Series Analysis: forecasting and control* (revised Ed.) Holden-Day, Oakland.

Box, G.E.P., Jenkins, G.M. & Reinsel, G.C. (2008). *Time Series Analysis: forecasting and control* (4th ed.) John Wiley and Sons, Oxford.

Cryer, J. D. (1986). *Time series analysis*. Duxbury Press.

Fosler-Lussier, E. (1998). *Markov models and hidden Markov models: A brief tutorial, TR-98-041*. International Computer Science Institute.

Grimmett, G., & Stirzaker, D. (2004). *Introduction to probability and random processes* (3rd ed.). Oxford University Press.

Chapter 17
Odds, Odds Ratios, and Comparing Proportions

Abstract Odds and odds ratios can be helpful when comparing the likelihoods to an event under two different conditions. This chapter emphasizes the use of resampling methods for comparing likelihoods of discrete events.

17.1 Difference Between Proportions and Odds Ratios

Comparing two proportions seems like it ought to be fairly simple. In fact, it is not. There is an approximate test for the hypothesis that two probabilities are equal, namely:

$$H_0 : P_1 = P_2$$

The test statistic is based on a normal approximation (Fleiss et al., 2003). The statistic is:

$$Z_{\text{diff}} = \frac{p_1 - p_2}{\sqrt{\frac{p_1(1-p_1)}{n_1} + \frac{p_2(1-p_2)}{n_2}}}$$

where p_1 is the sample proportion of "successes" in group 1 (size n_1), and p_2 is the sample proportion of "successes" in group 2 (size n_2).

If Z_{diff} is outside the interval $(z_{\alpha/2}, z_{1-\alpha/2})$, then H_0 is rejected, that is, if it is either in the $100(\alpha/2)$ or $100(1 - \alpha/2)$ percentiles of a standard normal distribution, then reject the null hypothesis, H_0. This test is based on the assumption that the statistic Z_{diff} is at least approximately normally distributed, which may be violated.

Another way to compare proportions is by odds and odds ratios. If P is the probability of a "success" event, then the odds of the event are:

$$O = \frac{P}{1-P}$$

S. A. Pardo, *Statistical Methods and Analyses for Medical Devices*,
https://doi.org/10.1007/978-3-031-26139-8_17

The odds give the likelihood of the event relative to the likelihood of its compliment, e.g., the event not occurring or not observed. You may recall previously that the logarithm of odds is called log odds and was useful in formulating the logistic regression model. In any case, the odds have the interpretation that if the odds are greater than 1, then there is a greater chance of the "success" event happening (or being observed) than not. If the odds are less than 1, then there is a lesser chance of the "success" event happening (or being observed) than not. If the odds are exactly 1, then it is just as likely for the event to occur as not.

Some people prefer to ask the question, "What percent is the event of interest more (or less) likely to occur than to not occur?" To answer that question, use the following transformation of odds:

$$\pi\% = \left(1 - \frac{1}{O}\right) 100\%$$

For example, suppose 99 out of 100 items conform to specifications. Then the odds of conformance are:

$$O_{\text{conform}} = \frac{99/100}{1/100} = 99$$

So, conformance is 99 times more likely than not. Therefore:

$$\pi\% = \left(1 - \frac{1}{O_{\text{conform}}}\right) 100\% = \left(1 - \frac{1}{99}\right) 100\% \approx 98.99\%$$

And you can say that it is (approximately) 98.99% more likely that the item will be conforming as not. Similarly, if the odds are less than 1, then the event is less likely to occur. So, for example, if 1 out of 100 items is nonconforming, then:

$$O_{\text{non-conform}} = \frac{1/100}{99/100} \approx 0.0101$$

Thus:

$$\pi\% = \left(1 - \frac{1}{O_{\text{non-conform}}}\right) 100\% \approx \left(1 - \frac{1}{0.0101}\right) 100\% \approx -98.01\%$$

In other words, it is 98.01% less likely to be nonconforming.

For the most part, in this book, we will just refer to odds as the *number* of times the event is more likely to occur.

The ratio of odds can be used to provide a comparison of two proportions. Suppose that O_1 represents the odds of a particular device performing adequately and O_2 the odds of adequate performance for a similar device. The odds ratio (OR) is:

$$\mathrm{OR} = \frac{O_1}{O_2}$$

If OR $>$ 1, then the first device (call it device 1) is more likely to perform adequately compared to the other (device 2).

As in the case of the difference of two proportions, the sampling distribution of OR is fairly intractable. Fleiss et al. (2003) give an approximate confidence interval formula for OR. A simpler method is via bootstrap. The code for computing bootstrap confidence intervals for both OR and the difference in proportions, $p_1 - p_2$, was given in Fig. 17.1.

Permutation methods can be used to test the hypotheses:

$$H_0 : p_1 - p_2 = 0$$

or:

$$H_0 : \frac{O_1}{O_2} = 1$$

Figure 17.1 presents an R code for performing permutation tests for both OR and $p_1 - p_2$. Figure 17.2 shows the null permutation distribution for the difference in proportions and Fig. 17.3 the null permutation distribution for OR.

In the example data, $n_1 = n_2 = 80$, and $x_1 = 78$, $x_2 = 60$. That is to say, both groups had a sample size of 80, group 1 had 78 "successes" out of 80, whereas group 2 had only 60. In both OR and $p_1 - p_2$, the test statistics were beyond the extreme tails of the null permutation distributions, so in both cases H_0 is rejected.

An example from chemistry provides an example of how odds ratios might be used. A molecule is called chiral if it has two mirror-image versions (called R and S enantiomers) that cannot be geometrically transformed to be identical to each other; it is said that the two enantiomers cannot be "superimposed" on each other. Often in chemical synthesis, a chiral molecule is the desired product, but only one of the enantiomers is desirable. However, the synthesis process may produce both enantiomers. In the case of medical chemistry, the enantiomer is critical. Such was the case of thalidomide, a chiral molecule. The R-thalidomide enantiomer aids in curing indigestion; S-thalidomide causes devastating and deadly birth defects.

Suppose that two synthetic chemical processes produce the same chiral molecule, but it is not known which one produces more of the R-enantiomer, which is the desired chiral partner. So, chemists perform and experiment, producing a batch of the chiral molecule with each of the two processes. They assay each batch to determine the proportions of the R-enantiomer. In the batch from process A, 87%

```
options(na.action=na.omit)
#
# setwd tells R which folder to use for this script
#

setwd("<your path here>")
df1 <- read.csv("20220613 Ch 17 Two Proportions Summary Data.csv")
#
# Variables:
#  Group ( 1 and 2)
#  X
#  N
#
attach(df1)
OR <- c()
ODDS1 <- c()
ODDS2 <- c()
pDiff <- c()
samp1 <- c()
samp2 <- c()
one.zero.stacked <- c()
#
# The following are for equivalence limits,
# not necessary unless doing an equivalence test
#
P1 <- 0.99
P2 <- 0.95
Del <- abs(P1 - P2)
maxOR <- (P1/(1-P1)) / (P2/(1-P2))
minOR <- 0.90
#
# Now expand the summary data into vectors of 0's and 1's
#
one.1 <- rep(x=1,times=X[1])
zero.1 <- rep(x=0,times=N[1] - X[1])
one.zero.1 <- c(one.1,zero.1)
```

Fig. 17.1 Permutation test script for OR and $p_1 - p_2$

```
#
# This shuffles the original (expanded) data
#
rand.1 <- runif(length(one.zero.1),0,1)
rand.one.zero.1 <- one.zero.1[order(rand.1)]
#
one.2 <- rep(x=1,times=X[2])
zero.2 <- rep(x=0,times=N[2] - X[2])
one.zero.2 <- c(one.2,zero.2)
#
# This shuffles the original (expanded) data
#
rand.2 <- runif(length(one.zero.2),0,1)
rand.one.zero.2 <- one.zero.2[order(rand.2)]
#
one.zero.stacked <- c(one.zero.1,one.zero.2)

#
reps <- 10000

alpha <- 0.05
set.seed(26)
#

#
sampP1 <- X[1] / N[1]
sampP2 <- X[2] / N[2]
sampODDS1 <- sampP1 / ( 1 - sampP1)
sampODDS2 <- sampP2 / (1 - sampP2)
sampOR <- sampODDS1 / sampODDS2
sampDiff <- sampP1 - sampP2

#

for(i in 1:reps) {
  samp1 <- sample(x=one.zero.stacked,size=N[1],replace=TRUE)
  samp2 <- sample(x=one.zero.stacked,size=N[2],replace=TRUE)
```

Fig. 17.1 (continued)

```
permute1 <- mean(samp1)
permute2 <- mean(samp2)
if (permute1==1){
 ODDS1[i] <- 0
 }
else {
 ODDS1[i] <- permute1 / (1 - permute1)
 }
if (permute2==1) {
 ODDS2[i] <- 0
 }
else {
 ODDS2[i] <- permute2 / (1 - permute2)
 }
if (ODDS1[i]==0 | ODDS2[i]==0) {
 OR[i] <- NA
 }
else {
 OR[i] <- ODDS1[i] / ODDS2[i]
 }
pDiff[i] <- permute1 - permute2
}

#
# these are critical values for permutation tests
#
CL.pDiff <- quantile(pDiff,probs=c(alpha/2,1-alpha/2),type=8,na.rm=TRUE)
CL.OR <- quantile(OR,probs=c(alpha/2,1-alpha/2),type=8,na.rm=TRUE)
#
# CL = Critical Limits
#
LCL.pDiff <- CL.pDiff[1]
UCL.pDiff <- CL.pDiff[2]
LCL.OR <- CL.OR[1]
UCL.OR <- CL.OR[2]

#
# Histograms of Permutation "Null" Distributions
```

Fig. 17.1 (continued)

```
#
hist.OR <- hist(OR,plot=FALSE)

hist.diff <- hist(pDiff,plot=FALSE)
#
hist.OR$density <- 100*hist.OR$counts / sum(hist.OR$counts)
hist.diff$density <- 100*hist.diff$counts / sum(hist.diff$counts)
F.diff <- ecdf(pDiff)
F.OR <- ecdf(OR)
#
# Compute p-values
#
sides.test <- 2
#
if (sampDiff < 0) {
p.val.diff <- sides.test*F.diff(sampDiff)
 } else {
 p.val.diff <- sides.test*(1 - F.diff(sampDiff))
 }

#
if (sampOR < 1) {
 p.val.OR <- sides.test*(F.OR(sampOR))
 } else {
 p.val.OR <- sides.test*(1 - F.OR(sampOR))
 }

plot(hist.OR,freq=FALSE,col="light grey",main="Permutation Distribution of
OR",xaxt="n",yaxt="n",xlim=c(0,15),xlab="OR")
axis(side=1,at=seq(from=0,to=15,by=1)) #controls x-axis
axis(side=2,at=seq(from=0,to=40,by=1)) #controls y-axis
abline(v=sampOR,lty=1)
abline(v=LCL.OR,lty=2)
abline(v=UCL.OR,lty=2)
#abline(v=minOR,lty=3) # equivalence lower limit
#abline(v=maxOR,lty=3) # equivalence upper limit
legend(x=6,y=27,legend=c("Sample OR","Crit.Values"),lty=c(1,2))
```

Fig. 17.1 (continued)

```
#
# Use the following legend if performing equivalence test:
#
#legend(x=6,y=27,legend=c("Sample OR","Crit. Values","Equiv. Limit"),lty=c(1,2,3))

dev.new()
plot(hist.diff,freq=FALSE,col="light grey",main="Permutation Distribution of p1 -
p2",xaxt="n",yaxt="n",xlim=c(-0.50,0.40),xlab="p1-p2")
axis(side=1,at=seq(from=-0.50,to=0.40,by=0.01)) #controls x-axis
axis(side=2,at=seq(from=0,to=40,by=1)) #controls y-axis
abline(v=sampDiff,lty=1)
abline(v=LCL.pDiff,lty=2)
abline(v=UCL.pDiff,lty=2)
#abline(v=-Del,lty=3) # equivalence lower limit
#abline(v=Del,lty=3)  # equivalence upper limit

legend(x=-0.40,y=27,legend=c("Sample Diff","Crit. Values"),lty=c(1,2))
#
# Use the following legend if performing equivalence test:
#
#legend(x=0.062,y=27,legend=c("Sample Diff","Conf.Limit","Equiv.
Limit"),lty=c(1,2,3))

#
df2 <- cbind(OR,pDiff) # these are bootstrap samples

write.csv(df2,"20220510 Bootstrap Samples OR and pDiff.csv")
#
LCL.pDiff
UCL.pDiff
LCL.OR
UCL.OR
sampDiff
sampOR
p.val.diff
p.val.OR

#####
```

Fig. 17.1 (continued)

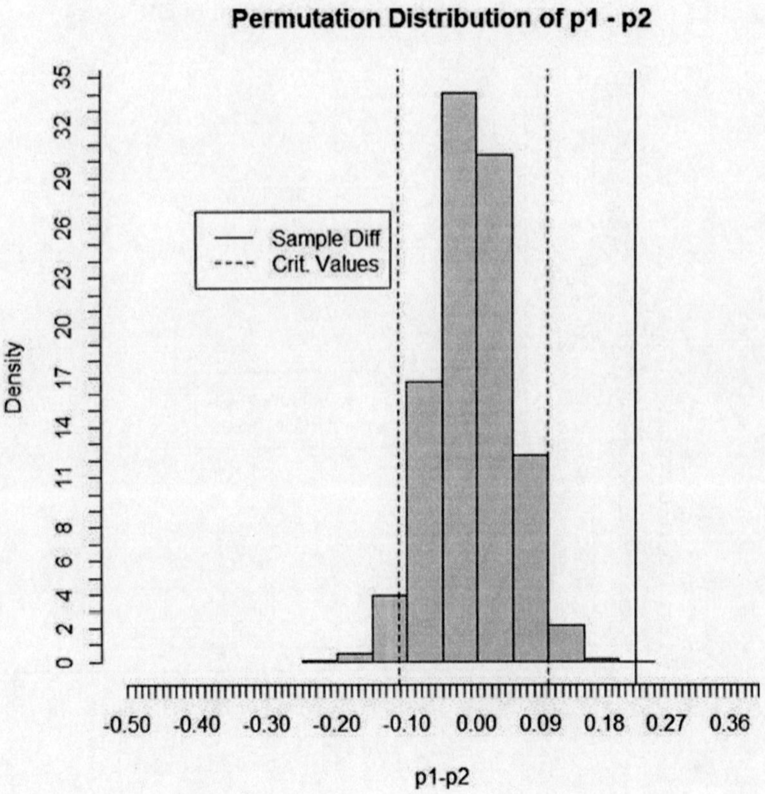

Fig. 17.2 Permutation distribution for $p_1 - p_2$

of the chiral molecule produced was the R-enantiomer. In the process B batch, 80% was R-enantiomer. The OR for process A to process B is:

$$\mathrm{OR} = \frac{O_1}{O_2} = \frac{p_1/(1-p_1)}{p_2/(1-p_2)} = \frac{0.87/(1-0.87)}{0.80/(1-0.80)} \approx \frac{6.692}{4.000} \approx 1.673$$

Note that if p = proportion of R-enantiomer found, then $1-p$ = proportion of S-enantiomer in the mixture.

In purity assays, the proportion of the analyte of interest (in this case, R-enantiomer) is computed by dividing the area under the peak of the target analyte by the areas of all peaks found in the assay. A peak is a local maximum of the assay response. Thus, it does not lend itself to computations requiring a discrete sample size, n. We would like to know whether we should believe that process 1 repeatably has at the very least a better chance of producing the R-enantiomer. Suppose that the run of each process produced 2500 g of the chiral substance. Of these, 2175 g were the R-enantiomer (87%) with process 1, and 2000 g (0.80) were the R-enantiomer

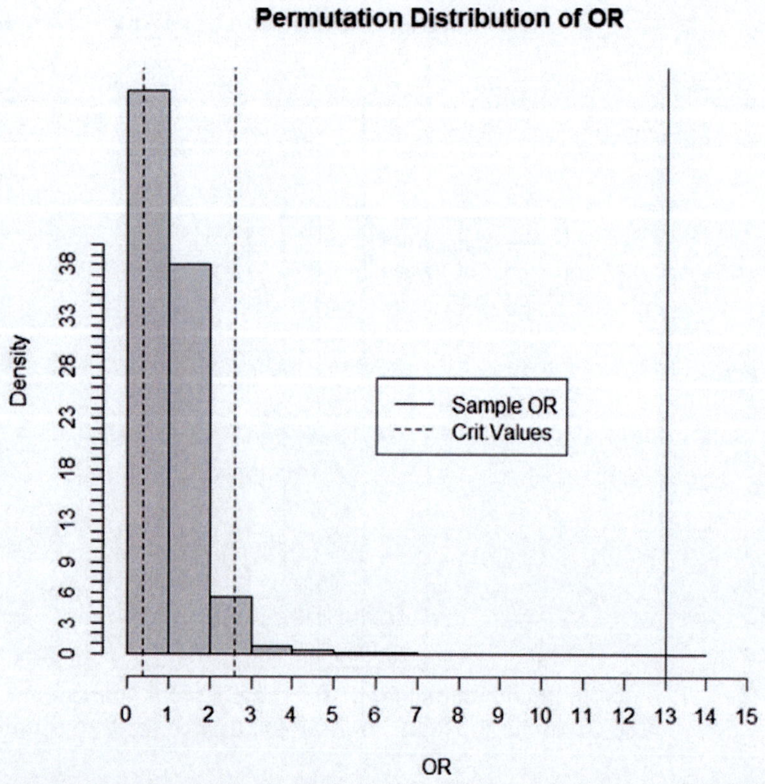

Fig. 17.3 Permutation distribution for OR

for process 2. Thus, the "sample size" could be considered as 2500, with number of "successes" equal to 2175 for process 1 and 2000 for process 2.

The code for computing bootstrap confidence intervals for both OR and difference of proportions is given in Fig. 17.4. Histograms of bootstrap OR and difference of proportions with confidence limits are shown in Figs. 17.5 and 17.6.

Table 17.1 shows the point estimates and confidence intervals.

17.2 Logistic Regression, Again

Logistic regression is a means of relating explanatory variables to a binary response. Earlier, the notion of odds ratio was introduced in the context of a logistic regression model.

```
options(na.action=na.omit)
#
# setwd tells R which folder to use for this script
#

setwd("<your path here>")

df1 <- read.csv("20220614 Ch 17 Two Proportions Summary Data R-enantiomer
Assay.csv")
#
# Variables:
#  Process ( 1 and 2)
#  X
#  N
#
attach(df1)
OR <- c()
ODDS1 <- c()
ODDS2 <- c()
pDiff <- c()
samp1 <- c()
samp2 <- c()
#
# Now expand the summary data into vectors of 0's and 1's
#
one.1 <- rep(x=1,times=X[1])
zero.1 <- rep(x=0,times=N[1] - X[1])
one.zero.1 <- c(one.1,zero.1)
#
# This shuffles the original (expanded) data
#
rand.1 <- runif(length(one.zero.1),0,1)
rand.one.zero.1 <- one.zero.1[order(rand.1)]
#
one.2 <- rep(x=1,times=X[2])
zero.2 <- rep(x=0,times=N[2] - X[2])
one.zero.2 <- c(one.2,zero.2)
```

Fig. 17.4 Bootstrap confidence intervals for OR and $p_1 - p_2$

```
#
# This shuffles the original (expanded) data
#
rand.2 <- runif(length(one.zero.2),0,1)
rand.one.zero.2 <- one.zero.2[order(rand.2)]

#
bootn <- 2000

alpha <- 0.05
set.seed(26)
#

#
sampP1 <- X[1] / N[1]
sampP2 <- X[2] / N[2]
sampODDS1 <- sampP1 / ( 1 - sampP1)
sampODDS2 <- sampP2 / (1 - sampP2)
sampOR <- sampODDS1 / sampODDS2
sampDiff <- sampP1 - sampP2
#

for(i in 1:bootn) {
 samp1 <- sample(x=rand.one.zero.1,size=N[1],replace=TRUE)
 samp2 <- sample(x=rand.one.zero.2,size=N[2],replace=TRUE)
permute1 <- mean(samp1)
permute2 <- mean(samp2)
if (permute1==1){
 ODDS1[i] <- 0
 }
else {
 ODDS1[i] <- permute1 / (1 - permute1)
 }
if (permute2==1) {
 ODDS2[i] <- 0
 }
```

Fig. 17.4 (continued)

```
  else {
    ODDS2[i] <- permute2 / (1 - permute2)
    }
  if (ODDS1[i]==0 | ODDS2[i]==0) {
    OR[i] <- NA
    }
  else {
    OR[i] <- ODDS1[i] / ODDS2[i]
    }
  pDiff[i] <- permute1 - permute2
  }
#
# this is a percentile-method bootstrap confidence interval
#
CL.pDiff <- quantile(pDiff,probs=c(0.05,0.95),type=8,na.rm=TRUE)
CL.OR <- quantile(OR,probs=c(0.05,0.95),type=8,na.rm=TRUE)
LCL.pDiff <- CL.pDiff[1]
UCL.pDiff <- CL.pDiff[2]
LCL.OR <- CL.OR[1]
UCL.OR <- CL.OR[2]

#
hist.OR <- hist(OR,plot=FALSE)

hist.diff <- hist(pDiff,plot=FALSE)
#
hist.OR$density <- 100*hist.OR$counts / sum(hist.OR$counts)
hist.diff$density <- 100*hist.diff$counts / sum(hist.diff$counts)


plot(hist.OR,freq=FALSE,col="light grey",main="Histogram of
OR",xaxt="n",yaxt="n",xlim=c(0.5,2.5),xlab="OR",ylab="Percent")
axis(side=1,at=seq(from=0.5,to=2.5,by=0.10)) #controls x-axis
axis(side=2,at=seq(from=0,to=50,by=1)) #controls y-axis
abline(v=sampOR,lty=1)
abline(v=LCL.OR,lty=2)
abline(v=UCL.OR,lty=2)
#abline(v=minOR,lty=3) # equivalence lower limit
#abline(v=maxOR,lty=3) # equivalence upper limit
legend(x=0.51,y=20,legend=c("Sample OR","Conf.Limit"),lty=c(1,2))
```

Fig. 17.4 (continued)

```
#
# Use the following legend if performing equivalence test:
#
#legend(x=6,y=27,legend=c("Sample OR","Conf.Limit","Equiv. Limit"),lty=c(1,2,3))
dev.new()
plot(hist.diff,freq=FALSE,col="light grey",main="Histogram of
Diff",xaxt="n",yaxt="n",xlim=c(-0.01,0.13),xlab="p1-p2",ylab="Percent")
axis(side=1,at=seq(from=-0.01,to=0.13,by=0.01)) #controls x-axis
axis(side=2,at=seq(from=0,to=50,by=1)) #controls y-axis
abline(v=sampDiff,lty=1)
abline(v=LCL.pDiff,lty=2)
abline(v=UCL.pDiff,lty=2)
#abline(v=-Del,lty=3) # equivalence lower limit
#abline(v=Del,lty=3)  # equivalence upper limit
legend(x=0.0,y=15,legend=c("Sample Diff","Conf.Limit"),lty=c(1,2))
#
# Use the following legend if performing equivalence test:
#
#legend(x=0.062,y=27,legend=c("Sample Diff","Conf.Limit","Equiv.
Limit"),lty=c(1,2,3))

#
df2 <- cbind(OR,pDiff) # these are bootstrap samples

write.csv(df2,"20220614 Bootstrap Samples OR and pDiff R-enantiomer.csv")
#
sampDiff
LCL.pDiff
UCL.pDiff
sampOR
LCL.OR
UCL.OR
df3 <- rbind(sampDiff,LCL.pDiff,UCL.pDiff,sampOR,LCL.OR,UCL.OR)
```

Fig. 17.4 (continued)

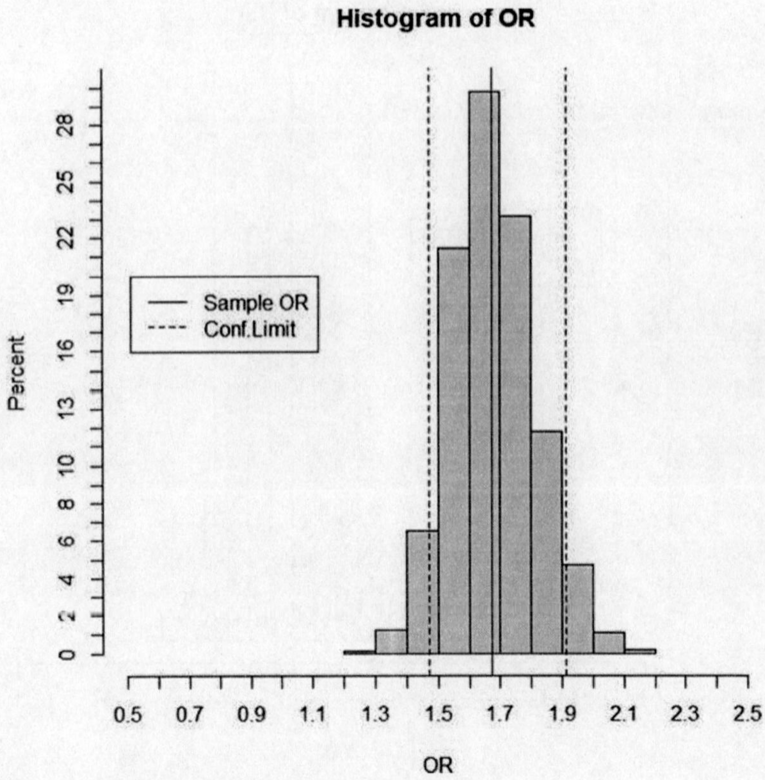

Fig. 17.5 Histogram of bootstrap OR with 95% confidence limits

The response variable is binary:

$$Y = \begin{cases} 1 \text{ if state is} + \\ 0 \text{ if state is} - \end{cases}$$

Then a set of p regressors can potentially affect the probability that $Y = 1$. The model is:

$$\Pr\{Y = 1 | x_1, x_2, \ldots, x_p\} = h(x_1, x_2, \ldots, x_p) = \frac{\exp\left(\beta_0 + \sum_{i=1}^{p} \beta_i x_i\right)}{1 + \exp\left(\beta_0 + \sum_{i=1}^{p} \beta_i x_i\right)}$$

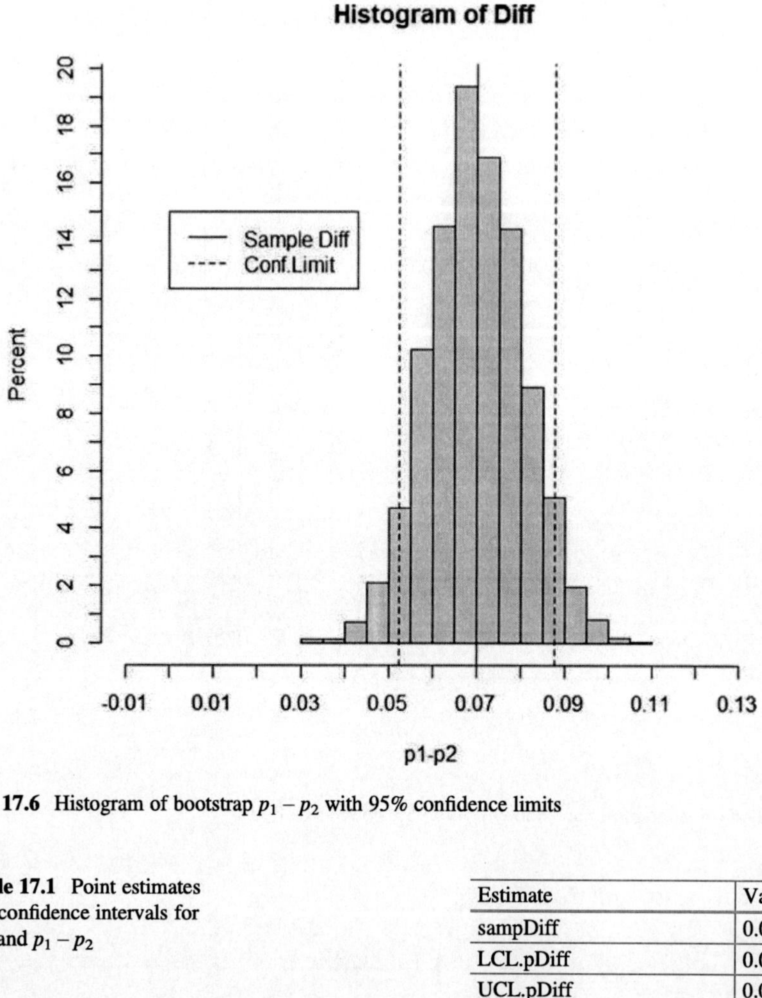

Fig. 17.6 Histogram of bootstrap $p_1 - p_2$ with 95% confidence limits

Table 17.1 Point estimates and confidence intervals for OR and $p_1 - p_2$

Estimate	Value
sampDiff	0.070
LCL.pDiff	0.053
UCL.pDiff	0.088
sampOR	1.673
LCL.OR	1.471
UCL.OR	1.910

We noted that:

$$1 - h(x, x, \ldots, x) = \Pr\{Y = 0 | x_1, x_2, \ldots, x_p\} = \frac{1}{1 + \exp\left(\beta_0 + \sum_{i=1}^{p} \beta_i x_i\right)}$$

So that:

$$r(x, x, \ldots x) = \ln\left(\frac{h(x_1, x_2, \ldots, x_p)}{1 - h(x_1, x_2, \ldots, x_p)}\right) = \ln\left(\exp\left(\beta_0 + \sum_{i=1}^{p} \beta_i x_i\right)\right) = \beta_0 + \sum_{i=1}^{p} \beta_i x_i$$

The function $r(x_1, x_2, \ldots, x_p)$ is called the log odds (Fleiss et al., 2003). Recall that the odds are:

$$O = \frac{P}{1 - P}$$

In the case of logistic regression:

$$P = h(x_1, x_2, \ldots, x_p) = \frac{\exp\left(\beta_0 + \sum_{i=1}^{p} \beta_i x_i\right)}{1 + \exp\left(\beta_0 + \sum_{i=1}^{p} \beta_i x_i\right)}$$

and:

$$1 - P = 1 - h(x, x, \ldots, x) = \frac{1}{1 + \exp\left(\beta_0 + \sum_{i=1}^{p} \beta_i x_i\right)}$$

The logarithm of the odds is also called a logit:

$$\ln(O) = \ln\left(\frac{P}{1 - P}\right)$$

The logistic regression model has the logit linearly related to the regressors:

$$r(x, x, \ldots x) = \ln\left(\frac{P}{1 - P}\right) = \ln\left(\exp\left(\beta_0 + \sum_{i=1}^{p} \beta_i x_i\right)\right) = \beta_0 + \sum_{i=1}^{p} \beta_i x_i$$

The parameters are estimated by maximizing the likelihood function:

$$L(\boldsymbol{\beta}) = \prod_{i=1}^{n} h^{y_i}(1 - h)^{1 - y_i}$$

where:

$$h = \frac{\exp\left(\beta_0 + \sum_{i=1}^{p} \beta_i x_i\right)}{1 + \exp\left(\beta_0 + \sum_{i=1}^{p} \beta_i x_i\right)} = \frac{1}{1 + \exp\left(\beta_0 + \sum_{i=1}^{p} \beta_i x_i\right)} = P$$
$$= \Pr\{Y = 1 | x_1, x_2, \ldots, x_p\}$$

It turns out the estimation process is simplified by using the logit transformation. Hosmer and Lemeshow (1989) give a detailed discussion.

The parameters of the logistic regression are actually log odds ratios. If the regressor has only two possible values (levels), then the interpretation of the model parameter is the log odds ratio between the two levels of the factor. As an example, consider two different microwelding systems, used to weld micro-cables onto the hybrid circuit boards for a pacemaker controller. In addition, there are two substrate materials that can be used for the boards. Two samples of circuit boards are obtained, with types of substrate: types 1 and 2. Then cables are welded, half of each sample using welder 1 and the other half welder 2. Each weld is subjected to a pull test. If the weld breaks, then that weld is a failure. If not, it is a "success."

Approximate confidence intervals for the coefficient estimates (log odds) can be computed using the standard errors of the estimates. That is to say, if $\widehat{\beta}$ represents a coefficient estimate, SE its standard error (computed by R), and $Z_{\alpha/2}$ the 100(1 − α/2) percentile of a standard normal distribution, then the two-sided confidence interval is given by:

$$\widehat{\beta} \pm Z_{\alpha/2}\mathrm{SE}$$

Exponentiating gives the interval for the odds ratio:

$$\exp\left(\widehat{\beta} \pm Z_{\alpha/2}\mathrm{SE}\right)$$

Figure 17.7 shows the R code for computing the logistic regression, extracting the coefficients and their standard errors, and computing confidence intervals for the odds ratios (ORs). Table 17.2 shows the estimates from R. Table 17.3 shows the numbers of successes with each factor level combination. Table 17.4 shows the numbers of successes for each factor level aggregated over the other factor's levels. This table also shows the proportions of successes (P), the odds (O), the odds ratios (OR), the reciprocal of the odds ratios (1/OR), and associated confidence limits. The confidence limits are based on maximum likelihood estimates and their standard errors from R. Note that R will compute the coefficients with the larger level's odds in the numerator. Recall that the levels were coded 1 and 2.

```
setwd("<your path here>")
df1 <- read.csv("20220613 Ch 17 Micro Welds.csv")
#
#
#
# VARIABLES:
# Substrate
# Welder
# test
#

attach(df1)
alpha <- 0.05
#

glm.model <- glm(test~Substrate + Welder,family=binomial(link="logit"),data=df1)
glm.predict <- 1 / (1 + exp(-predict(glm.model,df1)))
plot(x=Substrate,y=glm.predict,pch=1,xlab="Substrate",ylab="Predicted",main="Pred
icted Response vs. Substrate")
dev.new()
plot(x=Welder,y=glm.predict,pch=1,xlab="Welder",ylab="Predicted",main="Predicted
Response vs. Welder")

summary(glm.model)
coef.table <- coef(summary(glm.model))
#
intercept <- coef.table[1]
substrate.beta1 <- coef.table[2]
welder.beta2 <- coef.table[3]
se.intercept <- coef.table[4]
se.beta1 <- coef.table[5]
se.beta2 <- coef.table[6]
#
```

Fig. 17.7 Computing a logistic regression model

```
za <- qnorm(p=1-alpha/2,mean=0,sd=1)
substrate.OR <- exp(substrate.beta1)
welder.OR <- exp(welder.beta2)
lcl.intercept <- exp(intercept - za*se.intercept)
ucl.intercept <- exp(intercept + za*se.intercept)
lcl.substrate <- exp(substrate.beta1 - za*se.beta1)
ucl.substrate <- exp(substrate.beta1 + za*se.beta1)
lcl.welder <- exp(welder.beta2 - za*se.beta2)
ucl.welder <- exp(welder.beta2 + za*se.beta2)
#
df2 <- rbind(intercept,lcl.intercept,ucl.intercept,
        substrate.OR,lcl.substrate,ucl.substrate,
        welder.OR,lcl.welder,ucl.welder)

write.csv(df2,"20220613 Ch 17 Micro Welds Logistic Output.csv")
#####
```

Fig. 17.7 (continued)

Table 17.2 Parameter estimates (maximum likelihood) from R

Parameter	Result
intercept	0.715
lcl.intercept	0.228
ucl.intercept	18.316
substrate.OR	4.438
lcl.substrate	1.451
ucl.substrate	13.571
welder.OR	0.390
lcl.welder	0.136
ucl.welder	1.116

Table 17.3 Summary of sample weld test data

Substrate	Welder	*N*	*X*
1	1	25	19
1	2	25	15
2	1	25	24
2	2	25	21

Table 17.4 Aggregated weld test data with OR and 95% confidence limits

	N	X	P	O	OR	LCL(OR)	UCL(OR)	1/OR	LCL(1/OR)	UCL(1/OR)
Substrate 1	50	34	0.68	2.125	[redacted]					
Substrate 2	50	45	0.90	9.000	**4.235**	**1.451**	**13.571**	**0.236**	**0.074**	**0.689**
Welder 1	50	43	0.86	6.143	[redacted]					
Welder 2	50	36	0.72	2.571	**0.419**	**0.136**	**1.116**	**2.389**	**0.896**	**7.332**

The estimated OR (maximum likelihood) for substrate (type 2 over type 1) was 4.438. The 95% confidence interval was (1.451, 13.571). Since this interval does not contain the value 1, we conclude that substrate 2 is associated with a higher likelihood of passing the weld test. Clearly, substrate type 2 is better than type 1. Conversely, the estimated OR for welder was 0.390, with a 95% confidence interval of 0.136–1.116. Since this interval contains 1, we conclude that welder type does not make a repeatable difference in the probability of passing the weld test.

17.3 Summary

Two different methods were presented for comparing proportions. Both of them, the different between the proportions and the odds ratio, have approximate formulas for doing hypothesis tests or computing confidence intervals. They can easily be used for computing bootstrap confidence intervals or performing permutation tests.

Logistic regression can also be used to compare proportions and is particularly useful when there are two or more factors that may affect the probability of a "success." Especially in the cases where each factor has only two levels, the interpretation of the coefficient estimates is very simple; they are the log odds ratios. Using standard errors, approximate confidence intervals can be computed.

References

Hosmer, D. W., & Lemeshow, S. (1989). *Applied logistic regression*. Wiley.

Fleiss, J. L., Levin, B., & Paik, M. C. (2003). *Statistical Methods for Rates and Proportions* (3rd ed.). John Wiley and Sons, New York

Chapter 18
Afterword

Abstract A summary of the book, with some guidelines on where to look for solutions to particular problems is provided.

Both statistics and medical devices are very broad categories. The methods presented in this book are far from exhaustive. Hopefully, those methods will cover many applications that arise in the development of medical devices, from pacemakers to in vitro diagnostics. References have been provided, and the author strongly suggests the reader to investigate those references. There was an attempt to make the book self-contained. Hence, the two chapters on the basic notions or probability and inference were included, as well as the Appendix on some mathematical fundamentals. It is hoped that readers will endeavor to get more in-depth knowledge and understanding of any methods presented here. Furthermore, it is hoped that the readers will use some of the R code provided and modify the code as they require.

Table 18.1 shows a list of potential problems/applications, the recommended methods for solving those issues, and the chapter/section in which those methods can be found.

S. A. Pardo, *Statistical Methods and Analyses for Medical Devices*,
https://doi.org/10.1007/978-3-031-26139-8_18

Table 18.1 Some common problems and solution methods: a guide

Problem	Method(s)	Chapter/section
Confidence interval for proportion of PASSing product	Confidence interval for proportion	Chapter 3/Sect. 3.4
Comparing continuous variable between groups/treatments/designs	ANOVA	Chapter 5/Sects. 5.3 and 5.4; Chap. 6
Design experiment to assess effects of factors on a response variable	ANOVA	Chapter 6
Construct a predictive model with many regressors	Regression, nonlinear methods	Chapter 14
Determine optimal operating conditions for a process	Taguchi methods	Chapter 6/Sect. 6.10
Set up an acceptance test for final inspection	Acceptance sampling plan	Chapter 8
Bring a manufacturing process into statistical control	Process control charts	Chapter 7
Estimate variability relative to specification limits	Cp	Chapter 15/Sect. 15.4
Determine the "best" linear model	AIC or Mallows' C_p	Chapter 14/Sect. 14.2
Fit a predictive model to data observed at regular time intervals	Box and Jenkins time series analysis	Chapter 16
Decide which design is more likely to operate properly	Odds ratios	Chapter 17
Determine which factors have an effect on Pr{conformance}	Logistic regression	Chapter 9/Sect. 9.5; Chap. 14/Sect. 14.7; Chap. 17/Sect. 17.2
Use time-to-failure data to assess probability of failure within warranty	Reliability curve	Chapter 9/Sect. 9.2
Compute confidence interval when sampling distribution is unknown	Bootstrap	Chapter 4/Sect. 4.4; Chap. 11/Sect. 11.3; Chap. 12/Sect. 12.3
Assess equivalence of two systems	Equivalence testing	Chapter 11
Assess effects of factors with non-normal response	Rank transformation	Chapter 12
Incorporate prior information about potential ranges of parameters	MCMC	Chapter 13/Sect. 13.2
Using an accelerated life test to demonstrate reliability	Arrhenius Law	Chapter 9/Sect. 9.5

Appendix A: Some Mathematical Fundamentals

The Appendix provides a brief description of some mathematical methods used in the book.

A.1 Basics of Calculus

Limits
A limit is a sort of boundary to some sport of numerical progression. Clearly, some progressions are infinite, i.e., they have no bound. Their limit is said to be infinity ($+\infty$), or negative infinity, ($-\infty$). An example of a progression is:

$$f(k) = \frac{1}{k}, k = 1, 2, 3, \ldots$$

As k gets large (or, as we say, goes to $+\infty$), $f(k)$ gets smaller and smaller. Is it ever 0? No, but it gets closer and closer to 0 as k goes to $+\infty$. Thus, the limit of $f(k)$ is said to be 0.

Here is another example. The progression (or, as mathematicians like to call it, sequence):

$$f(k) = k, k = 1, 2, 3, \ldots$$

just keeps getting larger as k goes to $+\infty$. Thus, its limit is $+\infty$. When a sequence has as its limit either $\pm\infty$, it is said to have no limit, or its limit does not exist, or it is unbounded.

The above examples were sequences of positive integers, but you could have sequences of real numbers, too. It is just that you can't write the sequence of the "input" values by separating numbers with commas. Why? Because between any

S. A. Pardo, *Statistical Methods and Analyses for Medical Devices*,
https://doi.org/10.1007/978-3-031-26139-8

two real numbers, no matter how close they are to each other, there is always another real number. So, to represent the sequence of input values when they are real numbers, we use inequality signs. For example:

$$f(x) = \frac{1}{x}, x > 0$$

Guess what is the limit of the sequence $f(x)$? Did you say 0? Give the player a cupie doll!

The two important ideas of calculus, derivatives and integrals, are based on limits of real-valued sequences.

A.2 Derivatives and Differentiation

Take a nice, smooth function of real numbers; call it $f(x)$. Suppose you pick a value of x, and then another value a little to the right of x—call it $x + h$ (h is some fixed amount). The difference in the function from x to $x + h$ is:

$$d(x, h) = f(x+h) - f(x)$$

So, d is the amount the function changed when moving from x to $x + h$. The thing is, if h is very big, you might get a different amount of change in f than if h is very small. So, we like to talk about the change in f per unit of x. This could be expressed as:

$$\delta(x, h) = \frac{f(x+h) - f(x)}{x+h-x} = \frac{f(x+h) - f(x)}{h}$$

Note that h is in the same units as x.

The function $\delta(x,h)$ might have a limit as h is made smaller and smaller (closer and closer to 0, but not 0 itself). This limit is called the derivative of f at x. It turns out that many common functions have formulas that describe these limits, or derivatives. The derivative of f at x tells you the "slope" of the function, or how fast it is changing when the input value is "near" x. In one case, the derivative is the same number no matter what the value of x; this special class of functions can be described as:

$$f(x) = a + bx, \text{ where } a \text{ and } b \text{ are two constants.}$$

That is correct; this is a straight-line function with intercept a and slope b. The slope, b, is the value of the derivative no matter what the value of x is.

There are several notations for expressing the derivative of a function. If the function is $f(x)$, then its derivative (which is in turn a function itself) is expressed as:

$$\frac{df(x)}{dx}$$

or:

$$f'(x)$$

or:

$$\dot{f}(x)$$

Those are the three most common notations.

It turns out that certain common functions have formulas for their derivatives. Here are a few:

$f(x)$	$\frac{df}{dx}$
ax^n	anx^{n-1}
$\frac{1}{x^n}$	$\frac{-n}{x^{n+1}}$
$\sin(x)$	$\cos(x)$
$\cos(x)$	$-\sin(x)$
e^{ax}	ae^{ax}
$\log_e(x)$	$\frac{1}{x}$

For a special case of ax^n, consider:

$$f(x) = ax$$

Then:

$$\frac{d}{dx}(ax) = ax^{1-1} = ax^0 = a * 1 = a$$

The most important rules about derivatives are linearity and the chain rule:

$$\frac{d}{dx}(af(x) + bg(x)) = a\frac{df}{dx} + b\frac{dg}{dx}$$

and:

$$\frac{d}{dx}f(g(x)) = f'(g(x))g'(x)$$

Notice how I changed notations from describing the linearity rule to the chain rule. It was just more convenient.

Example of linearity:

$$\frac{d}{dx}\left(ax^2 + bx^3\right) = 2ax^{2-1} + 3bx^{3-1} = 2ax + 3bx^2$$

Example of chain rule:

$$\frac{d}{dx}\sin(ax^n) = \cos(ax^n) * anx^{n-1}$$

A special case of linearity:

$$\frac{d}{dx}(af(x)) = a\frac{df}{dx}$$

Also, if a is a constant, and:

$$f(x) = a$$

regardless of the value of x, then:

$$\frac{d}{dx}(f(x)) = \frac{d}{dx}(a) = 0$$

Note that the derivative of a function is itself a function. As a result, its value can be computed for a given value of x. So, for example, suppose:

$$f(x) = 3x^2$$

So its derivative would be:

$$f'(x) = 2 * 3x^{2-1} = 6x$$

Thus, the value of the derivative, $f'(x)$, at $x = 1$ is:

$$f'(1) = 6(1) = 6$$

When a function has more than one "x" variable, then it has derivatives for each of the x's, independent of the other x's. So, suppose f is a function of two variables, call them x_1 and x_2.

Then there is a derivative of f with respect to x_1 and another derivative of f with respect to x_2. The notation commonly used is:

$$\frac{\partial f(x_1, x_2)}{\partial x_1}$$

and:

$$\frac{\partial f(x_1, x_2)}{\partial x_2}$$

When forming these "partial derivatives," the variables not being included in the differentiation process (the process of forming the derivative) are treated like they were constants. A special but important case would arise in multiple linear regression. Suppose the function was:

$$f(x_1, x_2) = \beta_1 x_1 + \beta_2 x_2$$

Then:

$$\frac{\partial f(x_1, x_2)}{\partial x_1} = \beta_1$$

and:

$$\frac{\partial f(x_1, x_2)}{\partial x_2} = \beta_2$$

This is just included so that you will be aware of the notation and have some familiarity with it.

Higher-Order Derivatives
If there is a function, *f*(*x*), that has a derivative, then the derivative is also a function in the variable *x*. Thus, the derivative of *f* can also have a derivative. This derivative of the derivative of *f* is called the second derivative of *f*. This second derivative in turn can have a derivative, which is now the third derivative of *f*. This can go on forever. Keep in mind that the derivative of a function that is equal to a constant (just a single number) is 0, regardless of the value of the constant. Of course, if the constant is 0, then the derivative is still 0.

Here is an example. Suppose:

$$f(x) = 2 + 3x^4$$

Its derivative is:

$$f'(x) = \frac{df}{dx} = 3 * 4x^{4-1} = 12x^3$$

This function in turn has a derivative:

$$f''(x) = \frac{d^2f}{dx^2} = 3 * 12x^{3-1} = 36x^2$$

We use the superscript "2" to indicate that this is the second derivative of the original function, *f*.

We can continue taking derivatives of the derivatives:

$$f'''(x) = \frac{d^3 f}{dx^3} = 2 * 36x^{2-1} = 72x^1$$

$$f''''(x) = \frac{d^4 f}{dx^4} = 1 * 72x^{1-1} = 72x^0 = 72$$

$$f'''''(x) = \frac{d^5 f}{dx^5} = 0 * 72 = 0$$

Once the *n*th order derivative (in this case, fifth) is 0, all subsequent derivatives will be 0.

A.3 Derivatives and Optima

A function of a single variable is at its maximum or minimum when its first derivative is equal to 0. That means if there is a value of the variable that makes the first derivative equal to 0, then that is the value of the variable where the original function is either maximum or minimum.

Without even graphing the original function, you can determine whether the point at which its first derivative is 0 is either a maximum or minimum. You do this by taking the derivative of the derivative (this is the second derivative). If at the value of the variable making the first derivative 0, the second derivative is negative, then the function is at its maximum at that value of the variable. If the second derivative is positive, then this is a minimum for the original function.

We will illustrate this with a funny function:

$$f(p, q) = pq^2$$

This is a function of two variables, but suppose $q = 1 - p$. Then the function becomes:

$$f(p) = p(1-p)^2 = p(p^2 - 2p + 1)$$
$$= p^3 - 2p^2 + \mathrm{p}$$

The derivative of this function is:

$$\frac{df}{dp} = 3p^2 - 4p + 1$$

To find the minimum or maximum of the original function, set the derivative equal to 0 and solve:

$$3p^2 - 4p + 1 = 0$$
$$(3p - 1)(p - 1) = 0$$

So, either $p = 1/3$ or $p = 1$. First, find out of these two values of p make the second derivative negative or positive.

The second derivative is:

$$\frac{d^2f}{dp^2} = 6p - 4$$

Using $p = 1/3$ yields $6(1/3) - 4 = -2$. With $p = 1$, $6(1) - 4 = 2$. At what value is the second derivative in between being negative or positive? When it equals 0:

$$6p - 4 = 0$$
$$p = 4/6 = 2/3$$

So, if $p < 2/3$, the second derivative is negative, so that at $p = 1/3$, the original function has a "local maximum" (a bump). If $p > 2/3$, the second derivative is positive, so at $p = 1$, the original function has a "local minimum" (a trough). At $p = 2/3$, the original function is changing from going up to going down (an inflection point).

Look at the plot:

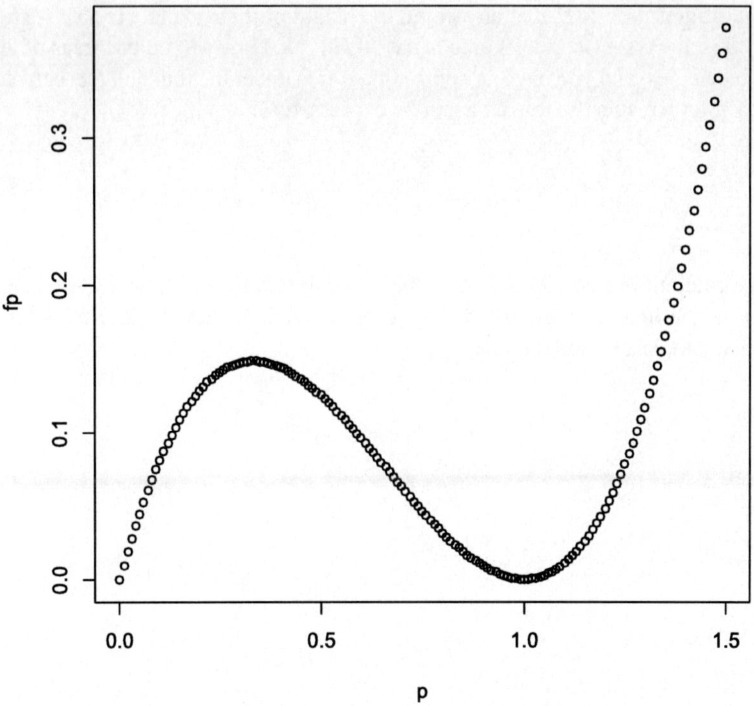

Note that the "trough" is at $p = 1$. If the variable p is restricted to the open "unit interval" (0, 1), then we may only be interested in the "local maximum" at $p = 1/3$. But the interval of interest depends on whether the function is a "model" describing some real phenomenon.

A.4 Integrals and Integration

Suppose there was some function, $f(x)$, and you wanted to know how much area there was under it (assume for the moment that the function values are always non-negative) between two values on the x-axis (say x_1 and x_2, with $x_1 < x_2$). For now, don't ask why on earth you would want to know this area; just take for granted that there will be some good reasons. Suppose you broke up the range of values from x_1 to x_2 in the following fashion:

$$x_1, x_1 + h, x_1 + 2h, x_1 + 3h, \ldots, x_2$$

The intervals between each consecutive value in this progression all have the same length, h. The function has values at each of the x's in the progression:

$$f(x_1), f(x_1 + h), f(x_1 + 2h), \ldots, f(x_2)$$

Suppose that we choose a value k such that $x_2 = x_1 + kh$. The area of a rectangle is width times height. Consider the rectangle starting at x_1 with "base" width h and with height $f(x_1)$. Its area would be $f(x_1) * h$. Then consider the next rectangle from $x_1 + h$ to $x_1 + 2\,h$, with height $f(x_1 + h)$. Its area would be $f(x_1 + h)*h$. Go through the whole progression, so that at any point, the height of the rectangle is $f(x_1 + kh)$ and its with is always h, so that its area is $f(x_1 + kh) * h$. Now add up the areas of all the little rectangles, and you have an approximate area under the function $f(x)$ between x_1 and x_2. Mathematically, this area can be expressed as:

$$L = \sum_{j=0}^{k-1} f(x_1 + jh)h$$

We could have also approximated the area a little differently. Instead of using the height of the first rectangle to be $f(x_1)$, suppose we had started with $f(x_1 + h)$. The approximate area would then be:

$$U = \sum_{j=1}^{k} f(x_1 + jh)h$$

Suppose we try to find the limit of L as h gets smaller and smaller, and the limit of U as h gets smaller and smaller. If these two limits are not infinite, and they equal each other, then the limit is called the (Reimann) integral of $f(x)$ between x_1 and x_2. The usual notation for the integral of f between values x_1 and x_2 is:

$$A = \int_{x_1}^{x_2} f(\xi)d\xi$$

The Greek letter ξ acts like an index variable in summation notation. When the integral has a formula, call it $F(x)$, then the area, A, is computed as the difference $F(b) - F(a)$, which is sometimes denoted as:

$$F(b) - F(a) = F(x)|_a^b$$

As in the case of derivatives, the integrals of some common functions have easy-to-compute forms. The table gives some examples:

$f(x)$	$\int_a^b f(x)dx$
anx^{n-1}	$ax^n\|_a^b$
ae^{ax}	$e^{ax}\|_a^b$
$\cos(x)$	$\sin(x)\|_a^b$
$-\sin(x)$	$\cos(x)\|_a^b$
$\frac{1}{x}$	$\log_e(x)\|_a^b$
$\frac{-n}{x^{n+1}}$	$\frac{1}{x^n}\|_a^b$

You may have noticed from this table and the table of derivative formulas that when a derivative has a formula, the integral is the original function that gave rise to the derivative. Sometimes the integral is referred to as the anti-derivative. This relationship is useful in solving a class of problems called differential equations. Differential equations are equations in which relationships between derivatives are known but the original function is unknown. The integral is useful for finding the original function that satisfies the differential equation. As a simple case, suppose that we know:

$$\frac{df(x)}{dx} = \cos(x)$$

Then the original function must be:

$$\int \frac{df(x)}{dx} = \int \cos(x)dx = \sin(x)$$

We have not specified any limits on the integral, since we were just trying to illustrate the idea that the integral as antiderivative is useful in solving differential equations. Such integrals are referred to as indefinite; often the resulting formula is written with the letter C as an added term:

$$\int \cos(x)dx = \sin(x) + C$$

The integration process has some very useful rules. Possibly the most useful is, like the derivative, linearity. Specifically:

$$\int af(x) + bg(x)dx = a\int f(x)dx + b\int g(x)dx$$

A.5 Matrix and Vector Algebra

A matrix is a rectangular array of numbers:

$$\boldsymbol{M} = \begin{bmatrix} a_{11} & a_{12} & \dots & a_{1n} \\ \vdots & \vdots & \ddots & \vdots \\ a_{m1} & a_{m2} & \dots & a_{mn} \end{bmatrix}$$

The "entries," a_{ij}, can be real or complex numbers, but we will only deal with real matrices and vectors.

A vector is a special matrix that has only either one row or one column. For the most part, we will consider vectors to be a column of numbers:

$$\boldsymbol{v} = \begin{bmatrix} b_1 \\ \vdots \\ b_m \end{bmatrix}$$

We will represent the names of matrices with bold-face, capital letters and the names of vectors with bold-face, lowercase letters. We will use the convention of labeling the entries, or components, with subscripts that represent the row number (first subscript) and column number (second subscript). In the case of vectors, the single subscript will refer to the row number for "column" vectors or the column number for "row" vectors. As we stated before, mostly when we refer to a vector, we will be thinking of it as a column. When we need to refer to a "row" vector, most of the time the name of the vector will have a superscript "T," which is an abbreviation for "transpose." A transpose vector or matrix is one where the rows become columns.

There are several operations involving matrices and vectors that will appear throughout the text. They are:

1. Scalar multiplication
2. Matrix and vector addition
3. Transposition
4. Matrix multiplication
5. "Dot" or scalar product of two vectors and vector length
6. Square matrices, the identity matrix, and matrix inverses
7. Determinants
8. Eigenvalues and eigenvectors
9. Diagonalization and powers of matrices

A.6 Scalar Multiplication

A scalar for our purposes is a number that is not a vector or matrix. In fact, the entries or components in all of our vectors and matrices will be scalars. To multiply a scalar by a matrix of vector, simply multiply that scalar by every entry in the matrix or vector. So:

$$\boldsymbol{A} = s\boldsymbol{M} = \boldsymbol{M}s = \begin{bmatrix} sa_{11} & sa_{12} & \dots & sa_{1n} \\ \vdots & \vdots & \ddots & \vdots \\ sa_{m1} & sa_{m2} & \dots & sa_{mn} \end{bmatrix}$$

or:

$$\boldsymbol{w} = s\boldsymbol{v} = \boldsymbol{v}s = \begin{bmatrix} sb_1 \\ \vdots \\ sb_m \end{bmatrix}$$

The order is not important for scalar multiplication. This is not the case of matrix multiplication.

A.7 Matrix and Vector Addition

Adding matrices and vectors is component-wise addition. This means that in order to add two matrices, they must have exactly the same number of rows and columns. So:

$$\boldsymbol{M} = \begin{bmatrix} a_{11} & a_{12} & \dots & a_{1n} \\ \vdots & \vdots & \ddots & \vdots \\ a_{m1} & a_{m2} & \dots & a_{mn} \end{bmatrix}$$

$$N = \begin{bmatrix} b_{11} & b_{12} & \dots & b_{1n} \\ \vdots & \vdots & \ddots & \vdots \\ b_{m1} & b_{m2} & \dots & b_{mn} \end{bmatrix}$$

$$M + N = N + M = \begin{bmatrix} a_{11} + b_{11} & a_{12} + b_{12} & \dots & a_{1n} + b_{1n} \\ \vdots & \vdots & \ddots & \vdots \\ a_{m1} + b_{m1} & a_{m2} + b_{m2} & \dots & a_{mn} + b_{mn} \end{bmatrix}$$

and:

$$v = \begin{bmatrix} a_1 \\ \vdots \\ a_m \end{bmatrix}$$

$$w = \begin{bmatrix} b_1 \\ \vdots \\ b_m \end{bmatrix}$$

$$v + w = w + v = \begin{bmatrix} a_1 + b_1 \\ \vdots \\ a_m + b_m \end{bmatrix}$$

A.8 Transposition

If:

$$M = \begin{bmatrix} a_{11} & a_{12} & \dots & a_{1n} \\ \vdots & \vdots & \ddots & \vdots \\ a_{m1} & a_{m2} & \dots & a_{mn} \end{bmatrix}$$

Then the transpose of matrix M, call it M^T, is:

$$M^T = \begin{bmatrix} a_{11} & \dots & a_{m1} \\ a_{12} & \dots & a_{m2} \\ \vdots & \ddots & \vdots \\ a_{1n} & \dots & a_{mn} \end{bmatrix}$$

To illustrate the point, the subscripts on the entries in M^T have not been changed from the original, although our convention is to have the row number first followed by the column number. As a more concrete example, if:

$$M = \begin{bmatrix} 1 & 2 & 3 & 4 \\ 5 & 6 & 7 & 8 \\ 9 & 10 & 11 & 12 \end{bmatrix}$$

then its transpose would be:

$$M^T = \begin{bmatrix} 1 & 5 & 9 \\ 2 & 6 & 10 \\ 3 & 7 & 11 \\ 4 & 8 & 12 \end{bmatrix}$$

A transposed column vector becomes a row vector and vice versa:

$$v = \begin{bmatrix} a_1 \\ \vdots \\ a_m \end{bmatrix}$$

$$v^T = \begin{bmatrix} a_1 & \dots & a_m \end{bmatrix}$$

A.9 Matrix Multiplication

Not all pairs of matrices can be multiplied. Furthermore, matrix multiplication is not commutative, as was matrix addition. In order for matrix M and N to be "multipliable" (usually called conformable with respect to multiplication), to obtain the product MN, the number of columns of M must equal the number of rows in matrix N. Symbolize the entries of matrix M as a_{ij}, and the entries of matrix N as b_{jk}. If M has m rows and n columns, and N has n rows and r columns, then the product matrix:

$$P = MN$$

will have m rows and r columns. The ith row and jth column of P will be the sum:

$$p_{ij} = a_{i1} * b_{1j} + a_{i2} * b_{2j} + \dots + a_{in} * b_{nj}$$

As an example, consider:

$$M = \begin{bmatrix} 1 & 2 \\ 3 & 4 \end{bmatrix}$$

$$N = \begin{bmatrix} 5 & 6 & 7 \\ 8 & 9 & 10 \end{bmatrix}$$

Then:

$$P = MN = \begin{bmatrix} 1*5+2*8 & 1*6+2*9 & 1*7+2*10 \\ 3*5+4*8 & 3*6+4*9 & 3*7+4*10 \end{bmatrix} = \begin{bmatrix} 21 & 24 & 27 \\ 47 & 54 & 61 \end{bmatrix}$$

The product NM is not defined.

A very important special case of matrix multiplication is between an $m \times n$ matrix and an $n \times 1$ vector (which is a very special kind of matrix). That is:

$$\boldsymbol{Ax} = \begin{bmatrix} a_{11} & a_{12} & \dots & a_{1n} \\ \vdots & \dots & \ddots & \vdots \\ a_{m1} & a_{m2} & \dots & a_{mn} \end{bmatrix} \begin{bmatrix} x_1 \\ x_2 \\ \vdots \\ x_n \end{bmatrix} = \begin{bmatrix} b_1 \\ \vdots \\ b_m \end{bmatrix} = \boldsymbol{b}$$

A.10 Dot or Scalar Product

Two vectors can be multiplied in a fashion that yields a scalar. In order to obtain this scalar product, one vector must be a row vector and the other a column vector, and they both must have equal numbers of components. The "dot" product is symbolized with a "•". If:

$$\boldsymbol{v}^{\boldsymbol{T}} = [a_1 \quad \dots \quad a_n]$$

$$\boldsymbol{w} = \begin{bmatrix} b_1 \\ \vdots \\ b_n \end{bmatrix}$$

then the dot product $\boldsymbol{v}^{\boldsymbol{T}} \bullet \boldsymbol{w}$ is:

$$\boldsymbol{v}^{\boldsymbol{T}} \bullet \boldsymbol{w} = [a_1 \quad \dots \quad a_n] \begin{bmatrix} b_1 \\ \vdots \\ b_n \end{bmatrix} = a_1 * b_1 + \dots + a_n * b_n.$$

On the other hand, the product $\boldsymbol{w}\boldsymbol{v}^{\boldsymbol{T}}$ is actually a matrix. Suppose:

$$\boldsymbol{w} = \begin{bmatrix} b_1 \\ \vdots \\ b_m \end{bmatrix}$$

with $m \neq n$.

The vector $\boldsymbol{w}$ can be thought of as a matrix with m rows and 1 column, and $\boldsymbol{v}^{\boldsymbol{T}}$ can be thought of as a matrix with 1 row and n columns. Thus, the product $\boldsymbol{w}\boldsymbol{v}^{\boldsymbol{T}}$ is a matrix with m rows and n columns.

A vector has "length," which we will designate as $\|\boldsymbol{v}\|$, and is defined to be:

$$\|\boldsymbol{v}\| = \sqrt{\sum\nolimits_{i=1}^{n} a_i^2} = \left(\boldsymbol{v}^{\boldsymbol{T}} \bullet \boldsymbol{v}\right)^{1/2}$$

Inasmuch as this length is a scalar, another vector that has length = 1 can be constructed as:

$$\tilde{v} = \frac{1}{\|v\|}v$$

A vector of all 0s has length = 0.

A.11 Square Matrices, the Identity Matrix, and Matrix Inverses

A square matrix is one that has the same number of rows and columns. They turn out to be a very important case for a wide variety of applications. For one thing the product of a matrix and its transpose is always a square matrix. If $\boldsymbol{M}$ is an $m \times n$ matrix, then its transpose, $\boldsymbol{M}^{\boldsymbol{T}}$, is an $n \times m$ matrix, and:

$$\boldsymbol{S}_1 = \boldsymbol{M}^{\boldsymbol{T}}\boldsymbol{M}$$

is a square matrix with dimensions $n \times n$, and:

$$\boldsymbol{S}_2 = \boldsymbol{M}\boldsymbol{M}^{\boldsymbol{T}}$$

is a square matrix with dimensions $m \times m$.

So, going back to our least squares regression, the matrix $\boldsymbol{Z}$ is $n \times k$, so that $\boldsymbol{Z}^{\boldsymbol{T}}$ is $k \times n$, and therefore $\boldsymbol{Z}^{\boldsymbol{T}}\boldsymbol{Z}$ is a $k \times k$ square matrix.

An identity matrix is a special square matrix where most of the entries are 0, except for the "diagonal entries (a_{ii}) which are all 1. In general, the $n \times n$ matrix, $\boldsymbol{I}_n$ is:

$$\boldsymbol{I}_n = \begin{bmatrix} 1 & 0 & \dots & 0 \\ 0 & 1 & \dots & 0 \\ \vdots & 0 & \ddots & \vdots \\ 0 & 0 & \dots & 1 \end{bmatrix}$$

It turns out that for any $n \times n$ square matrix, $\boldsymbol{S}_n$, the products with the identity matrix of corresponding dimensions are the matrix itself:

$$\boldsymbol{S}_n\boldsymbol{I} = \boldsymbol{I}\boldsymbol{S}_n = \boldsymbol{S}_n$$

This begs the question, can we find another matrix so that its product with square matrix $\boldsymbol{S}_n$ is the $n \times n$ identity? The answer is, sometimes. This matrix, if it exists, is called the inverse of $\boldsymbol{S}_n$. When a square matrix has an inverse, which we denote with a superscript "−1," it has the property that:

$$\boldsymbol{S}_n\boldsymbol{S}_n^{-1} = \boldsymbol{S}_n^{-1}\boldsymbol{S}_n = \boldsymbol{I}_n$$

Table A.1 Perfect linear relationship example

X_1	Y
1	7
2	9
3	11
4	13
5	15
6	17
7	19
8	21
9	23

In the case of multiple regression as described earlier, the answer to the existence of the inverse matrix $[\mathbf{Z}^T\mathbf{Z}]^{-1}$ is a resounding: almost always. Oh. The usual case for regression is that the matrix:

$$\mathbf{S}_k = \mathbf{Z}^T\mathbf{Z}$$

does in fact have an inverse matrix:

$$\mathbf{S}_k^{-1} = \left[\mathbf{Z}^T\mathbf{Z}\right]^{-1}$$

Is it possible to know whether this matrix exists? Generally, software will tell you if you try to fit a multiple regression and the matrix $\mathbf{Z}^T\mathbf{Z}$ is not invertible. This condition will occur if any column of the regressor matrix, **Z**, is in fact a linear function of other columns. Such a condition is called collinearity. Sometimes a column is very close to being a linear function of other columns, but not quite. Such a condition is called ill conditioning. Ill-conditioned matrices may have an inverse, but it may be too hard or at least very difficult to compute it (we have not discussed how to actually find an inverse at this point; suffice it to say that there is a numerical procedure for doing so). As an example, consider the data in Table A.1. In this case, X_1 is just the numbers from 1 to 9, and $Y = 2 * X_1 + 5$ exactly. In the R-code shown in Fig. A.1, the variable X_2 is X_1 plus a little noise (rnorm generates random normal variable values).

The output is shown in Fig. A.2

The first regression (linreg1) was perfectly fine. The intercept was close to 5 and the slope parameter (coefficient for X_2) was close to 2. Remember that we added some random noise, so that the fit would not be "perfect."

The second regression had the collinear term X_3 included. The *lm* function could not estimate its associated coefficient. The message "(1 not defined because of singularities)" indicates that there is some collinearity with respect to variable X_3, hence the term "singularities."

As to how a square matrix' inverse is computed, it is accomplished through some variant of a numerical procedure called Gauss-Jordan elimination. The details of this process may be found in any book on linear algebra (see, e.g., Strang, 2016).

```
setwd("<your path here>")

df1 <- read.csv("20230216 Appendix A.6.1.csv")

#

# Variables:

#  X1

#  Y

#  Note: Y = 2X + 5, without noise

#

#

attach(df1)

size <- length(X1)

X2 <- X1 + rnorm(n=size,mean=0,sd=.75)

X3 <- 2.5*X2

lnreg0 <- lm(Y ~ X1)

lnreg1 <- lm(Y ~ X2)

lnreg2 <- lm(Y ~ X2 + X3)


summary(lnreg1)

summary(lnreg2)
```

Fig. A.1 R code for linear regressions with noisy *X* and collinear *X*

A.12 Determinants

Square matrices have a number associated with them called a determinant. Determinants in general are computed using the Gauss-Jordan elimination process and are in fact related to a square matrix's inverse. Perhaps the most important property of the determinant is that if it is 0, then the associated square matrix does not have an inverse. A determinant can either be negative, 0, or positive. The general definition of the process by which a determinant is computed is beyond the scope of this discussion. It is important to realize that such a quantity exists. The determinant is often symbolized with either the functional notation det(***S***) or "absolute value" bars surrounding the name of the square matrix, e.g.:

$$\det(\boldsymbol{S}) = |\boldsymbol{S}|$$

However, a determinant can in fact be negative.

Fig. A.2 Output of linear regressions

```
> summary(lnreg1)
Call:
lm(formula = Y ~ X2)
Residuals:
   Min     1Q  Median    3Q    Max
-2.2965 -1.6409 -0.3524 1.0603 3.1833
Coefficients:
            Estimate Std. Error t value Pr(>|t|)
(Intercept)    5.407      1.513   3.574 0.009046 **
X2             1.974      0.278   7.102 0.000193 ***
---
Signif. codes: 0 '***' 0.001 '**' 0.01 '*' 0.05 '.' 0.1 ' ' 1
Residual standard error: 2.044 on 7 degrees of freedom
Multiple R-squared: 0.8781, Adjusted R-squared: 0.8607
F-statistic: 50.44 on 1 and 7 DF, p-value: 0.0001933
> summary(lnreg2)
Call:
lm(formula = Y ~ X2 + X3)
Residuals:
   Min     1Q  Median    3Q    Max
-2.2965 -1.6409 -0.3524 1.0603 3.1833
Coefficients: (1 not defined because of singularities)
            Estimate Std. Error t value Pr(>|t|)
(Intercept)    5.407      1.513   3.574 0.009046 **
X2             1.974      0.278   7.102 0.000193 ***
X3                NA         NA      NA       NA
---
Signif. codes: 0 '***' 0.001 '**' 0.01 '*' 0.05 '.' 0.1 ' ' 1
Residual standard error: 2.044 on 7 degrees of freedom
Multiple R-squared: 0.8781, Adjusted R-squared: 0.8607
F-statistic: 50.44 on 1 and 7 DF, p-value: 0.0001933
```

R has a built-in determinant function, det(.), where the argument is a square matrix. Consider this example:

```
> col1 <- c(1,2,3)
> col2 <- c(4,5,6)
> col3 <- c(9,7,8)
> mat1 <- cbind(col1,col2,col3)
> det(mat1)
[1] -9
```

Although no simple formula exists for determinants in general, a 2 × 2 square matrix has a fairly simple determinant formula. For a 2 × 2 matrix:

$$\boldsymbol{A} = \begin{bmatrix} a & b \\ c & d \end{bmatrix}$$

the determinant is:

$$\det(\boldsymbol{A}) = ad - bc$$

Another special case of determinant is when the square matrix is symmetric across its diagonal. That is to say, if a_{ij} represents the element of the matrix in the ith row and jth column, then the matrix is symmetric, also referred to as triangular, if $a_{ij} = a_{ji}$, $i \neq j$. Then the determinant is equal to the product of the diagonal elements:

$$\det(\boldsymbol{A}) = \prod_{i=1}^{n} a_{ii}$$

The matrix A is $n \times n$. This fact is most useful for a special matrix, called the variance-covariance matrix, or sometimes called the covariance matrix. If X_1 and X_2 are random variables with means μ_1 and μ_2, respectively, then the covariance of X_1 and X_2 is:

$$\text{Cov}(X_1, X_2) = \text{Cov}(X_2, X_1) = \sigma_{12} = \sigma_{21} = \int_{-\infty}^{+\infty}\int_{-\infty}^{+\infty} (x_1 - \mu_1) \times (x_2 - \mu_2) f(x_1, x_2) dx_1 dx_2$$

The covariance between two variables X_1 and X_2 could be estimated from n observations with the formula:

$$C(X_1, X_2) = \frac{1}{n}\sum_{i=1}^{n} (x_{1i} - \overline{x}_1)(x_{2i} - \overline{x}_2)$$

where $\overline{x}$ is the arithmetic average.

The diagonal of the covariance matrix is just the variances of the respective variables. So, for example, the covariance matrix of X_1 and X_2 could be symbolized as:

$$\boldsymbol{\Sigma} = \begin{bmatrix} \sigma_1^2 & \sigma_{12} \\ \sigma_{21} & \sigma_2^2 \end{bmatrix}$$

It is very likely that there will never be a need to compute a determinant per se. Its most obvious application is in the area of multivariate analysis, i.e., situations where there are multiple "response" variables that are related to each other in some fashion. Many of the multivariate methods involve something called eigenvalues and eigenvectors, which are computed using equations with determinants.

A.13 Eigenvalues and Eigenvectors

Suppose $\boldsymbol{A}$ is a square matrix of dimensions $n \times n$:

$$\boldsymbol{A} = \begin{bmatrix} a_{11} & \cdots & a_{1n} \\ \vdots & \ddots & \vdots \\ a_{n1} & \cdots & a_{nn} \end{bmatrix}$$

Sometimes, especially when the components of matrix $\boldsymbol{A}$ are pairwise correlations between a set of response variables, it is of interest to find a scalar, λ, such that:

$$\boldsymbol{Ax} = \lambda \boldsymbol{x}$$

Since multiplying the identity matrix by a vector yields the original vector:

$$\boldsymbol{Ix} = \boldsymbol{x}$$

then the previous equation can be expressed as:

$$\boldsymbol{Ax} = \lambda \boldsymbol{Ix}$$

Because the two sides of this equation both have the same dimensions ($n \times 1$), the right-hand side can be subtracted from the left:

$$\boldsymbol{Ax} - \lambda \boldsymbol{Ix} = \boldsymbol{0}$$

where $\boldsymbol{0}$ is an $n \times 1$ vector of all zeros. The vector x can be "factored" to yield:

$$(\boldsymbol{A} - \lambda \boldsymbol{I})\boldsymbol{x} = \boldsymbol{0}$$

It turns out that the equation only has a solution if:

$$|\boldsymbol{A} - \lambda \boldsymbol{I}| = |\boldsymbol{0}| = 0$$

The determinant of the $n \times 1$ vector of all zeros is 0.

It turns out that there are potentially n different values of λ that satisfy the determinant equation, called the characteristic equation for matrix $\boldsymbol{A}$. That is to say, the determinant:

$$|\boldsymbol{A} - \lambda \boldsymbol{I}|$$

actually is a polynomial of order n, so setting it equal to 0 allows for the possibility of n solutions, or roots. The roots of the characteristic equation are called, believe it or not, characteristic roots, or characteristic values, or eigenvalues. Each root, λ_i, then corresponds to a vector, $\boldsymbol{x}_i$, that satisfies the equation:

$$\boldsymbol{A}\boldsymbol{x}_i = \lambda_i \boldsymbol{x}_i$$

The solution vectors, $\boldsymbol{x}_i$, $i = 1, n$, are called characteristic vectors, or eigenvectors.

A.14 Diagonalization and Powers of Matrices

A diagonal matrix is a square ($n \times n$) matrix with diagonal entries, a_{ii} not necessarily 0, but all other entries necessarily 0 ($a_{ij} = 0, i \neq j$). Of particular interest is a diagonal matrix, $\boldsymbol{\Lambda}$, whose diagonal entries are the eigenvalues of another n x n matrix, $\boldsymbol{A}$. If matrix $\boldsymbol{S}$ is the matrix of eigenvectors of A, then it turns out that (Strang, 2016):

$$\boldsymbol{A} = \boldsymbol{S}\boldsymbol{\Lambda}\boldsymbol{S}^{-1}$$

While there are many applications and mathematical reasons why this relation is important, we will focus here on one fact, relating to powers of a matrix. The kth power of matrix $\boldsymbol{A}$ is simply $\boldsymbol{A}$ multiplied by itself k times. It turns out that for a diagonal matrix, say our matrix of eigenvalues, $\boldsymbol{\Lambda}$, the kth power is simply a diagonal matrix whose diagonal elements are a_{ii}^k. This is very convenient, inasmuch as it makes computing the kth power of $\boldsymbol{A}$ fairly easy:

$$\boldsymbol{A}^k = \boldsymbol{S}\boldsymbol{\Lambda}^k\boldsymbol{S}^{-1} = \boldsymbol{S}\begin{bmatrix} a_{11}^k & \cdots & 0 \\ \vdots & \ddots & \vdots \\ 0 & \cdots & a_{nn}^k \end{bmatrix}\boldsymbol{S}^{-1}$$

All but the first $\boldsymbol{S}$ and last $\boldsymbol{S}^{-1}$ will "cancel" each other, since $\boldsymbol{S}\boldsymbol{S}^{-1} = \boldsymbol{I}$, the identity matrix.

The R function *eigen()* can be used to extract both eigenvalues and eigenvectors from a square matrix. Figure A.3 shows some code and associated output for using

```
setwd("<your path here>")
dfl <- read.csv("20170220 Example A.9 4x4 Matrix.csv")
#
#
# VARIABLES:
# s1
# s2
# s3
# s4
#
attach(dfl)
Amat <- as.matrix(dfl)
Smat <- eigen(Amat)
Smat$values
Smat$vectors
Lamda <- diag(Smat$values)
Eig <- as.matrix(Smat$vectors)
EigInv <- solve(Eig)
Arecover <- Eig%*%Lamda%*%EigInv
Lamda
Eig
Amat
Arecover
#####
```

OUTPUT

```
> Lamda

     [,1]      [,2]       [,3]          [,4]

[1,]    1  0.0000000 0.0000000  0.000000e+00

[2,]    0 -0.4422144 0.0000000  0.000000e+00
```

Fig. A.3 Extracting eigenvalues and eigenvectors and diagonalization

```
[3,]    0  0.0000000 0.1922144  0.000000e+00

[4,]    0  0.0000000 0.0000000 -1.289329e-17

> Eig

     [,1]       [,2]          [,3]         [,4]

[1,]  0.5  0.7026433 -0.1512474 -0.07142857

[2,]  0.5 -0.1993420  0.8144128  0.50000000

[3,]  0.5 -0.3825925  0.1353958 -0.78571429

[4,]  0.5 -0.5658429 -0.5436212  0.35714286

> Amat

      a1   a2   a3    a4

[1,] 0.1 0.2 0.30 0.40

[2,] 0.4 0.3 0.20 0.10

[3,] 0.5 0.2 0.15 0.15

[4,] 0.6 0.1 0.10 0.20

> Arecover

     [,1]  [,2] [,3] [,4]

[1,]  0.1  0.2 0.30 0.40

[2,]  0.4  0.3 0.20 0.10

[3,]  0.5  0.2 0.15 0.15

[4,]  0.6  0.1 0.10 0.20

>
```

Fig. A.3 (continued)

eigen() with a 4 × 4 matrix. The code also shows how the original matrix can be "recovered" using the eigenvalues and eigenvectors. The function *diag*() is used to generate the diagonal matrix from the vector of eigenvalues.

A.15 Least Squares and Variants

Solving Linear Equations
When there are k equations in k unknowns, the problem of solving for the unknowns can be stated in matrix/vector form:

$$\mathbf{y} = \mathbf{X}\boldsymbol{\beta}$$

$\mathbf{y}$ is a $k \times 1$ column vector of known values. $\mathbf{X}$ is a $k \times k$ matrix of coefficients. Finally, $\boldsymbol{\beta}$ is a $k \times 1$ vector of unknowns, for which we would like to solve. Assuming that $\mathbf{X}$ is an invertible matrix, and its inverse is $\mathbf{X}^{-1}$, the solution is simply:

$$\boldsymbol{\beta} = \mathbf{X}^{-1}\mathbf{y}$$

What if X is not a square matrix? Suppose there are more equations, n, than unknowns, k. Then X cannot be simply inverted. However, the problem can still be solved using something called a generalized, or Moore-Penrose inverse (Strang, 1980):

$$\mathbf{X}^* = \left[\mathbf{X}^T\mathbf{X}\right]^{-1}\mathbf{X}^T$$

The matrix $\mathbf{X}^T$ is the transpose of $\mathbf{X}$. So $\mathbf{X}^T\mathbf{X}$ is a $k \times k$ square matrix. The number of columns in $\boldsymbol{X}$ is $k < n$. A set of columns are said to be linearly dependent if one of them can be computed as a linear function of the others. Statistically speaking, two columns in the $\mathbf{X}$ matrix are linearly dependent if the two variables they represent are correlated with each other. If the set of columns are not linearly dependent, they are called linearly independent. The rank of a matrix is the number of its linearly independent columns, also called its dimension. The number of linearly independent columns in $\boldsymbol{X}$ is k, and the number of linearly independent rows of $\boldsymbol{X}^T$ is also k. A square matrix is invertible if the number of columns is its rank. This means that the rank of $\boldsymbol{X}^T\boldsymbol{X}$ is also k, so that it is invertible. Thus:

$$\mathbf{X}^*\mathbf{X} = \left[\mathbf{X}^T\mathbf{X}\right]^{-1}\mathbf{X}^T\mathbf{X} = \mathbf{I}_k$$

The matrix $\mathbf{I}_k$ is the $k \times k$ identity matrix. So finally:

$$\left[\mathbf{X}^T\mathbf{X}\right]^{-1}\mathbf{X}^T\mathbf{y} = \left[\mathbf{X}^T\mathbf{X}\right]^{-1}\mathbf{X}^T\mathbf{X}\boldsymbol{\beta} = \mathbf{I}_k\boldsymbol{\beta} = \boldsymbol{\beta}$$

It turns out that this is actually the least squares solution to the regression model:

$$\mathbf{y} = \mathbf{X}\boldsymbol{\beta} + \boldsymbol{\varepsilon}$$

where $\boldsymbol{\varepsilon}$ is a $k \times 1$ vector of random "errors."

So, as long as the number of observations, i.e., the length of vector **y**, n, is greater than or equal to k, the number of regression parameters, the least squares solution exists. What happens if $n < k$? Consider a simplistic example, with $n = 2$ and $k = 3$. Then the matrix equation is still:

$$\mathbf{y} = \mathbf{X}\boldsymbol{\beta}$$

(for now, forget the error vector, $\boldsymbol{\varepsilon}$).

Express the equation in a more expanded form:

$$\begin{bmatrix} y_1 \\ y_2 \end{bmatrix} = \begin{bmatrix} x_{11} & x_{12} & x_{13} \\ x_{21} & x_{22} & x_{23} \end{bmatrix} \begin{bmatrix} \beta_1 \\ \beta_2 \\ \beta_3 \end{bmatrix}$$

Note that the product $\mathbf{X}^T\mathbf{X}$ would be a 3×3 matrix, and $\mathbf{X}^T\mathbf{y}$ would be 3×1 vector:

$$\mathbf{X}^T\mathbf{y} = \begin{bmatrix} x_{11} & x_{21} \\ x_{12} & x_{22} \\ x_{13} & x_{23} \end{bmatrix} \begin{bmatrix} y_1 \\ y_2 \end{bmatrix}$$

So, it seems that we could multiply both sides of the original equation by $\mathbf{X}^T$:

$$\mathbf{X}^T\mathbf{y} = \mathbf{X}^T\mathbf{X}\boldsymbol{\beta}$$

The problem is that when $k > n$, the matrix $\mathbf{X}^T\mathbf{X}$ cannot be inverted, even though it is a $k \times k$ square matrix. This is because the **X** matrix, having more columns than rows, has some columns that can be expressed as linear functions of the other columns, i.e., they are linearly dependent. The rank of a transpose matrix is equal to the number of linearly independent rows, which is the number of linearly independent columns of the original matrix. Thus, if a matrix, **X**, is $n \times k$ with $k > n$, then its rank is $k - n$ (assuming there are $k - n$ linearly independent columns in **X**) and its transpose, $\mathbf{X}^T$, also had rank $k - n$. The product $\mathbf{X}^T\mathbf{X}$ is a $k \times k$ matrix. It turns out that the product of two matrices is a matrix whose rank is the minimum of the two factor matrices, so rank of $\mathbf{X}^T\mathbf{X}$ is $k - n$. However, $\mathbf{X}^T\mathbf{X}$ has k rows and k columns. Since its rank $(k - n)$ is less than the number of columns (and rows), it is not invertible. Therefore, there is no least squares solution to a problem with k regressors and n observations where $k > n$.

A.16 Taylor Series Expansions

Suppose you have a function, $f(x)$, that is nice and smooth, so that it has derivatives. The value of f can be approximated if you know the values of its derivatives (somehow you knew the values of its derivatives but not the function itself).

So, it turns out that the Taylor series approximation for $f(x)$ is:

$$f(x) \approx f(0) + f'(0)x + \frac{1}{2!}f''(0)x^2 + \frac{1}{3!}f'''(0)x^3 + \frac{1}{4!}f''''(0)x^4 \ldots$$

So, if you know the function's value when $x = 0$, call it $f(0)$, and the value of some arbitrary number of its derivatives at $x = 0$, you could approximate the value of the function at any arbitrary value of x. The more terms you add, the better the approximation.

As an example, suppose:

$$f(x) = 5\sin(2x)$$

That means:

$$f'(x) = 2 * 5\cos(2x)$$
$$f''(x) = -2 * 2 * 5\sin(2x)$$
$$f'''(x) = -2 * 2 * 2 * 5\cos(2x)$$
$$f''''(x) = 2 * 2 * 2 * 2 * 5\sin(2x)$$

We also can compute:

$$f(0) = 5\sin(0) = 0$$
$$f'(0) = 2 * 5\cos(0) = 10$$
$$f''(0) = -2 * 2 * 5\sin(0) = 0$$
$$f'''(0) = -2 * 2 * 2 * 5\cos(0) = -40$$
$$f''''(x) = 2 * 2 * 2 * 2 * 5\sin(0) = 0$$

So, an approximation of the function:

$$f(x) = 5\sin(2x)$$

is:

$$f(x) \approx f(0) + f'(0)x + \frac{1}{2!}f''(0)x^2 + \frac{1}{3!}f'''(0)x^3 + \frac{1}{4!}f''''(0)x^4$$

$$f(x) \approx 0 + 10x + 0 + \frac{1}{3!}(-40)x^3 + 0$$

So, we know that $\sin\left(2 * \frac{\pi}{2}\right) = 0$, and $f(x) = 5\sin\left(2\frac{\pi}{2}\right) = 0$. Our approximation is:

$$f\left(\frac{\pi}{2}\right) \approx 0 + 10\left(\frac{\pi}{2}\right) + 0 + \frac{1}{3!}(-40)\left(\frac{\pi}{2}\right)^3 + 0 \approx -10.1306$$

OK, so it's not a very good approximation. I guess we would need to use more terms to get a better approximation.

A.17 Quadratic Solution and Vertices

The solution to the equation:

$$ax^2 + bx + c = 0$$

is:

$$x = \frac{-b \pm \sqrt{b^2 - 4ac}}{2a}$$

That is to say, the value of x that makes the quadratic equal to zero is either:

$$x_1 = \frac{-b - \sqrt{b^2 - 4ac}}{2a}$$

or:

$$x_2 = \frac{-b + \sqrt{b^2 - 4ac}}{2a}$$

Sometimes these values are complex numbers.
As an example, suppose we have the quadratic expression:

$$2x^2 + 1x - 6$$

The solution to the equation:

$$2x^2 + 1x - 6 = 0$$

is:

$$x = \frac{-1 \pm \sqrt{1^2 - 4(2)(-6)}}{2(2)} = \frac{-1 \pm \sqrt{49}}{4} = \left(\frac{-8}{4}, \frac{6}{4}\right)$$

If you put either −8/4 or 6/4 into the expression, you will get 0.

A.18 Vertex

The vertex of a parabola (which is described by a quadratic expression) can be derived using derivatives.

Let:

$$f(x) = ax^2 + bx + c$$

From the rules of differentiation, we get:

$$f'(x) = 2ax + b$$

To find the value of x at which the original function is a maximum (for down-pointing parabolas) or minimum (for up-pointing parabolas), set the derivative (also called the first derivative) to zero and solve for x:

$$f'(x_v) = 2ax_v + b = 0$$

$$x_v = \frac{-b}{2a}$$

To find the value of the parabola at the vertex, simply put the value of x_v back in the original expression.

For example, suppose:

$$f(x) = x^2 - 2x + 7$$

Then:

$$f'(x) = 2x - 2$$

Setting this equal to 0 and solving:

$$2x_v - 2 = 0$$

$$x_v = \frac{2}{2} = 1$$

The vertex is at $x_v = 1$. The parabola's height at $x_v = 1$ is:

$$f(1) = 1^2 - 2(1) + 7 = 6$$

The way to know if the parabola is pointing up (so that the vertex is its minimum) or pointing down (so that the vertex is its maximum) is simple. If $a > 0$, the parabola is pointing up. If $a < 0$, the parabola is pointing down. So, in this example, since $a = 1 > 0$, the value 6 is the parabola's minimum.

A.19 Inequalities and Absolute Value Expressions

An inequality expression is just like an equation, with the "=" sign replaced with one of the following:

$<$ (strictly less than)
$>$ (strictly greater than)
$\leq$ (less than or equal to)
$\geq$ (greater than or equal to)

So, the expressions:

A $<$ B means "A is strictly less than B"
A $>$ B means "A is strictly greater than B"
A $\leq$ B means "A is either less than or equal to B"
A $\geq$ B means "A is either greater than or equal to B"

OK, so those are the elementary definitions. It gets a little more complicated when A or B are algebraic expressions in terms of some unknown variables, say x or y. Sometimes we are asked to "solve" an inequality for the variable. The solving process is very similar to solving equalities, with one very important exception. If you multiply or divide both sides of the inequality by a negative number, the direction of the inequality changes. This is true whether the inequality sign is strictly less than, less than or equal, strictly greater than, or greater than or equal. Furthermore, the "solution" is often an infinite set of numbers, instead of one (in linear equalities) or two (quadratic equalities) numbers.

For example, consider the inequality:

$$3x–6 < 0$$

If this were the equality:

$$3x - 6 = 0$$

We would solve for x by adding six to both sides and dividing both sides by 3:

$$3x = 6$$
$$x = 6/3 = 2$$

Do the same thing for the inequality:

$$3x < 6$$
$$x < 2$$

So, any value of x that is strictly less than 2 will satisfy the inequality:

$$3x - 6 < 0$$

The biggest difference in solving inequalities is when multiplying or dividing both sides by a negative number. In that case, the direction of the inequality is reversed. For example:

$$-3x \geq 6$$

Dividing both sides by −3 gives you:

$$x \leq 6/(-3)$$

or

$$x \leq -2$$

Notice that it does not matter if the inequality is strict ($<$ or $>$) or not strict ($\leq$ or $\geq$).

So far, the inequalities only involved linear terms. What about quadratic expressions?

Suppose you had:

$$3x^2 - 27 > 0$$

We could use the quadratic formula, but this expression is too simple. Add 27 to both sides and divide by 3:

$$x^2 > 9$$

You can take square roots of both sides (this will *not* change the direction of the inequality), but keep in mind that the solution can be either negative or positive:

$$x > \pm\sqrt{9} = \pm 3$$

OK, this does not make sense. Suppose, for example, $x = -2 > -3$. Well: $x^2 = (-2)^2 = 4$ which is certainly not >9. How do we resolve this? The answer is absolute value.

If $x^2 > 9$, then clearly $x > 3$ makes the statement true, but so does $x < -3$. In other words, the solution is:

$$|x| > 3$$

So, whenever you have a squared variable, think of it as:

$$|x|^2$$

As another example, suppose:

$$5x^2 > 125$$

First translate this expression into something with an absolute value:

$$5|x|^2 > 125$$

Then divide both sides by 5:

$$|x|^2 > 25$$

Now take square roots of both sides:

$$|x| > 5$$

So this means any value of x that is either greater than +5 or less than −5 will make the inequality true.

Notice that the direction of the inequality did not change, since we never multiplied or divided both sides by a negative number.

If the original inequality were < instead of >:

$$5x^2 < 125$$

The solution would be:

$$|x| < 5$$

This means that if x is anywhere between -5 and $+5$, the inequality is true. Since this is a strict inequality, the values $x = -5$ or $x = +5$ do not make the inequality true (i.e., -5 and $+5$ do not satisfy the inequality).

A.20 Expressions with Absolute Values

Sometimes you may encounter an expression that is linear in terms of the variable but also has an absolute value for the expression. For example:

$$|2x-8| < 10$$

Such inequalities actually are describing two separate expressions:

$$2x-8 < 10$$

and

$$-(2x-8) < 10$$

This is because

$$|-(2x-8)| = |2x-8|$$

So, get a complete answer to the question:
What values of x will make the inequality true?
You must solve both inequalities.
The first one is fairly straightforward:

$$2x < 10 + 8 = 18$$
$$x < 18/2$$

so, $x < 9$.

The other inequality requires multiplication by a negative number (-1):

$$-(2x-8) < 10$$

$2x - 8 > -10$ (notice the change of direction in the inequality sign)

$$2x > -10 + 8 = -2$$

$x > -1$ (notice that the direction of the inequality did *not* change when dividing by 2)

So, the complete answer is that $x > -1$ and $x < 9$. To see if this is correct, substitute for x any number between -1 and 9 (but don't include either -1 or 9, since this is a strict inequality.)

Try $x = 0$:

$$|2(0)-8| = 8 < 10$$

Or try $x = 5$:

$$|2(5)-8| = 2 < 10.$$

One more; try $x = -1/2$:

$$|2(-1/2)-8| = |-1-8| = |-9| = 9 < 10$$

The case of absolute values with quadratic expressions is very similar in nature, namely, that the absolute value of the expression generates two inequalities to solve. For example:

$$|x^2 - 2| \leq 1$$

The two inequalities are:

$$x^2 - 2 \leq 1$$

and

$$-(x^2 - 2) \leq 1$$

In the case of the first, the solution is fairly simple to obtain:

$$x^2 - 2 \leq 1$$
$$x^2 \leq 1 + 2 = 3$$

Now that we have:

$$|x|^2 \leq 3$$

we must recall how we solved inequalities with squared variables:

$$|x|^2 \leq 3$$
$$|x| \leq \sqrt{3}$$

So either:

$$x \leq \sqrt{3}$$

or:

$$-x \leq \sqrt{3}$$

If $-x \leq \sqrt{3}$, then $x \geq -\sqrt{3}$

So, for the first part of the original inequality, either $x \geq -\sqrt{3}$ or $x \leq +\sqrt{3}$.

The second part proceeds similarly.

$$-(x^2 - 2) \leq 1$$
$$x^2 - 2 \geq -1$$
$$x^2 \geq -1 + 2 = 1$$

so:

$$|x|^2 \geq 1$$

This means that either:

$$x \geq 1$$

or:

$$-x \geq 1 (\text{i.e.}, x \leq -1)$$

So, in summary, we now have that in order for the original inequality to be true, then:

$$x \geq -\sqrt{3} \text{ or } x \leq +\sqrt{3}$$

and:

$$x \geq 1 \text{ or } x \leq -1$$

Well, if $x = -\sqrt{4}$, it is certainly ≤ -1, but it is *not* $\geq -\sqrt{3}$. Similarly, if $x = +\sqrt{4}$, then it is ≥ 1 but *not* $\leq +\sqrt{3}$.

As a check, let $x = -\sqrt{4}$ in the original inequality:

$$\left|(-\sqrt{4})^2 - 2\right| = |4-2| = 2 > 1, \text{so} -\sqrt{4} \text{ is not a solution.}$$

Conversely, if $x = -\sqrt{3}$, then:

$$\left|(-\sqrt{3})^2 - 2\right| = |3-2| = 1$$

Since the original inequality is not strict, $-\sqrt{3}$ is in fact a valid solution. In this case, the second part of solving the inequality does not result in any more solutions.

In summary:

1. Inequalities are for the most part solved just like equalities, with one exception.
2. The exception is that when multiplying or dividing both sides by a negative number, the direction of the inequality changes.
3. When the expression has a squared variable, then before solving, convert the squared variable into the absolute value of the squared variable; x^2 becomes $|x|^2$.
4. When an absolute value of the variable is in the expression, then it creates two separate expressions that both must be solved. That is because $|x| = |-x|$.
5. Expressions with absolute values of quadratic expressions may have up to four solutions.

A.21 Complex Numbers and Variables

The imaginary number is symbolized by:

$$j = \sqrt{-1}$$

A general imaginary number is a real number, y multiplied by j:

$$jy$$

A complex number is a sort of "sum" of a real number and an imaginary number:

$$z = x + jy$$

The "+" is not quite the same as the "+" used to add real numbers, that is, $x + jy$ does not mean the same thing as $x + y$.

There are rules for combining complex numbers in such ways that will be called "addition" and "multiplication." When adding or multiplying complex numbers, you always get another complex number. Real numbers are special cases of complex numbers:

$$z = x + j0 = x$$

Typically, x is referred to as the real part, and y the imaginary part, of the complex number:

$$z = x + jy$$

So, if $z_1 = x_1 + jy_1$ and $z_2 = x_2 + jy_2$, then:

$$z_1 * z_2 = x_1x_2 + jx_1y_2 + jx_2y_1 - y_1y_2 = (x_1x_2 - y_1y_2) + j(x_1y_2 + x_2y_1)$$

$$z_1 \oplus z_2 = x_1 + x_2 + j(y_1 + y_2)$$

The symbols used for multiplication (*) and addition ($\oplus$) are different to indicate that these are for complex numbers. Note that adding or multiplying two complex numbers results in another complex number. Recall that real numbers can be thought of complex numbers with $y = 0$, and imaginary numbers can be thought of as complex numbers with $x = 0$. So, in that sense complex arithmetic is a generalization of real arithmetic.

A point in Cartesian coordinates (x, y) can be represented in polar coordinates as:

$$x = r\cos\theta$$

$$y = r\sin\theta$$

The polar expression of the point is (r, θ). The value of r, called the modulus, is:

$$r = \sqrt{x^2 + y^2}$$

The value of θ is the angle that a ray from the origin (0, 0) to the point (x, y) makes with the x-axis. It can be computed if you know (x, y) as:

$$\theta = \tan^{-1}\frac{y}{x}$$

Of course, as long as $x \neq 0$. When $x = 0$, $\theta = \frac{\pi}{2} = 90^\circ$.

A complex number, $z = x + jy$, can be expressed in terms of (r, θ) by substituting the polar representations for x and y:

$$z = x + jy = r(\cos\theta + j\sin\theta) = re^{j\theta}$$

The exponential form is based on Euler's relation:

$$e^{j\theta} = \cos\theta + j\sin\theta$$

It turns out that the "polar" form of a complex number is very useful when performing operations and computing functions of complex numbers.

An important concept is the complex conjugate. If:

$$z = x + jy$$

then its complex conjugate is:

$$\bar{z} = x - jy = re^{-jy}$$

It seems simple enough. Here is an interest set of facts:

$$z\bar{z} = \bar{z}z = (x+jy)(x-jy) = x^2 + y^2 = r^2$$

$$\frac{z}{\bar{z}} = \frac{re^{j\theta}}{re^{-j\theta}} = e^{2j\theta}$$

$$\frac{\bar{z}}{z} = \frac{re^{-j\theta}}{re^{j\theta}} = e^{-2j\theta}$$

$$z^k = \left(re^{j\theta}\right)^k = r^k e^{jk\theta}$$

A.22 Fourier Series

Any piece-wise continuously valued, periodic function, with period T, can be represented by a series:

$$f(t) = \frac{a_0}{2} + \sum_{i=1}^{\infty} a_i \cos \omega_i t + b_i \sin \omega_i t$$

The values a_0, a_i, b_i are called "Fourier coefficients" and the:

$$\omega_i = \frac{2\pi i}{T}$$

are called frequencies. The value:

$$\omega_1 = \frac{2\pi}{T}$$

is sometimes referred to as the principal frequency. Generally, the higher the value of the frequency, the less impact it has on the value of the function $f(t)$.

The series representation of $f(t)$ is called the Fourier series of $f(t)$. The values ω_i are called "frequencies." Using Euler's relation, the Fourier series can also be expressed as:

$$f(t) = \sum_{k=-\infty}^{+\infty} c_k e^{2\pi j \omega_k t}$$

The coefficients c_k are related to a_k and b_k as:

$$c_k = \frac{a_k - jb_k}{2}$$

Typically, c_{-k} is the complex conjugate of c_k:

$$c_{-k} = \frac{a_k + jb_k}{2}$$

The values c_k are also called "Fourier coefficients," but they have complex values. If you know two values of c_k, say c_1 and c_{-1}, you can solve for a_k and b_k.

It turns out that the Fourier coefficients can be estimated using data and least squares. Such analyses fall under the general topic of time series analysis. In particular, using a time sequence of data, $z_1, z_2, \ldots, z_t, \ldots z_T$, the Fourier coefficients can be estimated up to $\frac{T}{2}$ terms, which, for large enough T, should give a pretty good estimate of the function $z(t)$. In this context, we are using the letter z to represent real numbers (sorry for the confusion).

A.23 Spectral Analysis of Time Series

Suppose a series of values (e.g., signal amplitudes) are observed at discrete times, $t = 1, 2, \ldots, T$. Represent these values with $z_1, z_2, \ldots, z_T$.

The Fourier model for this series can be expressed as:

$$z_t = \alpha_0 + \sum_{i=1}^{q} \alpha_i \cos(2\pi f_i t) + \beta_i \sin(2\pi f_i t) + \epsilon_t$$

For $t = 1, 2, \ldots, T$

$$q = \begin{cases} \frac{T-1}{2}, T \text{ odd} \\ \frac{T}{2}, T \text{ even} \end{cases}$$

$$f_i = \frac{i}{T}$$

The values α_0, α_i, and β_i are unknown constants (e.g., to be estimated with data via least squares), and ϵ_t is Gaussian white noise (i.e., noise distributed as $N(0, \sigma^2)$). The f_i are referred to as the ith harmonic of the fundamental frequency, $\frac{1}{T}$.

The least squares estimates for the unknown constants (called Fourier coefficients) are given by the formulas:

$$\widehat{\alpha}_0 = \bar{z} = \frac{1}{T}\sum_{t=1}^{T} z_t$$

$$\widehat{\alpha}_i = \frac{2}{T}\sum_{t=1}^{T} z_t \cos(2\pi f_i t)$$

$$\widehat{\beta}_i = \frac{2}{T}\sum_{t=1}^{T} z_t \sin(2\pi f_i t)$$

$i = 1, q$.

If T is even, the estimates for the last coefficients ($i = q$) are a little funny. They are:

$$\widehat{\alpha}_q = \sum_{t=1}^{T} (-1)^t z_t$$

and:

$$\widehat{\beta}_q = 0$$

Note that in the Fourier model, the sum is over the harmonics, f_i, $i = 1, q$, but in the least squares formulas, the sum is over the T observations, z_t, $t = 1, T$. That is to say, there are $q + 1$ parameters, or coefficients, to estimate, and each one is estimated using all T observations.

Now compute the function, called the sample intensity at f_i, as:

$$\widehat{I}(f_i) = \frac{T}{2}\left(\widehat{\alpha}_i^2 + \widehat{\beta}_i^2\right), i = 1, q$$

The plot of $\widehat{I}(i)$ on the y-axis and f_i on the x-axis, $i = 1, q$, is called the *periodogram*, or *sample spectrum*. If instead of computing the estimates of the coefficients using f_i, you compute them with an arbitrary sequence of values for f between 0 and 0.5, then the plot is called the *spectrum* instead of the *periodogram*.

It turns out that the sample intensity and spectrum have a nice relationship to something called the autocovariance (or autocorrelation) function of the series z_t, $t = 1, T$.

Suppose that z_t is "centered," by subtracting the mean value of the time series y_t:

$$z_t = y_t - E(y_t)$$

Then the autocovariance function for z_t is:

$$\gamma(h) = E(z_t z_{t-h})$$

Note that you could use z_{t+h}, or just make sure that $t - h \geq 0$, if you like dealing with nonnegative time subscripts.

So, the spectrum is the Fourier transform of the autocovariance. More typically, we use the "cosine" Fourier transform:

$$I(f) = \frac{1}{2\pi}\left(\gamma(0) + \sum_{h=1}^{\infty}\gamma(h)\cos(hf)\right)$$

When using sample data, assume that T values of the series have been observed so that:

$$\widehat{I}(f_i) = \frac{1}{2\pi}\left(\widehat{\gamma}(0) + \sum_{h=1}^{T/2}\widehat{\gamma}(h)\cos(hf_i)\right)$$

The values of f_i are taken to be values between 0 and 0.5. The values can either be arbitrarily assigned with uniform increments or computed as described earlier:

$$f_i = \frac{i}{T},\quad i = 1, \frac{T}{2}$$

A.24 A Note on Importing Data

You can either use the reads.csv() function from Base R, or the read_excel() function in library readxl. The structure of the data is the same, namely, the first row is column headings (variable names), and each row thereafter is a record.

Reference

Strang, G. (1980). *Linear Algebra and Its Applications* (2nd ed.). Academic Press, New York

Index

S. A. Pardo, *Statistical Methods and Analyses for Medical Devices*,
https://doi.org/10.1007/978-3-031-26139-8